AA002870

Eurotherm Seminar 102: Thermal Management of Electronic Systems 2014

Journal of Physics: Conference Series Volume 525

Limerick, Ireland
18 – 20 June 2014

ISBN: 978-1-63439-306-5
ISSN: 1742-6588

Printed from e-media with permission by:

Curran Associates, Inc.
57 Morehouse Lane
Red Hook, NY 12571

Some format issues inherent in the e-media version may also appear in this print version.

Copyright© by the Institute of Physics
All rights reserved.

Printed by Curran Associates, Inc. (2014)

For permission requests, please contact the Institute of Physics
at the address below.

Institute of Physics
Dirac House, Temple Back
Bristol BS1 6BE UK

Phone: 44 1 17 929 7481
Fax: 44 1 17 920 0979

techtracking@iop.org

Additional copies of this publication are available from:

Curran Associates, Inc.
57 Morehouse Lane
Red Hook, NY 12571 USA
Phone: 845-758-0400
Fax: 845-758-2634
Email: curran@proceedings.com
Web: www.proceedings.com

TABLE OF CONTENTS

MATERIALS & INTERFACES

HEAT SINKS & THERMAL CONTROL TECHNIQUES

LIQUID COOLING

AIR MOVERS: JETS AND FANS II

Preface

About EUROTHERM

The aim of the EUROTHERM Committee (www.eurothermcommittee.eu) is to promote and foster European cooperation in Thermal Sciences and Heat Transfer by gathering together scientists and engineers working in specialized areas. The Committee consists of members representing and appointed by national bodies in the EU countries.

The current President of EUROTHERM is Professor Anton van Steenhoven from the University of Eindhoven (The Netherlands).

The Committee organizes and coordinates European scientific events such as the EUROTHERM Seminars (about 4 per year) and the European Thermal Sciences Conference (every 4 years).

About EUROTHERM Seminar 102 (www.eurothermseminar102.com)

This seminar, part of the long-running series of European seminars on the thermal sciences, took place in June 2014 at the University of Limerick in Limerick, Ireland.

The seminar addressed the topic of 'Thermal Management of Electronic Systems', a critical contemporary application area which represents a vibrant challenge for practitioners of the thermal sciences.

We convey special thanks to the reviewers who have evaluated these papers. We also thank the scientific committee, consisting of internationally recognized experts. Their role has been to manage the evaluation of abstracts and the papers selection process as co-coordinators for specific topics.

This seminar was hosted by the Stokes Institute at the University of Limerick. It could not have been organized without the efficient help of our administrators and technicians for IT support.

This volume of *Journal of Physics: Conference Series* includes 27 articles presented at the seminar.

Dr. Jeff Punch, Chair

Stokes Institute, University of Limerick, Limerick, Ireland
Email: jeff.punch@ul.ie

Prof. Edmond Walsh, Co-Chair

Associate Professor, Osney Laboratories, Department of Engineering Science, University of Oxford, UK
Email: edmond.walsh@bnc.ox.ac.uk

Content from this work may be used under the terms of the Creative Commons Attribution 3.0 licence. Any further distribution of this work must maintain attribution to the author(s) and the title of the work, journal citation and DOI.
Published under licence by IOP Publishing Ltd

Peer review statement

All papers published in this volume of *Journal of Physics: Conference Series* have been peer reviewed through processes administered by the proceedings Editors. Reviews were conducted by expert referees to the professional and scientific standards expected of a proceedings journal published by IOP Publishing.

Content from this work may be used under the terms of the Creative Commons Attribution 3.0 licence. Any further distribution of this work must maintain attribution to the author(s) and the title of the work, journal citation and DOI.
Published under licence by IOP Publishing Ltd

Journal of Physics: Conference Series **525** (2014) 012001 doi:10.1088/1742-6596/525/1/012001

Use of Thermal Zones as a Technique to Optimize Evaluation of a High-Temperature Bidirectional Converter

Ian Wilson[1]; Senior Mechanical Engineer, Microsemi Corporation
Mikaël Mohaupt[2]; Project Engineer , Euro Heat Pipes (EHP)
Laurent Barremaecker[2]; Operating Director, Euro Heat Pipes (EHP)

[1]Microsemi Corporation; 14930 East Alondra Blvd. La Mirada, CA 90638-5752
[2]Euro Heat Pipes (EHP); Rue de l'industrie 24, Nivelles, Belgium

Corresponding author: ian.wilson@microsemi.com

Abstract. Current electrical component technology makes it impossible for entire systems to operate at the extreme temperatures required for many applications. Therefore, thermal zoning is a straightforward method to maximize the temperature capability of complex systems. The technique involves simple thermal analyses to determine variable temperature limits, design requirements, and cooling techniques for an electrical system. This paper discusses how the thermal zoning technique is used during program development in order to maximize the demonstration temperature of a 50 kW bidirectional converter.

1. Introduction

Due to the complex interaction between cooling and power dissipation within an electrical assembly it can be difficult to find the limiting factor for a thermal design. In many cases attempts to improve component capabilities can be detrimental to the overall design. The thermal zoning technique is a straightforward eight step procedure, outlined in Table 1, which involves simple designs, thermal analyses, and a little research.

Table 1: 8-Step Thermal Zoning Technique

Step	Definition
1	Create a preliminary design
2	Divide the assembly into thermal zones
3	Define the thermal circuit
4	Review zones
5	Solve circuit
6	Determine capability of zones
7	Iterate zone capability based upon limits
8	Check results

A 50kW bidirectional converter is under development to demonstrate continuous functional capability with cold source temperatures in excess of 200°C, and a goal of achieving 250°C. As this 50kW bidirectional converter program is a technology demonstrator (TRL3) the intended goal is gaining practical experience with the components, interconnections, cooling technologies,

Content from this work may be used under the terms of the Creative Commons Attribution 3.0 licence. Any further distribution of this work must maintain attribution to the author(s) and the title of the work, journal citation and DOI.
Published under licence by IOP Publishing Ltd

manufacturing techniques, and materials required for operation at extreme temperatures. A block diagram of the system is shown in Figure 1.

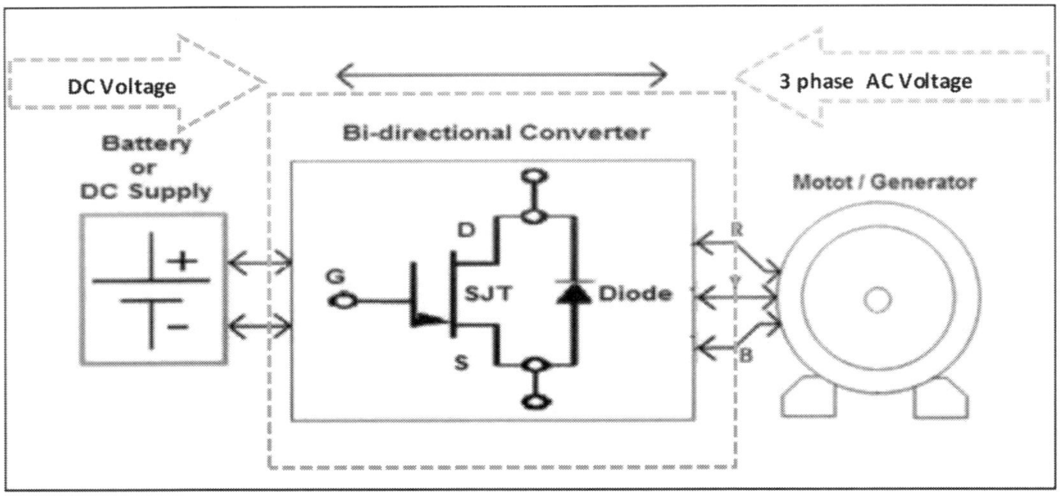

Figure 1: Block Diagram of a 50 kW Bidirectional Converter

2. Bidirectional Converter System

The first two steps in applying the Thermal Zoning Technique are creating a preliminary design of the system and dividing it into a set of thermal zones. A preliminary interconnect for the bidirectional converter is shown in Figure 2. The bidirectional converter system's five temperature zones: High Temperature Ambient, Power Bridge, Power Board, Control Board, and Low Temperature Ambient, are based upon the mechanical and electrical partitioning used to define the interconnects. The Power Bridge zone is a set of electrical assemblies used for electrical conversion. The Power Board zone is a printed wiring assembly used to filter, measure, and condition the electrical power. Control of the system is completed in the Control Board zone on a printed wiring assembly. The High Temperature Ambient zone is the air space around the Power Board and Power Bridge, but outside the Temperature Controlled Enclosure. Similarly, the Low Temperature Ambient zone is the general lab area where the system will be installed.

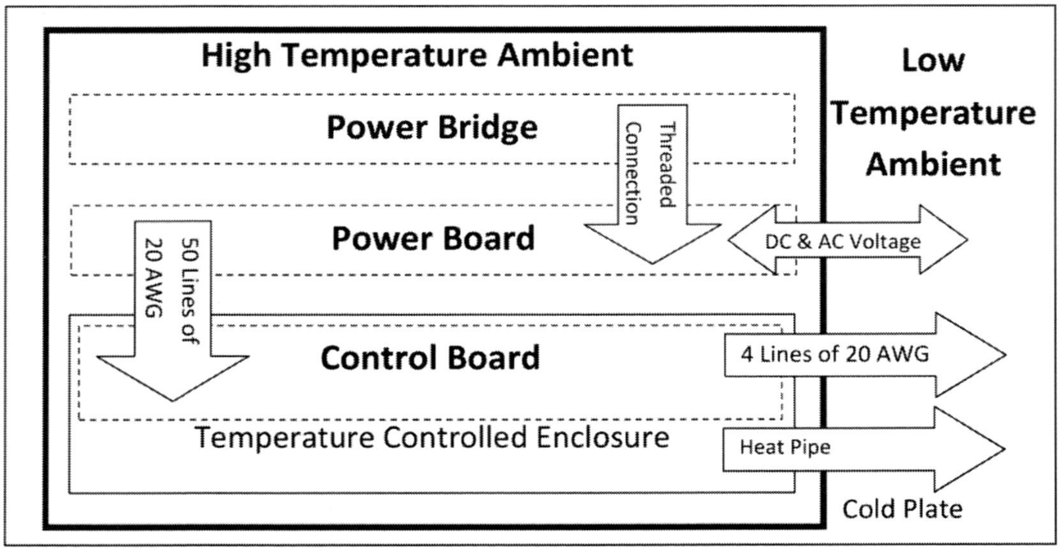

Figure 2: Representation of the Thermal Zones and Interconnections

Figure 3 represents the initial physical layout for the bidirectional converter assembly. In this design the Power Bridge is split into four identical assemblies which are mounted to a piece of fin stock for improved cooling. Each Power Bridge is mounted to the Power Board with five steel threaded connections. Fifty 20AWG wires form the control cable used to carry electrical signals between the Control Board and the Power Board. Heavy gauge wire will bring the AC and DC Voltage to the Power Board via the power cable. Due to limited temperature capability of the intricate devices used on the Controller Board, this zone represents the greatest thermal challenge. Therefore, a temperature controlled enclosure thermally isolates the Control Board from the rest of the system. Nevertheless, the enclosure has access for the control cable running between the Control and Power Boards. Two additional cutouts are included in the temperature controlled enclosure to allow a 4-wire signal cable and the loop heat pipe (LHP) to exit the thermal chamber.

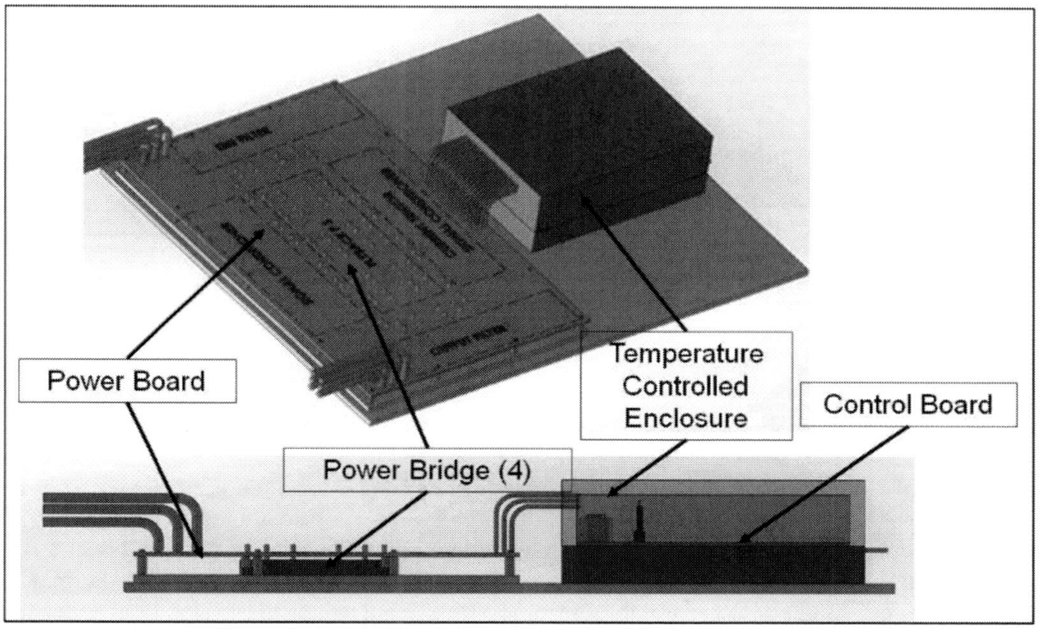

Figure 3: Bidirectional Converter System Mechanical Concept

3. Thermal Analysis

Step 3 of the Thermal Zoning Technique is a simple thermal analysis. Two cold sources are available for cooling the bidirectional converter system. The air within the thermal oven, referred to as the High Temperature Ambient zone, will be the cold source for the Power Bridge and the Power Board. The expected heat flow for these two assemblies under normal operating conditions is between 1.75 and 3 kW. The Control Board will also be mounted within the High Temperature Ambient zone, but unlike the Power Board and Power Bridge, insulation will be used to keep the High Temperature Ambient from heating the Control Board. The Low Temperature Ambient zone will be coupled via LHP to the Control Board and used as the cold source. This low temperature zone is external to the thermal chamber and in the general lab area.

Figure 4 is a circuit representing the thermal interconnections between the Power Bridge, Power Board, and High Temperature Ambient thermal zones. In this figure the power bridge dissipations are represented as current sources "B1," "B2," "B3," and "B4" while "P" represents another current source equal to the power dissipation on the Power Board. Each of the six resistor reference designators identifies a specific path available for heat transfer. Finally, the High Temperature Ambient zone is represented as ground in this circuit. Since forced convection will be the dominant source of system cooling the initial thermal analysis ignores the effects of radiation. This is not considered to be a significant risk to the program because radiant cooling will increase the energy transfer from the system and improve its maximum operating temperature. Nevertheless, should a designer wish to include the effects of radiant cooling explicitly in the analysis this thermal path would also be included as a resistor.

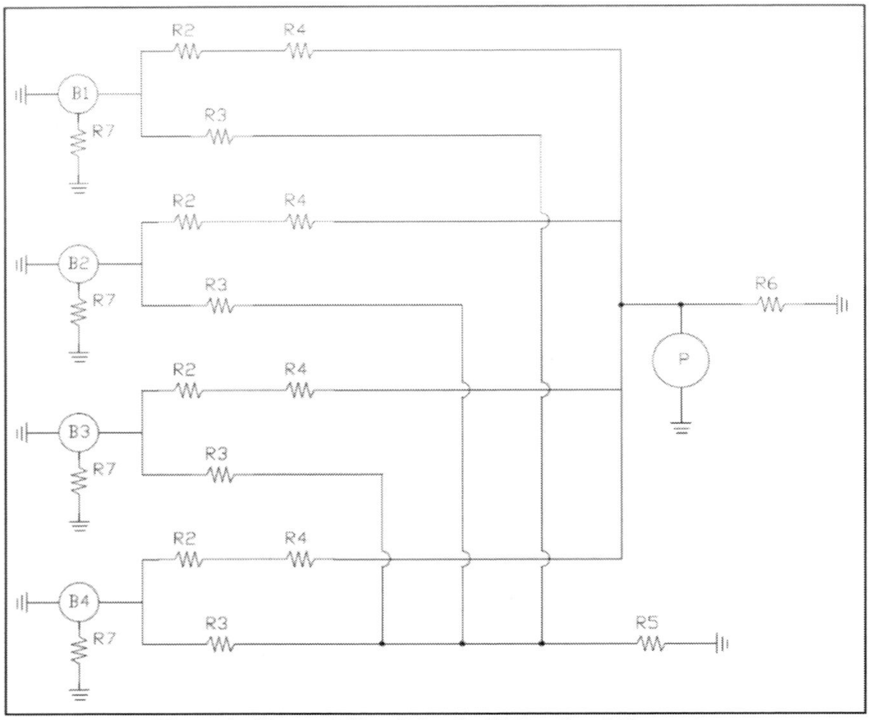

Figure 4: Thermal Circuit for the Bidirectional Converter

The Control Board has a minimal number of powered components, and is isolated from the High Temperature Ambient and the Power Board. Although it does represent another parallel path for heat rejection, the control cable is assumed to be insulated and of sufficient length to conduct a negligible amount of heat. Thus, it has been removed from the Power Board and Power Bridge thermal analysis. The heat transfer along the control cable becomes significant when the system is viewed from the perspective of the Control Board. For this reason, a detailed discussion of the Controller Board and Temperature Controlled Enclosure study is included in Section 4.5. Nevertheless, simplifying the system to remove any impact of the Control Board and temperature controlled enclosure should be marginally conservative for the remaining thermal zones.

With a thermal resistance values, power levels, and a fully developed thermal circuit the resultant temperature increases can be calculated. Software such as PSPICE, Simetrix, Orcad, or other programs supplied by component manufacturers will quickly solve the circuit. Power levels shown in Table 2 are based upon the 96% minimum system efficiency expected for the converter. The Power Bridge represents the largest source of waste heat in the bidirectional converter; therefore, approximately 95% of the power dissipation has been allocated to the four Power Bridges. Junction to case thermal resistances, included as R1, are a consequence of the particular die packaging used during component manufacturing. As this is a finer detail than required by a preliminary zoning study it has been neglected. The remaining thermal resistances are calculated from the interface resistivity in [1] and the material properties in [2]. This analysis assumes a constant temperature in each thermal zone. Therefore, to ensure that localized heating does not create failures, when determining the appropriateness of a component or material for use in a particular zone design factors must be used.

Although it was not included in the preliminary design represented in Figure 3, a resistance value for direct convection off the Power Bridge and into the High Temperature Ambient was included in the thermal analysis. This additional thermal path allowed for a change in the mounting configuration if temperatures on the Power Bridge became too great. However, results in Table 2 show that power dissipated directly from the Power Bridge and into the High Temperature Ambient is negligible. As a result, the board surface area is better used for cooling the Power Board and the mounting of components as originally designed.

Table 2: Thermal Circuit Solution

Resistor	Resistance	Power	ΔT	Notes
(#)	(ºC/W)	(W)	(ºC)	
1	0	472.5	0.00	Conduction from component to Power Bridge (θ_{JC})
2	5.25	8.25	43.31	Conduction along Power Bridge to Power Board interface
3	0.026	460.3	12.04	Conduction from Power Bridge and into Fin Stock
4	0.034	8.25	0.28	Conduction from Power Bridge and into Power Board
5	0.032	1841.0	59.67	Convection off of Fin Stock
6	0.206	133.0	27.40	Convection off of Power Board
7	17.71	4.00	70.86	Convection directly off of the Power Bridge

4. Capability of the Thermal Zones

The Thermal Zoning Technique's sixth step is determining capabilities for each thermal zone. Understanding what limiting factors exist for a particular thermal zone allows developers to avoid paying for unnecessary capability in other areas of the assembly. With the calculated temperature deltas in Table 2, it is possible to iterate the selection of materials, components, and cold source temperatures until the weakest link in the system is determined. This iteration is Step 7 in the Thermal Zoning Technique.

4.1. Low Temperature Ambient

Test and support equipment required for the bidirectional converter to operate will be located in the Low Temperature Ambient zone. This is a large general lab area which will have active temperature control using facility heating, venting, and cooling (HVAC) systems to maintain a roughly 20°C ambient. Heating in this zone is expected to be the result of less than 100 W; this represents control power dissipation, wild heat from the thermal chamber, and power losses from the support equipment. This zone does not represent a limiting factor on the bidirectional test temperature.

4.2. High Temperature Ambient

The High Temperature Ambient zone will be created by a thermal chamber and represents the primary cold sink for the bidirectional converter assembly. The thermal chamber has an active control system which causes the temperature of the High Temperature Ambient to be constant and known. Therefore, cooling capability for the Power Bridge and the Power Board is not limited by the capacitance of the air. Expected heat flow in the High Temperature Ambient zone is approximately 2 kW, but may be as large as 3 kW. A limitation of the thermal chamber means that the High Temperature Ambient zone cannot exceed 200 °C and still be actively controlled [3]. However, the temperature of the fin stock used for cooling the Power Bridge will be notably higher due the heat flow from the power electronics. This zone is not expected to represent a limit on the bidirectional test temperature.

4.3. Power Bridge

The zone referred to as the Power Bridge is an assembly which contains the 3-Phase electrical bridge assemblies, diodes, and inductors. With expected power dissipation levels just below 475 W per bridge, at 96% system efficiency, the total surface heat flux exceeds 12 W/cm^2, and heat flux levels for the powered components approach 50 W/cm^2. Nevertheless, as they are self-heating, powered component must be robust for extreme thermal environments.

A breakdown of the maximum operational temperatures for components, materials, and interconnects used in the Power Bridge are shown in Table 3. Based upon program development testing, the use of silicon-carbide diodes and super junction transistors [4] increases the operational temperature range significantly above the operational limits of commercially available silicon components [5]. Similarly, reconfiguring the design to move inductors out of the Power Bridge and onto the Power Board also improves the thermal capability of the Power Bridge. Finally, eutectic gold-tin solder (Au80Sn20) can be used for most interconnections within the Power Bridge. Although the eutectic temperature for this solder is only 280 °C, the rework and operational limits can be much higher. After application to gold leaded components the reflow temperature of eutectic gold-tin solder increases to 320 °C which is sufficient to provide margin over expected Power Bridge temperatures [6]. Nevertheless, limited brazing and wire bonding will be used at locations such as the threaded inserts where high strength is required.

Table 3: Temperature Limits within the Power Bridge

Description	Max Temp (ºC)
Interconnection	
Eutectic gold-tin solder	320
Components	
SiC Diodes	250
SiC Super Junction Transistors	250
Temperature Sensors	280
Inductors	200

4.4. Power Board

The Power Board is the most complex portion of the bidirectional converter. In addition to interfacing directly with the Power Bridge assemblies, the Power Board will contain AC/DC current sensors, electrical filtering, link capacitors, auxiliary power supplies, inductors, and gate drivers. Cooling of the Power Board will occur via a small amount of forced convection produced by the thermal chamber control system and fan/blower. Depending upon the assumed geometry and the temperature differential, forced convection of air will result in a heat transfer coefficient between 25 and 250 W/m^{2}K [2]. Therefore, the convective heat transfer coefficient for the Power Board to High Temperature Ambient zone is assumed to be 25 W/m^{2}K. The high temperatures used within the thermal chamber and the complex geometries created by components mounted on the PWA justify using a heat convection coefficient value on the lowest end of the range defined for forced convection.

Component technology for the devices mounted on the Power Board varies greatly. Included in Table 4 is the breakdown of maximum operational temperatures for components, materials, and interconnects used in the Power Board. Operational temperature ranges for capacitors [7], resistors [9], thermal sensors, and laminates [10] can vary by several hundred degrees. However, the temperature capability of passive devices, such as inductors and other magnetics, are limited to 150°C [5] which can be increased slightly by modifying the materials and manufacturing processes used in assembly. Furthermore, by carefully laying out the board with these less capable components placed further from the Power Bridge connections and powered devices the temperature capability of the Power Board can be improved. Furthermore, limited cooling technologies, such as heat sinks, may be implemented to locally decrease device temperatures where local heating or hot spots lead to inadequate operational temperature potential. Finally, despite a high temperature tin-silver solder having an apparently acceptable temperature limit for the Power Bridge [11], local heating and hot spots make gold-tin solder preferable.

Table 4: Temperature Limits within the Power Board

Description	Max Temp (ºC)
Board Material	
Glass reinforced hydrocarbon/ceramic laminates	280
Interconnection	
Eutectic gold-tin solder	320
High temperature tin-silver solder	221
Components	
Inductors & other magnetics	200
Capacitors	200
Temperature Sensors	280
CAN Driver	210
Gate Driver	230
Logic Integrated Circuits	225

4.5. Controller Board

Signal conditioning circuitry, power supplies, voltage sensors, connectors, and integrated circuits will all be mounted on the Controller Board. Limited temperature capability of the intricate devices used on the Controller Board makes this zone the greatest thermal challenge. A temperature controlled enclosure isolates the Control Board from the rest of the system to keep the device temperatures below 80ºC. The enclosure does have access for the Control Cable between the Control and Power Boards. The control of temperature inside this box has been chosen to be a two-phase device using heat pipe technology.

Heat pipe technology consists of the transport of thermal power from a hot dissipating point to a remote cold source through the latent heat characteristic of a working fluid. As classical heat pipes would fail to operate in very severe conditions and on geometrically complex solution, LHP technology allows overcoming these limiting aspects thanks to the use of very fine and efficient porous media as capillary pump ([12], [13]). This operating principle is described briefly on Figure 5.

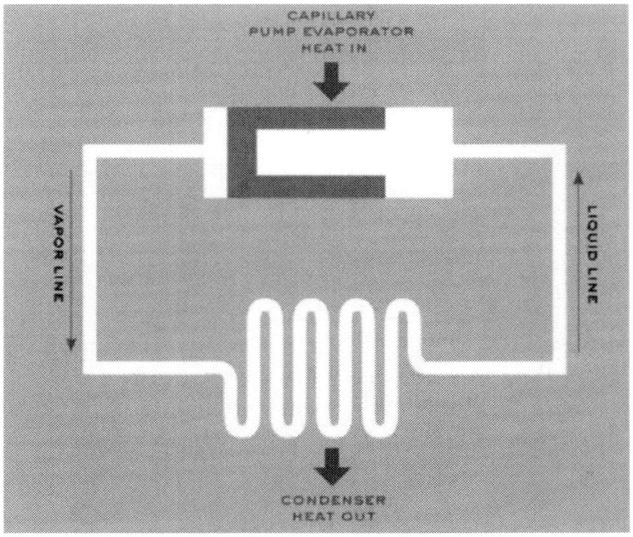

Figure 5: LHP operating principle

Multiple motivations raise the choice of heat pipe technology to cool down the Control Board. First of all, an efficient cooling solution is needed to maintain the insulated box temperature under 80°C according to available cold sources (ambient air at lab condition or cold plate). A heat pipe reduces to its lowest value the thermal resistance between the cold source and the Control Board, reducing at the minimum the Board temperature. In addition, this lab test set-up shall demonstrate the ability of heat pipe technology to manage severe ambient condition which could disturb, significantly or not, the operating of the fluid loop. Finally the two-phase technology offers a passive way to bring the heat to the cold source and so improves reliability and reduces the operating costs (no active pump or fan).

The greatest challenge lie in dealing with all possible parasitic heat fluxes which should disturb the insulated box thermal balance: the convective and radiative heat flux coming from the High ambient temperature zone, the heat flux coming from connection wires and the parasitic heat on the path of the heat pipe through the high ambient zone. Finally, the sum of all parasitic heat fluxes will be largely greater than the heat to extract from the Control Board. The insulation of critical parts of the two-phase device as well as the whole temperature controlled enclosure is in development to propose a solution based on a fine thermal analysis of the system composed of the LHP and the temperature controlled box.

5. Conclusion

Using the thermal zoning technique, the Control Board was identified as the initial thermal challenge on the 50 kW bidirectional converter under design. The issue of very low temperature capability for components mounted on the Control Board via the concept of a LHP coupled to the Low Temperature Ambient zone. The next weak link becomes the Power Board. Particular areas of concern for the Power Board are passive components such as magnetics and capacitors.

As shown in Table 5, based upon the assumed 96% efficiency and the limiting temperatures of each zone, the demonstration test can be completed at 225°C provided the High Temperature Ambient is set at 150°C. On the other hand, the large power dissipation into the Fin Stock causes the Power Bridge to see an effective cold source temperature of approximately 209°C. These parameters achieve the stated demonstration goal. Unfortunately, demonstrating functionality at higher cold source temperatures will place components on the Power Bridge at risk.

Table 5: Resultant Temperatures by Zone for 96% Efficiency

Zone	Power (W)	Temp (ºC)
High Temperature Ambient (Air)	0	150.0
High Temperature Ambient (Finstock)	0	209.2
Low Temperature Ambient	0	25.0
Control Board	10	80.0
Power Bridge	1890	221.2
Power Board	100	180.2

A bidirectional converter with an overall efficiency at 94%, but maintaining the 100 W Power Board dissipation level will increase the Fin Stock temperature without significantly increasing the Power Board temperature. As a result, when the High Temperature Ambient air is at 150°C components mounted on the Power Bridge will be exposed to mounting surface temperatures near 250°C. The Power Board temperature in this configuration will remain between 185 to 220°C depending upon localized heating and hot spots. These results are shown in Table 6.

Table 6: Resultant Temperatures by Zone for 94% Efficiency

Zone	Power (W)	Temp (ºC)
High Temperature Ambient (Air)	0	150.0
High Temperature Ambient (Finstock)	0	240.6
Low Temperature Ambient	0	25.0
Control Board	10	80.0
Power Bridge	2890	258.8
Power Board	100	185.3

Current electrical component technology makes it impossible for the entire bidirectional converter to operate with a 200 to 250°C temperature cold source. However, it is difficult to predict the weakest link in the system. Thermal zoning is a straightforward technique which involves simple thermal analyses to determine variable temperature limits, design requirements, and cooling techniques for an electrical system. The technique prioritizes problems for solutions and may lead to such counter intuitive results as less efficient Power Bridges are the most effective path forward for a technology demonstrator.

References

[1] Charles A. Harper, *Handbook of Electronic Packaging* (San Francisco: McGraw-Hill 1969)
[2] Frank P. Incropera and David P. DeWitt, *Fundamentals of Heat and Mass Transfer 4th Edition* (New York: John Wiley & Sons 1996)
[3] TestEquity, *Specification TestEquity 1007C Temperature Chamber*, http://www.testequity.com/products/598/
[4] Microsemi Corporation, *Datasheet Zero Recovery Silicon Carbide Schottky Diode APT20SCD120BHB* (USA: 2013)
[5] Microsemi Corporation, *Datasheet ThinkeyTM Silicon Schottky Diode LDS-0297* (Lawrence: 2013)
[6] Indium Corporation, *Application Note Gold Tin – The Unique Eutectic Solder Alloy, Form No. 98051 R1* (USA: Unknown)
[7] Vishay Intertechnology, Inc. *Interactive data book VSE-DB0059-1201e* (Yankton: 2011)
[8] Presidio Components, Inc. *Ceramic Capacitors Catalog 3500 Rev. L* (San Diego: 2014)
[9] Kemet Corporation, *Datasheet High Temperature 200°C, C0G Dielectric, 10 – 200 VDC c1001_C0G_SMD* (Greenville: 2014)
[10] Rogers Corporation, *Datasheet RO4000® Series High Frequency Circuit Materials* (Chandler: 2013)
[11] Kester, *Alloy Temperature Chart* (Itasca: 2010)
[12] Maidanik J F Vershinin S V Kholodov V F and Dolgirev J E 1985 Heat Transfer Apparatus U.S. Patent No. 4515209
[13] Stenger F J 1966 Experimental Feasibility Study of Water-Filled Capillary-Pumped Heat Transfer Loops *NASA Technical Memorandum* NASA TM-X-1310

Journal of Physics: Conference Series **525** (2014) 012002 doi:10.1088/1742-6596/525/1/012002

Low Reynolds turbulence model CFD simulation for complex electronic system: an industrial point of view

M.Giannuzzi

Prompt Engineering srl - Via Livorno 60, 10144 Torino, IT

E-mail: michele.giannuzzi@prompt-engineering.com

Abstract. In electronic systems the presence of bluff bodies, sharp corners and bends are the cause of flow separation and large recirculation bubbles. Since the recirculation vortices develop they encapsulate the heat from an electronic component becoming one of the major contributors of malfunction. Going in depth in this, some numerical simulations of conjugate heat transfer for a heat wall-mounted cube have been performed using the commercial CFD code scSTREAM V11 by Software Cradle Co, Ltd. It is well known that the reliability of CFD analysis depends heavily on the turbulent model employed together with the wall functions implemented. The three low-Reynolds $k-\varepsilon$ turbulent models developed by Abe-Nagano-Kondoh have been validated against experimental data consisting mainly of velocity profiles and surface temperature distributions provided in literature. The performed validation shows a satisfactory agreement between the measured and simulated data. The turbulent model chosen is then used for the CFD simulation of a complex electronic system.

1. Introduction

Nowadays the electronics industry sees a constant and ever more stringent technological progress. The adoption of new packaging and miniaturization processes increases the problems associated with the thermal load dissipation. Due to the presence of multi-property materials (such as printed circuit boards and integrated circuits), multiple heat sources and complicated internal geometries, the electronics systems are characterized by complex heat transfer behavior.

In [1] was addressed the importance of combining experimental and numerical approaches to understand the thermal behaviour of complex electronic devices. Testing campaigns were meant to give all the technical information necessary to reproduce a real system by means of numerical model. In this paper, time was spent analyzing hypotheses and guidelines modeling useful to simplify the virtual model maintaining high standard of result's reliability.

Under this framework when commercial numerical software is used, the unquestionable importance of a validation step for the numerical model is needed in order to assure the effectiveness and reliability of the results carried-out. Although CFD simulations are accepted as a design tool, they are still subject of modeling errors. It is well known that the reliability of CFD analysis depends heavily on the turbulent model employed together with the wall function implemented. In this communication, time was spent in choosing the most suitable turbulence model useful for electronic cooling simulation. To this aim, a virtual test campaign over a channel with a wall-mounted cube was performed. The turbulence models used in this study

Content from this work may be used under the terms of the Creative Commons Attribution 3.0 licence. Any further distribution of this work must maintain attribution to the author(s) and the title of the work, journal citation and DOI.
Published under licence by IOP Publishing Ltd

have been validated with respect to the experimental data consisting mainly of velocity profiles and surface temperature distributions provided in literature [2; 3].

The commercial software scSTREAM V11 Thermo-fluid Analysis System by ©Software Cradle Co. Ltd. which is a structured/Cartesian mesh Finite Volume Method solver, has been used.

2. Which are the best turbulence models useful for electronic cooling simulation?

In electronic cooling systems off-design geometries such as bluff bodies, sharp corners and bends are mainly characterized by adverse pressure gradients for which the flow separates and large recirculation bubbles are generated [2]. Most electronics flows in the near-wall regions are characterized by strong viscous stress and wall blocking effects. Such separation and reattachment zones have a strong impact in temperature prediction profiles. By means of a literature overview, one can see that the high-Reynolds-number models are the most used in the field of electronic cooling simulations. However it is known that these models are unable to handle the near-wall region directly and they usually over-predict the heat transfer coefficient getting a lower surface temperature profile. From this, the need arises to exploit transitional and low-Reynolds number turbulent models. Moreover the importance of developing the technology to control turbulence and promote heat transfer is not a niche anymore, and in the last decades great effort on modeling near-wall turbulence has been spent. In this study, one of the latest low-Reynolds $k-\varepsilon$ model is used, *i.e.* the one developed by Abe-Nagano-Kondoh (AKN) [4; 5]. In scSTREAM V11 [6] several turbulence models can be found such as the standard $k-\varepsilon$, the RNG $k-\varepsilon$, the MP $k-\varepsilon$, the linear low-Reynolds-number $k-\varepsilon$, the non-linear low-Reynolds-number $k-\varepsilon$, the temperature field two-equation models (in which the air properties are temperature dependent) and the Large Eddy Simulation models. Since the interest is for electronic cooling problems only low-Re number models have been considered.

2.1. *Linear low-Reynolds Abe-Nagano-Kondoh model*

In Iaccarino [7] a low-Re model with the Launder and Sharma damping function [8] has been used for solving a steady flow in asymmetric, two-dimensional diffuser. The numerical predictions are in poor agreement with the experimental data since the model fails to respond correctly to the adverse pressure gradient and misses the separation and the recirculation completely.

By contrast, the linear AKN model allows accurate analysis of a wide range of flows, with Reynolds numbers ranging from low to high. It is able to achieve a significant improvement on conventional turbulence models in terms of accuracy of prediction of separation and reattachment flows. Moreover it is also valid for the analysis of the transition from laminar to turbulent flow, or the reverse transition from turbulent to laminar flow. The basic equations used in this low-Reynolds-number turbulence model are basically the same of those used for high-Reynolds-number turbulence models. The strong viscous stresses and the damping effects of turbulence are considered by adding some numerical terms: the most important is the damping function f_μ.

$$\mu_t = \rho C_\mu f_\mu \frac{k^2}{\varepsilon} \quad ; \quad f_\mu = \left(1 - e^{-\frac{y^+}{14}}\right)^2 \left[1 + \frac{5}{R_t^{3/4}} e^{-\left(\frac{R_t}{200}\right)^2}\right]$$

2.2. *Non-linear low-Reynolds AKN model*

In Klebanoff [9] has been demonstrated the importance of the region near the wall where the most turbulent energy is generated and the inadequacy of the concept of local isotropy. One of the limitations with Eddy Viscosity models is that they use an isotropic value which may not be appropriate and hence can increase the diffusion of the results obtained. This means that linear turbulence models are completely unsuited for the analysis of flow in which secondary eddies arise. By contrast, the non-linear turbulence model can be applied to the phenomenon of anisotropic turbulence by deriving the coefficient of non-linear eddy viscosity from the spatial gradient of mean velocity. More information about the mathematics of this model can be found in [4; 6].

2.3. *Two-equation heat transfer AKN model*

The two-equation heat transfer AKN model (known as thermal AKN model) was proposed by Abe et al. [5]. It is based on the low-Reynolds-number AKN model but the prediction accuracy for the turbulent heat transfer coefficients is now improved. To get this, eddy diffusivity for heat $\alpha_t = \frac{K_t}{\rho C_p}$ is represented by the temperature variance $\bar{t}^2$ and its dissipation rate ε_t instead of a constant turbulent Prandtl number and the eddy viscosity. In [10] it is pointed out that the turbulent Prandtl number, which is widely used as a conventional approach to solve turbulent heat transfer problems, is not constant at all. In general, the turbulent Prandtl number strongly depends on the distance from walls. Furthermore, it is strongly influenced not only by a kind of fluid, but also by the existence of secondary flow, pressure gradient, buoyancy and even by thermal boundary conditions on heated surface. The assumption of the constant turbulent Prandtl number is valid till that both the velocity and the thermal boundary layers have the same thickness and are developing downstream in the same way. However, one cannot always expect to find this similarity in the turbulent flows. More information about the mathematics of this model can be found in [5; 6].

2.4. *Analytical wall-function (AWF) for low-Reynolds number model*

It is known that low-Reynolds-number turbulence model always requires strict no-slip boundary conditions. Indeed a strong clustering of the grid points at the walls should be used so that the y^+ of the first grid point away from the wall is almost everywhere less than 1. However, it is quite difficult to assure this condition over the whole region due to the restriction of computer resources. Hence, the low-Reynolds-number turbulence model used in scSTREAM adopts a special boundary condition for which the strict no-slip condition is always imposed whenever the mesh size adjacent to the wall is satisfied by the condition $y^+ \leq 1$, while the conventional wall-function approach is applied to the region at $y^+ \geq 11.6 = y_v^+$. In the in-between region ($1 \leq y^+ \leq y_v^+$), both boundary conditions are connected with each other via the following formulations:

$$k = \left(\frac{y^+}{y_v^+}\right)^2 \frac{u^{*2}}{\sqrt{C_\mu}} \qquad y^+ \leq y_c^+$$

$$\varepsilon = \frac{2\,\nu\,k}{y^2} + \frac{y}{y_v}\left(\frac{u^{*3}}{k\;y_v} - \frac{2\,\nu\,k}{y_v^2}\right) \qquad y^+ \leq y_c^+$$

This wall boundary condition, named hybrid wall boundary (HWB) condition, allows the low-Reynolds-number turbulence model being used for simulating wall turbulent flows with coarse meshes near the wall. For industrial type applications this helps since it is computing saving (few mesh grids are needed near the wall) but it has also some drawbacks: firstly the y_v^+ is not always constant but it is strongly dependent by the flow nature; secondarily for some turbulent flow simulations this hybrid wall function is not adequate. For instance when separation and reattachment phenomena occur the damping function f_μ become zero since the $y^+ = \frac{u_\varepsilon y}{\nu}$ always becomes zero due to a zero friction velocity.

To overcome all this Craft *et al.*[11] developed an analytical wall-function (AWF) in a more general level than the log-law based schemes. In the AWF, the wall shear stress and heat flux are obtained through the analytical solution of simplified near-wall versions of the transport equation for the wall-parallel momentum and temperature, accounting for effect in smooth and rough wall turbulent flow [11; 12]. In particular, rather than using a conventional damping function it has been thought to shift the turbulent origin from the wall to the edge of the viscous sublayer. Hence the turbulent viscosity is taken as zero within the viscous layer, while for $y \geq y_v$ the same is assumed to increase linearly with the distance from the edge of sublayer:

$$\mu_t = max\{0, \alpha\,\mu\,(y^* - y_v^*)\}$$

where $y^* = \frac{y\sqrt{k}}{\nu}$ is the new non-dimensional distance, while $y_v^* = \frac{y_v\sqrt{k}}{\nu}$ is the edge of the viscous sublayer. The new non-dimensional wall-normal distance is quite effective in flow with separation and reattaching points because of the non-vanishing turbulence kinetic energy k at those points.

3. Numerical Simulation

3.1. Turbulence model validation

In this section the experimental results obtained by Meinders [2] are compared against numerical predictions. Benchmark criteria were based on the prediction of surface temperature profiles on some specified path and velocity profiles at different heights. The test channel is similar to the one proposed in [3]. Since the test layout allows a symmetry boundary condition along the central in-plane, half of the domain was considered. The whole dimensions are 23H x 3.4H x 4H in the streamwise (x), normal to the channel (y) and spanwise (z) directions, where H is the cube size (H = 15 [mm] as pointed out in [2]). All the computations have been performed with an Intel Xeon E5645 2.4GHz machine (2 processors and 12 cores).

The simulation consists of two sub-domains: firstly a channel domain used as driver region (in blue color) and secondarily a wall-mounted cube domain which is inside the first one (in black color). The channel domain is solved to get the transition from laminar to turbulent boundary layer. As addressed in [3] in cases with flow separation, free shear layers, laminar-turbulent transitional flows, the downstream flow is highly dependent on the condition specified at the inlet. On the contrary of what is performed in [1], where the heat conjunction problem of heated wall-mounted cube was resolved in a single analysis domain with a turbulent flow fully developed, in the present communication the resulting flow consists of developing turbulent boundary layer both on the floor and on the opposite wall. Although this flow is qualitatively comparable to the fully developed channel one, a noticeable difference exists in local heat transfer from the cube [2]. This might be seen comparing the surface temperature profiles obtained in this study with the ones shown in [1]. Since the resulting profiles are in accordance with the experiments they are used as inlet conditions for the simulation of the wall-mounted cube by means of the so called zooming function. In scSTREAM [6], the zooming function enables to apply an analysis result to initial conditions and/or boundary conditions in another analysis. An analysis result of upstream flow field can be used for inlet boundary condition of another analysis. Moreover zooming function enables to link any field parameters of different mesh sizes. So it can link global large size analysis and local small size analysis.

Channel domain: Turbulent characterization Channel flow simulations were carried out using the low-Reynolds turbulent models before introduced and the resulting flow fields are used as inlet conditions for the simulation of the wall-mounted cube.

The test layout allows a symmetry boundary condition along the central in-plane, half of the domain was considered. A non-conformal mesh was used with a computational grid set of 93 x 115 x 41 points in the x, y and z directions, respectively. A total number of 438,495 cells were used. A cluster of grid points at the walls has been used so that the y^+ is often less than 1. Symmetry, wall friction and free-stream conditions were applied at the domain boundaries. The outlet zone is set as fixed static pressure at 0 [Pa]. Since the inlet velocity is 4.47 [m/s] and the

Journal of Physics: Conference Series **525** (2014) 012002 doi:10.1088/1742-6596/525/1/012002

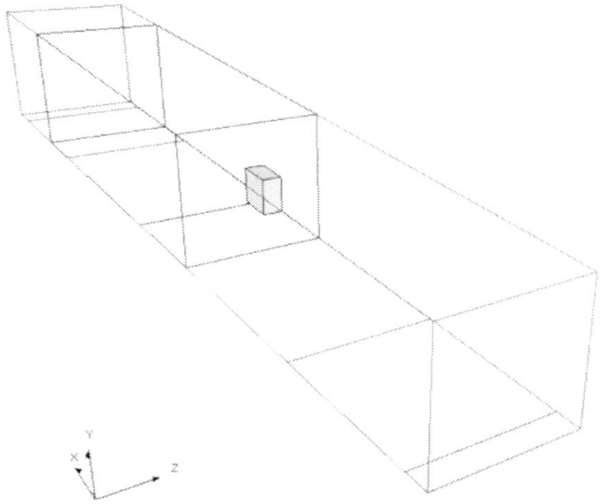

Figure 1: Whole computational domain

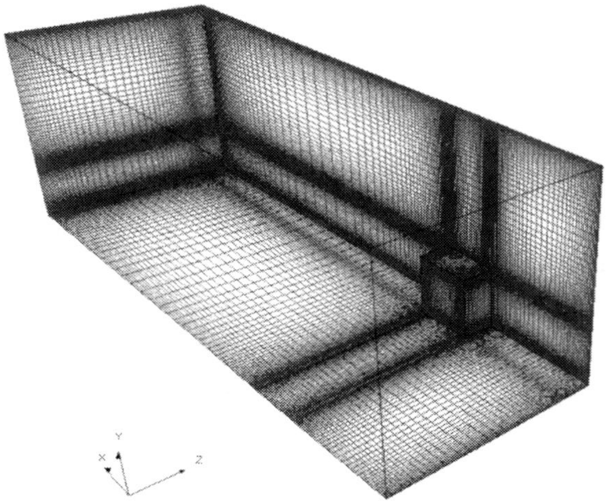

Figure 2: Wall-mounted cube domain mesh, 3D view

reference length is H, the Reynolds number is approximately 4440 [2]. The convergence level reached by the residual is at least 1e-6.

Wall-mounted cube domain: Validation study Since the origin of the axis system, the inlet and the outlet of computational domain are located at $x = -2|; H$ and $x = 4\ H$, respectively, while the cube is located between $0 \leq x \leq H$, $0 \leq y \leq H$, $-0.5\, H \leq z \leq 0$. The computational grid is set in 154 x 123 x 53 grid points in the x, y and z directions, respectively. A non-conformal mesh was used in the streamwise and wall-normal directions with grid points clustering near the channel and cube walls. The total number of cells is about one million. In comparison with the channel case the number of grid points has been doubled in a volume which is half of the whole one. The mesh of the wall-mounted cube domain is shown in 2.

The test channel base plate is constructed by phenol-formaldehyde which is 10 $[mm]$ in thickness and has a thermal conductivity from 0.33 $[\frac{W}{mK}]$. The structure of the cube can be decomposed into two materials: the core of the cube measuring 12 $[mm]$ is constructed from copper, with a thermal conductivity of 398 $[\frac{W}{mK}]$, which is kept at constant temperature of 75 $[C]$. The second material is an epoxy resin, with a thermal conductivity of 0.24 $[\frac{W}{mK}]$, which encapsulates the copper and has an uniform thickness of 1.5 $[mm]$. The ambient temperature was set at 21 $[C]$. In table 1 are addressed the numerical parameters used for the simulations. The convergence level reached by the residual is at least 1e-7.

Table 1: Numerical parameters used for the simulations. Legend: U,V,W= Mean Velocity Components, T=temperature, TQ=Turbulence Quantities,UW=upwind(1st order), QK=QUICK(2nd order)

	Spatial Discretization			Pressure Correction	Under-Relaxation			
	U,V,W	T	TQ		U,V,W	P	T	TQ
linear AKN	UW	QK	UW	SIMPLEC	0.9	1	0.9	0.6
nonlinear AKN	QK	QK	UW	SIMPLEC	0.9	1	0.9	0.2
thermal AKN	QK	QK	UW	SIMPLEC	0.6	1	0.9	0.6

Comparison Surface temperature and velocity fields are presented in Figures 3, 5 and 6. The experimental data used can be found in Meinders *et al.* [2].

In figure 3 the streamwise velocity profiles are shown at a location of $x/H = 18.7$ downstream the inlet at the centreline of the channel (i.e. at location of $x/H = 6.7$ far from the cube origin). It is evident that all the turbulence models match quite well the profile in the bulk flow, in the meanwhile all the models severely under-predict the wake region of the cube since they seem to still recover after the recirculation region. Similar results can be found in [13]. In figure 4 streamline plots in xy-plane and xz-plane are shown and compared with a DNS simulation and experimental visualization, respectively. It can be noted that all the features addressed in [2] are present: the horseshoe vortex at the base of front face and its saddle point, the arch vortex in the wake recirculation region and the vortex tube at the side. Among the three turbulence models presented the thermal AKN gives the best results. Indeed the locations of all the vortical patterns are in good accordance with respect to the experiment and the DNS visualizations. Analysing the velocity profiles at z/H=0 in figure 3, along path upstream and downstream the cube, again one can recognise that the thermal AKN model qualitatively represents better the experimental data, in particular it is able to individuate the vortex position upstream and downstream the cube by means the minimum of its curves. The nonlinear is the farthest from the experiments. It can be said also that for all the models it is evident that the shear layer separated at the top leading edge reattaches more farther than it should do.

Temperature experimental data were collected with an Infrared Thermography (IR) technique with accuracy within 0.4 [C]. More inaccuracies might be found at the corners of the cube, due to the restriction of such a data retrieval system. The surface temperature profiles are shown in figures 5 and 6. Looking at figure 5, for the leeward face of the cube (AB) it can be said that the AKN $k - \varepsilon$ turbulence model produces a good comparison against the experimental data, suggesting that the wake flow vortex resident in this region is accurately predicted. Also the nonlinear one shows good performances, while the thermal model over-estimates the range. Looking at the top face (BC) the linear AKN is compliant with the measured peak of the bound recirculation region while it fails to catch the thermal trend before it. The thermal AKN model behaves similarly to the linear AKN. The nonlinear over-estimates the whole range implying much more diffusion than the experiments. For the windward face (CD) since an over estimation of the turbulent kinetic energy generated near the stagnation point the numerical results under-estimate the measured values, so the thermal trend is predicted only in qualitative way by all the models.

The same comparison can be made over the in-plane path, see figure 6. For the lateral faces (AB) and (CD) linear and thermal AKN catch the peak where the recirculation happens, while the nonlinear completely over-estimates the data. For the windward face (BC) all the three models under-predict the experimental data. For the leeward face (DA) the linear AKN matches well the thermal behavior.

According to the comparison before seen and since the thermal AKN model is expensive in terms of computational time, it can be stated that the linear AKN is thought being the best choice to be used in the electronic cooling simulation over a real test case.

3.2. Infotainment and Navigation Systems (INS) thermal problem

To test the turbulence model chosen a real test case is taken into account. As known, 3D numerical models involve very expensive computations, both for hardware and time needed for obtaining reliable results. In [1] time was spent in analyzing hypotheses able to simplify the virtual model, maintaining the same standard of result's reliability. The simplification of physical

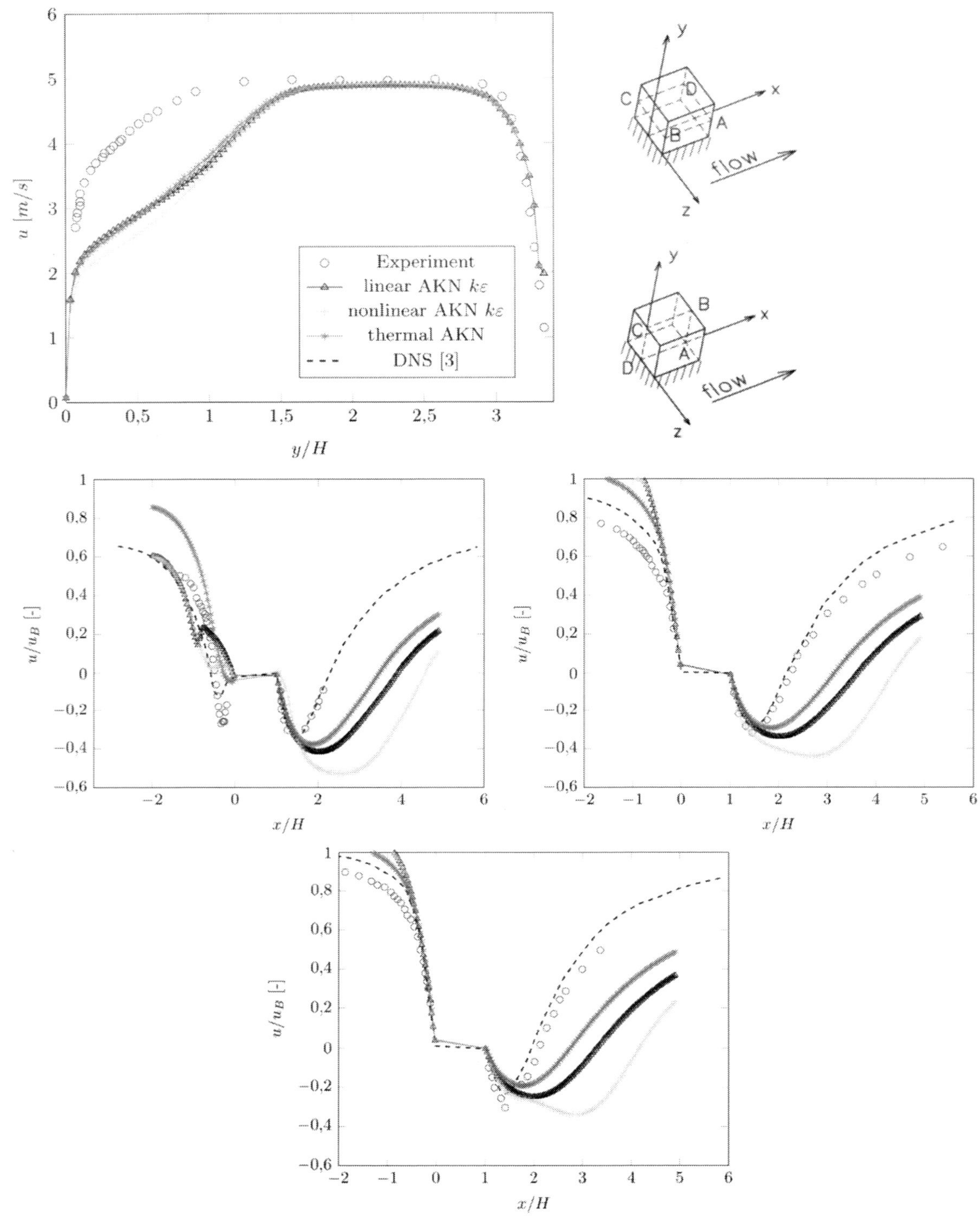

Figure 3: Streamwise velocity profiles at $x/H = 18.7$ (top-left). Cube's path ABCDA and ABCD for surface temperature profiles (top-right). Dimensionless streamwise velocity profiles (u_B=bulk velocity) at $z/H = 0$ for different heights: $y/H = 0.1$, $y/H = 0.3$ and $y/H = 0.5$

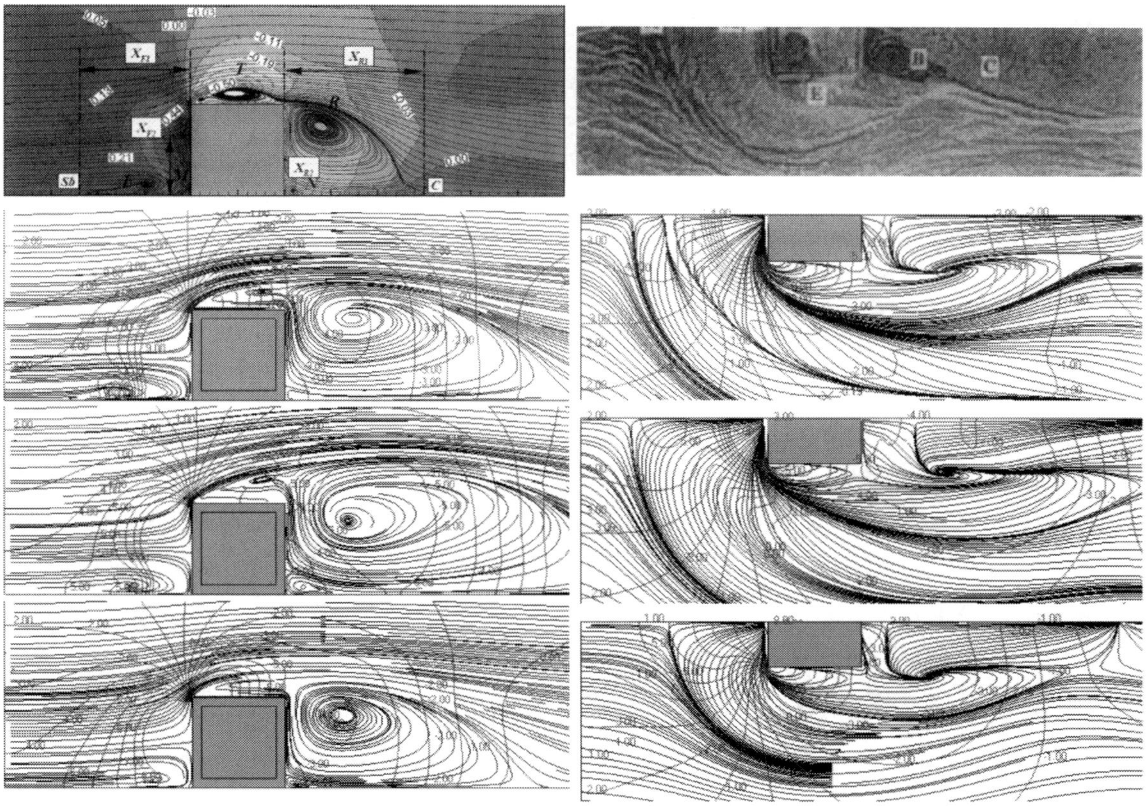

Figure 4: Visualizations of the surface flow and pressure distribution. from left to right and from top to bottom: DNS [3], Experiment [2], linear AKN, nonlinear AKN and Thermal AKN

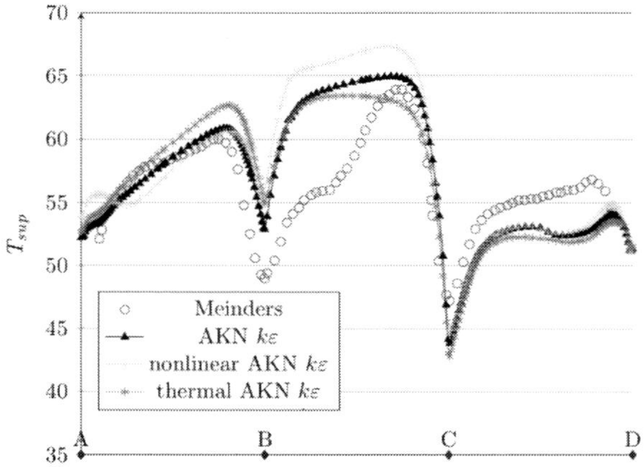

Figure 5: Surface temperature profile along path ABCD

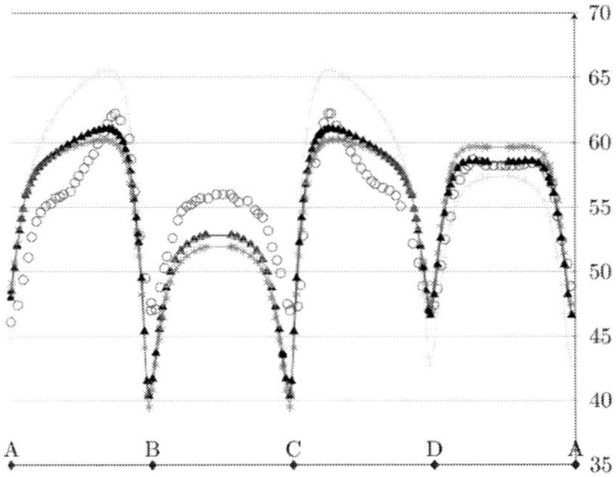

Figure 6: Surface temperature profile along path ABCDA

system allows drastically reducing degrees of freedom of the numerical model, so that memory space and computational time can be saved.

An infotainment and navigation system (INS) during operative conditions has been considered, see figure 7. The system mainly consists of three printed circuit boards: Main Board (MB), Daughter board (DB) and Power Board (PB). The electronics is all contained inside a metallic case, whose external dimensions are 195 x 165 x 53 mm. It presents several multimedia and connectivity devices such as Wi-Fi, Bluetooth, Cd-Rom etc. The electronics layout is suited to generate thermal interactions among components: it is an useful case to test very complex numerical models. The system presents an outlet fan working at 6200 RPM, with a pressure-flow rate curve from 0 to 5.75 $mmAq$ and flow rate from 0 to 0.28 m^3/min. During the idle condition the electronic system dissipates about 18 W, while during the operative test, with a sinusoidal waveform applied, it dissipates about 39 W (efficiency of the amplifier: 18%).

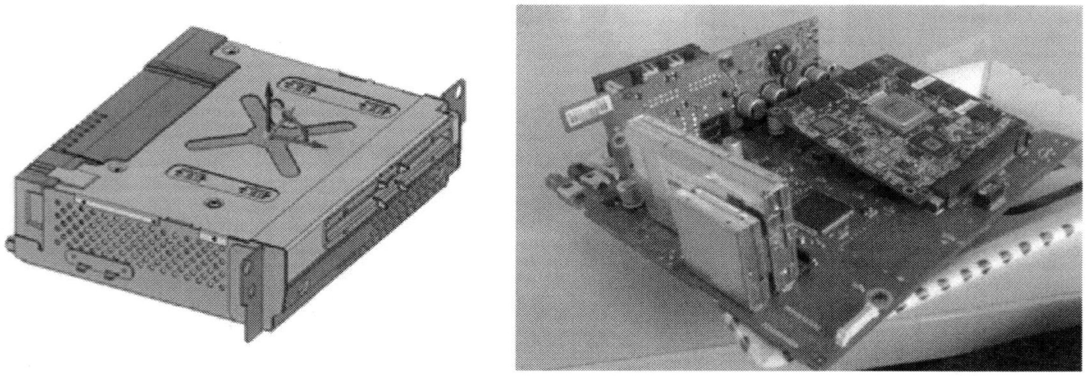

Figure 7: A real test case: infotainment and navigation system (INS)

Ambient temperature was set at 65 C. Since the INS was tested in a climatic chamber the boundary conditions imposed to the domain are: the friction and adiabatic wall for the bottom face, without introducing any errors the lateral faces were considered as free stream while a fixed total pressure was imposed at the top face.

Because of the absence of a characteristic length useful to describe the flow regime, the problem has been solved both in laminar and in turbulent regime. It is also reported the simulation with and without radiative heat exchange mode. Generally, when a forced ventilation is used, the radiative heat exchange might be considered negligible since the contribute in the heat exchange is minimal compared to the other two modes of heat transfer. In our case, however, the high velocity flow is confined only in a given area of the electronic device due to the design of some internal parts, so then all the remaining parts of the system also exchange in radiative way. This fact is well proven by the comparison between numerical and measured values, see table 2.

4. Conclusions

This study highlights the opportunity to exploit numerical approaches in order to simulate the thermo-fluid-dynamic behaviour of complex electronic devices. The study underlines the unquestionable importance of a validation step for the numerical model, strictly needed in order to assure effectiveness and reliability of the results carried-out. The accuracy of the numerical

Table 2: Prediction Discrepancy: measured vs. virtual temperatures

	DDR2	Daughter PCB	Heat spreader	Heat sink	Fan outlet air	Mother PBC
laminar flow + no-radiation	1.77%	1.36%	3.0%	3.94%	5.74%	5.14%
AKN model + radiation	1.13%	1.65%	0.1%	0.1%	2.49%	1.92%

models used is assessed against experimental measurements. The detected temperature values are well in accordance with experimental data.

References

[1] Giannuzzi M and Selci A 2013 Thermal predictions for an infotainment and navigation systems (ins). an experimental vs. numerical correlation *Proceedings of 31th UIT Heat Transfer Conference, ISBN 97888-6493-017-6* pp 329–338

[2] Meinders E R, Hanjalic K and Martinuzzi R J 1999 *Journal of Heat Transfer* **121** 564–573

[3] Yakhot A, Liu H and Nikitin N 2006 *International Journal of Heat and Fluid Flow* **27** 994 – 1009 ISSN 0142-727X

[4] Abe K, Kondoh T and Nagano Y 1994 *International Journal of Heat and Mass Transfer* **37** 139 – 151 ISSN 0017-9310

[5] Abe K, Kondoh T and Nagano Y 1995 *International Journal of Heat and Mass Transfer* **38** 1467 – 1481 ISSN 0017-9310

[6] Software Cradle Co., Osaka 530-0001 Japan 2011 *scSTREAM, Thermofluid Analysis System with Structured Mesh Generator*

[7] Iaccarino G 2001 *Journal of Fluids Engineering* **123** 819 ISSN 00982202

[8] Launder B and Sharma B 1974 *Letters in Heat and Mass Transfer* **1** 131 – 137 ISSN 0094-4548

[9] Klebanoff P 1955 Characteristics of turbulence in a boundary layer with zero pressure gradient Tech. rep. NASA

[10] Kays W M and Crawford M E 1993 *Convective Heat and Mass Transfer (Third Edition)* (McGraw-Hill) chap Chapter 13

[11] Craft T, Gerasimov A, Iacovides H and Launder B 2002 *International Journal of Heat and Fluid Flow* **23** 148–160 ISSN 0142727X

[12] Suga K, Craft T and Iacovides H 2006 *International Journal of Heat and Fluid Flow* **27** 852 – 866 ISSN 0142-727X special issue of the 6th International Symposium on Engineering Turbulence Modelling and Measurements

[13] Dhinsa K K, Bailey C J and Pericleous K A 2004 Turbulence modelling and its impact on cfd predictions for cooling of electronic components *In Thermomechanical Phenomena in Electronic Systems -Proceedings of the Ninth Intersociety Conference*

Journal of Physics: Conference Series **525** (2014) 012003 doi:10.1088/1742-6596/525/1/012003

Numerical Analysis of a Cold Plate for FM Radio Power Amplifiers

M Sabin[1], **S Piva**[2]

ENDIF, ENgineering Department In Ferrara, Università di Ferrara, via Saragat n.1, 44122 Ferrara

E-mail: stefano.piva@unife.it

Abstract. The results of a numerical investigation of heat and fluid flow in a liquid cold plate for FM radio power amplifiers are presented. The objective is to verify, by using a commercial CFD code, the performance of a blister cold plate designed to dissipate the heat generated by a known set of electronic components, in order to limit their maximum temperature during operations. Since in a blister cold-plate mainly the cover is thermally active, the cold-plate is simplified and lightened by using plastics in the base plate. A 3-D conjugate CFD approach, where thermal and fluid flow analyses are combined, is followed. Several design options for the cold plate are examined and the validity of the full 3-D CFD approach in the dimensioning of the cooling systems of electronic equipment is demonstrated.

1. Introduction

In the proper design of an electronic device it is necessary to ensure an adequate cooling of its active and passive components. Extensive reviews of the available techniques of industrial interest for electronics cooling are those of Çengel [1] and Kraus *et al.* [2]. Air cooling is the simplest and most common technique. It has several positive aspects, such as low costs and simplicity of implementation; on the other hand, it may show difficulties in dissipating high heat fluxes and often is the source of unacceptable levels of noise [3].

The active electronic components currently used can generate heat fluxes of up to 3 MW/m^2; for these high specific powers air cooling may be insufficient [3]. In this case, a cold plate is a practical solution frequently used (Incropera [4]). A cold plate consists of a plate, usually in aluminum, on which are fixed some electronic components and within which one or more hydraulic circuits are obtained for their cooling with water flowing in forced circulation. The design solutions are essentially two: "press-fitted" and "blistered". In a "press-fit" cold plate, the cooling circuits are manufactured with copper tubes pressed in special channels machined on the bottom surface of the plate. This technology is simple and cheap. However, the thermal resistance between tubes and plate could reduce the heat transfer. In a "blister" cold plate, the cooling channels are machined directly on the plate, which is then closed by a thin wall of high thermal conductivity (commonly aluminum). This technology allows better performance and greater design flexibility. On the other hand, the production costs are higher and therefore suitable for dedicated applications.

In the design of a cold plate, the main steps are: determination of number and layout of the cooling

---

[1] Actually in Altran Italy and detached to Maserati Spa.
[2] To whom any correspondence should be addressed.

Content from this work may be used under the terms of the Creative Commons Attribution 3.0 licence. Any further distribution of this work must maintain attribution to the author(s) and the title of the work, journal citation and DOI.
Published under licence by IOP Publishing Ltd

coils, diameter of the pipes and mass flow of refrigerant in the channels. In a preliminary stage, simplified one-dimensional calculations can be used. In the final design, such an approach seems to be not appropriate for the multidimensional effects of heat transfer, always present in these systems. These make the 3-D numerical approach the most appropriate strategy. Also, in a 3-D CFD approach, the analysis can be simplified, for instance by setting appropriate values of the heat transfer coefficient on the water side of the channels [5]. However, in this case, some flow details are lost.

In this paper, a 3-D conjugate CFD approach, where thermal and fluid flow analyses are combined, is preferred. The objective is to verify, by using a commercial CFD code, the performance of a blister cold plate designed to dissipate the heat generated by a known set of electronic components, in order to limit their maximum temperature during operations. Since in previous experiences [5], it was found that in a blister cold-plate mainly the cover is thermally active, it has been tried to simplify and lighten the cold-plate by using plastic in the base plate. The parameters taken into account in the discussion of the results are the thermal conductivity of the plate and the mass flow rate of refrigerant.

2. Problem statement

Figure 1 is a schematic representation of the cold plate of interest, consisting of a plastic base and of an aluminum cover housing the components to be cooled. In the base plate (21x302x372 mm) are machined cooling channels of square section (10x10 mm). The dimensions of the aluminum plate are 8x302x372 mm. The refrigerant is water. On the top of the cover plate are placed ten electronic components dissipating a known thermal power. In the model, these components are represented as blocks of copper (24.5x9.5x3 mm).

For this cold plate, a large number of simulations were performed to analyze the effect of different parameters. One of the tests is considered as the Reference test case. The thermophysical properties of the materials used in the Reference test case are reported in Table 1. Other significant parameters are:

- electric power dissipated by the electronic components: $P_{el} = 1465.17$ W;
- mass flow rate of the cooling water: $Q_m = 0.457$ kg/s;
- inlet temperature of the cooling water: $T_{in,water} = 25.8$ °C;
- ambient temperature: $T_\infty = 22.7$ °C;
- heat flow dissipated by the then electronic components and simulated as an uniform heat flux on their surface: $q = 629.5$ kW/m^2;
- thermal coupling between the cold plate and the environment, and modeled with a boundary condition of third type: $h_\infty = 10$ W/m^{2}K;
- to the contact thermal resistance due to the coupling between the cover plate and the electronic components, enhanced by a layer of thermal grease, was imposed a value from literature for metal-metal coupling (Moran et al.[7]): RTC = 0.7 10^{-5} m^{2}K/W.

Variants to the Reference case were analyzed to evaluate the effect of different parameters on the performance of the cold plate. In these variants, the following parameters were changed:

1. thermal conductivity of the base plate: a high value (60 W/mK), characteristic of an enhanced plastic added of a filler, and a low value (0.35 W/mK), characteristic of a standard plastic;
2. mass flow rate: 0.2285, 0.475 and 0.914 kg/s;
3. thermal conductivity of the electronic components: a high value typical of a component with prevailing copper (401 W/mK), and a low value, typical of a prevailing ceramic (30 W/mK).

3. Mathematical model

The numerical investigation is based on the following simplifying assumptions:

- incompressible Newtonian fluid;
- steady state;
- constant properties;
- negligible buoyancy forces;
- turbulent flow and k-ε model;
- negligible radiation heat transfer.

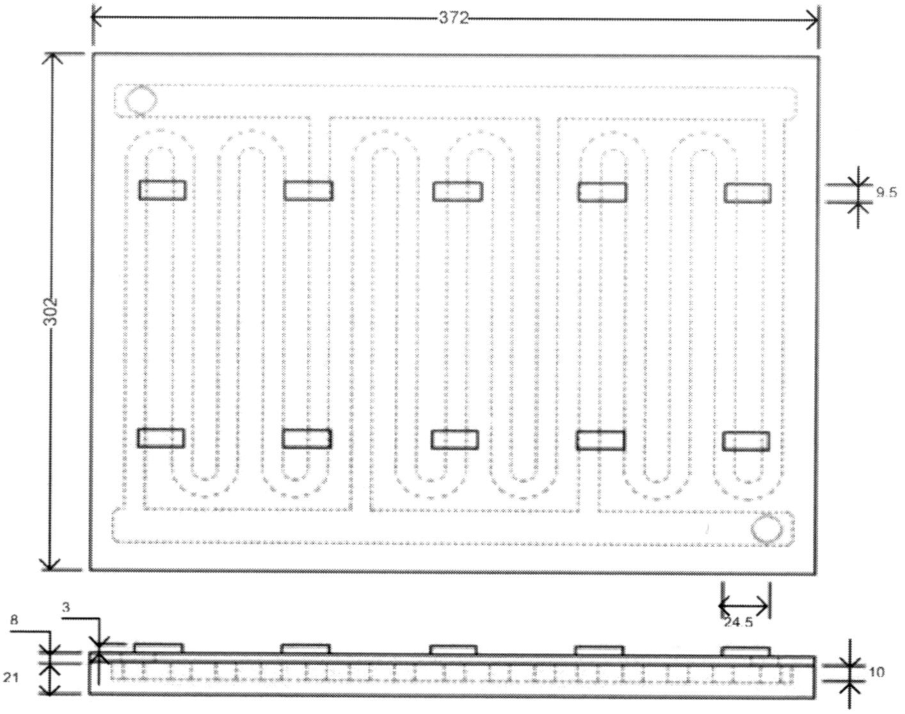

Figure 1. Sketch of the cold plate.

The equations to be solved are those assuring the conservation of mass, momentum and energy [6]:

$$\nabla\cdot(\rho\bar{u})=0 \tag{1}$$

$$\nabla\cdot(\rho\bar{u}\times\bar{u})=-\nabla p'+\nabla\cdot\left[\mu_{eff}\left(\nabla\bar{u}+\nabla\bar{u}^{t}\right)\right] \tag{2}$$

$$\nabla\cdot(\rho\bar{u}h_{tot})=\nabla\cdot\left[\left(\lambda+c\frac{\mu_t}{\sigma_H}\right)\nabla T\right] \tag{3}$$

Table 1. Reference Case: thermophysical properties.

Water		
ρ = 997 kg/m^3		c = 4181.7 J/kgK
λ = 0.6069 W/mK		μ= 8.899 10^{-4} Pa s
Plastic (base plate)		
ρ = 1100 kg/m^3	c= 2010 J/kgK	λ = 60 W/mK
Aluminium (cover plate)		
ρ = 2702 kg/m^3	c= 903 J/kgK	λ = 237 W/mK
Copper (electronic components)		
ρ = 8933 kg/m^3	c= 3.85 J/kgK	λ = 401 W/mK

where:

$$h_{tot} = h + \bar{u}^2/2 \tag{4}$$

$$p' = p + 2/3\rho k \tag{5}$$

$$\mu_{eff} = \mu + \mu_t \tag{6}$$

and the turbulent viscosity is calculated with a standard k-ε model [6].

The k-ε turbulence model is widely used in the literature because it is robust and efficient, and converging in an acceptable time. In this application, wall functions were specified near the walls. The technique of inflation was used with a scaling function such that the coordinate y^+ of the first grid point was always greater than 11.06:

$$y^+ = \max\left(11.06; y^+\right) \tag{7}$$

An advective High Resolution scheme was chosen for the equations of mass, momentum and energy conservation and for the turbulence equations.

The boundary conditions are:

- Adiabatic on lateral and bottom surfaces of the base plate, lateral surfaces of the cover plate and lateral surfaces of the electronic components;
- Boundary condition of the third kind on the upper surface of the cover plate:

$$-\lambda \left.\frac{\partial T}{\partial \bar{n}}\right|_w = U_\infty \left(T_w - T_\infty\right) \tag{8}$$

- Heat flow imposed on the upper surfaces of the electronic components;
- Mass flow rate and temperature assigned in the input section of the cooling channel;
- Thermal contact resistance in the interface between electronic components and cover plate.

For the generation of the computational grid the code, ANSYS Meshing was used. The grid is hybrid, composed of tetrahedron inside the domains and of five prismatic layers (inflated layers) near the walls, where the wall functions are implemented (Figure 2).

In Table 2, the geometrical properties of the grid are listed, by indicating for each domain the total number of elements (tetrahedrons and prisms), the number of nodes and the maximum size of the elements.

The size of the elements was chosen to have a maximum size of 1.5 mm in the fluid domain. The total thickness of the five layers of inflation at the wall is 1.5 mm, with a growth rate from the wall toward the center of the channel 1.2. In the other remaining solid domains, a maximum size of the elements of 6 mm was chosen.

The fluid flow and heat transfer characteristics of the cold plate were calculated with the finite volume commercial code ANSYS CFX 12.0. This release allows the simulation of the thermal contact resistances; this is an interesting opportunity when modeling the liquid cooling of electronic devices.

Table 2. Grid information.

Domain	Nodes	Elements	Max Elem. Size
Aluminium	128348	640105	20 mm
Fluid	346798	893001	1.5 mm
Plastic	19644	89526	20 mm
Electronic Components	728	1543	7 mm
All Domains	495518	1624175	

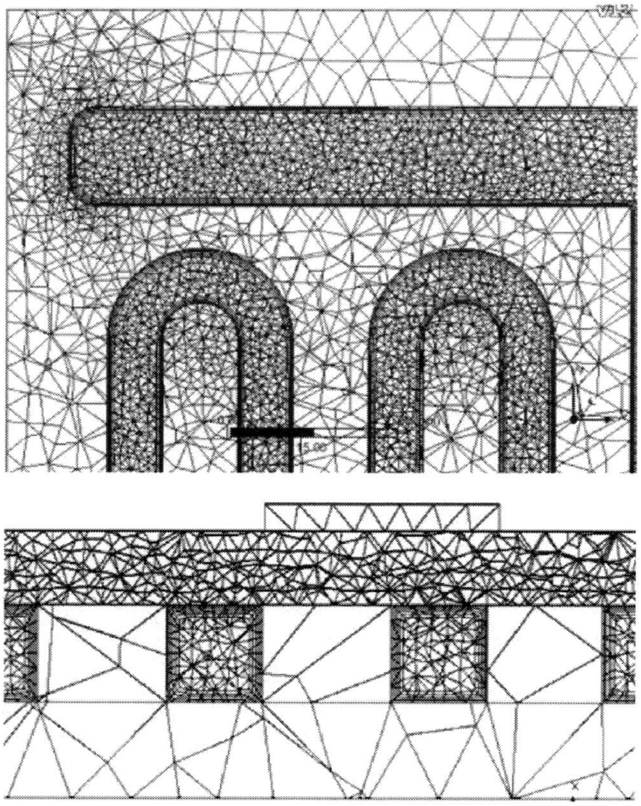

Figure 2. Details of the computational grid.

4. Numerical results

4.1. Reference Test Case

In Figures 3, 4, 5 and 6 are shown the results of the calculations for the Reference test case.

Specifically, Figure 3 shows the distribution of velocity inside the fluid domain. The mass flow is not equally distributed in the three coils; in particular near the outlet, on the left in Figure 3, the mass flow is higher, while near the inlet is lower. Near the entrance of the coils is observed the presence of recirculations. The extension of the area occupied by the recirculations shrinks when moving from inlet to outlet. This behavior is justified by the presence of sharp edges at the entrance. The existence of a possible correlation between the lower mass flow rate in the first coil and a higher temperature of the electronic components cooled by this coil is thence expected. In Figure 3, it may be observed that at the outlet the flow shows a pronounced swirl. This can be significant when more plates are connected in series.

In Figure 4 is shown the distribution of temperature on the surface of the aluminum cover of the blister. From Figure 4, it can be seen that the two resistances near the outlet and those near the inlet are the most disadvantaged (in particular, it is evident that there is a larger area interested by a disturbance in the distribution of temperature). This is due not only to the reduced flow, but also to the misalignment of the electronic components relative to the coils; these stressed components are not "strictly" placed on the coil and only a small portion of them is "wetted" by the fluid.

In Figure 5 are shown the temperature patterns on the surface of the ten electronic components. The two components near the inlet on the right show a temperature higher than the others, confirming the less intense heat transfer observed in Figure 4.

For an accurate analysis of the situation, in Table 3 are reported the mean values of temperature on the upper surface of the components. These values are ordered as the electronic components in Figures 3-6. Despite the coil near the inlet (left) is the most disadvantaged in terms of mass flow rate, the electronic components placed on the internal area of this coil are well cooled. This is due to the placement of the components well over the pipe and for this reason well "wetted" by the fluid.

In Figure 6 is shown the distribution of heat flux at the wall of the pipe, confirming the behavior observed in Figures 3-5.

Table 3. Average temperature on the surface of the electronic components.

Temperature (°C)				
55.64	56.19	56.30	56.70	57.39
56.02	55.54	56.22	55.79	56.84

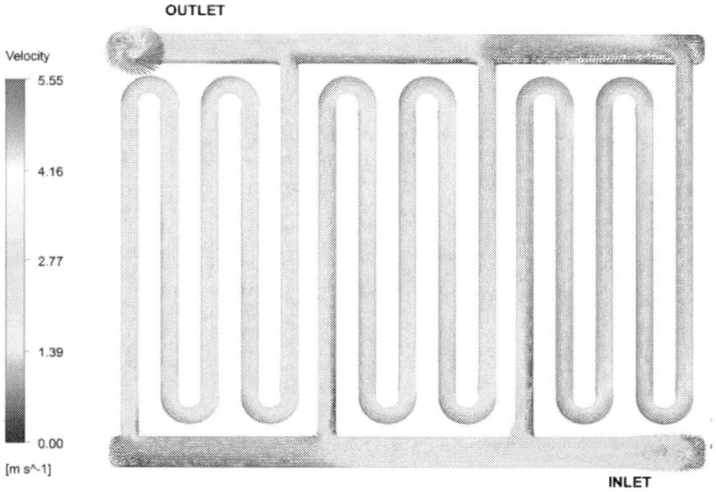

Figure 3. Distribution of velocity inside the fluid domain.

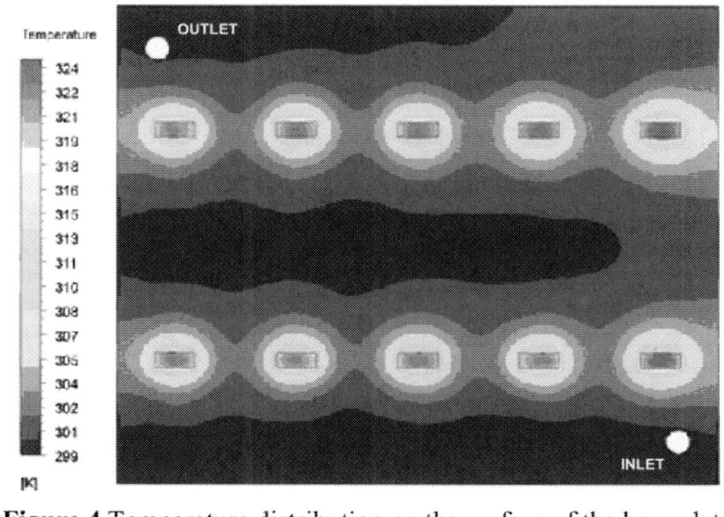

Figure 4 Temperature distribution on the surface of the base plate.

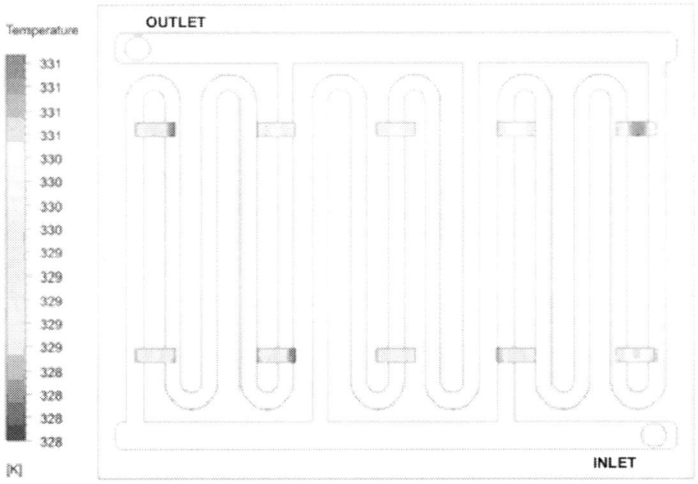

Figure 5. Temperature distribution and on the surface of the electronic components.

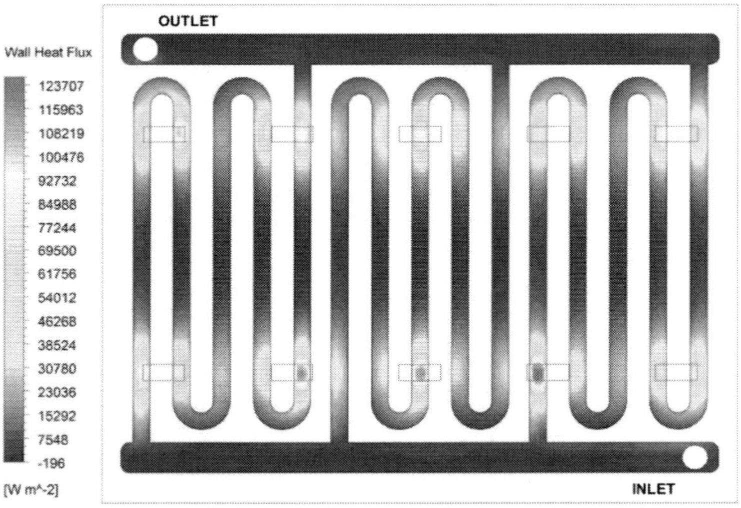

Figure 6. Heat flux on the walls of the coils.

4.2. Variants to the Reference Test Case

In Figure 7, for the variants to the Reference test case described in Section 2, are compared the results in terms of temperature distribution on the surface of the aluminum plate.

The values assumed by the significant parameters in these five cases are reported in Table 4, together with the values of the bulk temperature of the water at the outlet and the average temperature on the surface of the electronic components. The cases reported in Table 4 differ for a parameter at time. Case (a) is the Reference one; in case (b) the thermal conductivity of the plastic base is low; in case (c) the mass flow rate is half than in (a); in case (d) the mass flow is twice than in (a); in case (e) the thermal conductivity of the electronic components is low.

On the top surface of the cold plate, for a decrease of the thermal conductivity of the plastic base (b), the extent of the area showing high values of temperature grows and the local maximum of temperature increases.

When decreasing the mass flow rate (c), it is observed an increase of the extension of the areas with a high temperature; also the local maximum of temperature increases; the opposite happens when the mass flow rate is increased (d).

If the thermal conductivity of the base plate decreases (λ_{base} 60 → 0.35 W/m K as from case (a) to case (b)), the maximum local temperature on the plate grows slightly (~1 °C in this case).

In the last case (e), the mean surface temperature of the electronic components is approximately twice that observed for a high thermal conductivity.

In terms of average temperature on the electronic components (Table 4), for a low value of this parameter it is good, but not significant, a high thermal conductivity of the base plate and a high mass flow rate. On the opposite, the thermal characteristics of the electronic components is significant.

In Table 5 are shown the results of the pressure drop analysis for the three cases (b), (c) and (d), characterized by the same parameters and different the mass flow rates. In the range of flows of practical interest there is a slight dependence of the losses from the Reynolds number, due mainly to the inlet and outlet sections of the channels.

5. Concluding remarks

The detailed design of a cold plate requires the use of CFD, since its behavior is strongly 3D and the conjugate heat transfer is prevailing. The approach followed in this paper allows useful information to be obtained.

Special attention should be paid to the hydraulic aspects. It is not difficult to obtain unbalanced flows between the coils machined in the cold plate, resulting in a non uniform heat transfer.

Table 4. Average temperature on the surface of the components and bulk temperature in the outlet.

Case	Temperature (°C)				
a Q_m = 0.457 kg/s λ_{base} = 60 W/m K λ_{electr} = 401 W/m K	$T_{avg, out}$ = 27.39				
	55.64	56.19	56.30	56.70	57.39
	56.02	55.54	56.22	55.79	56.84
b Q_m = 0.457 kg/s λ_{base} = 0.35 W/m K λ_{electr} = 401 W/m K	$T_{avg, out}$ = 26.56				
	59.83	60.02	60.60	61.39	62.75
	60.26	60.21	60.22	59.66	61.91
c Q_m = 0.2215 kg/s λ_{base} = 0.35 W/m K λ_{electr} = 401 W/m K	$T_{avg ,out}$ = 27.33				
	64.97	65.11	65.90	67.09	69.22
	65.58	64.22	65.28	64.54	67.60
d Q_m = 0.914 kg/s λ_{base} = 0.35 W/m K λ_{electr} = 401 W/m K	$T_{avg, out}$ = 26.18				
	56.36	56.58	57.07	57.66	58.50
	56.81	55.94	56.77	56.39	57.99
e Q_m = 0.914 kg/s λ_{base} = 0.35 W/m K λ_{electr} = 30 W/m K	$T_{avg, out}$ = 26.18				
	114.85	114.97	115.28	116.13	117.03
	115.31	114.29	115.08	114.81	116.47

Table 5. Pressure drop analysis

Case	Q_m (kg/s)	Re	Δp (Pa)	β (-)
b	0.2215	6223	4317.4	5.04
c	0.457	12839	16664.9	4.87
d	0.914	25677	65615.3	4.79

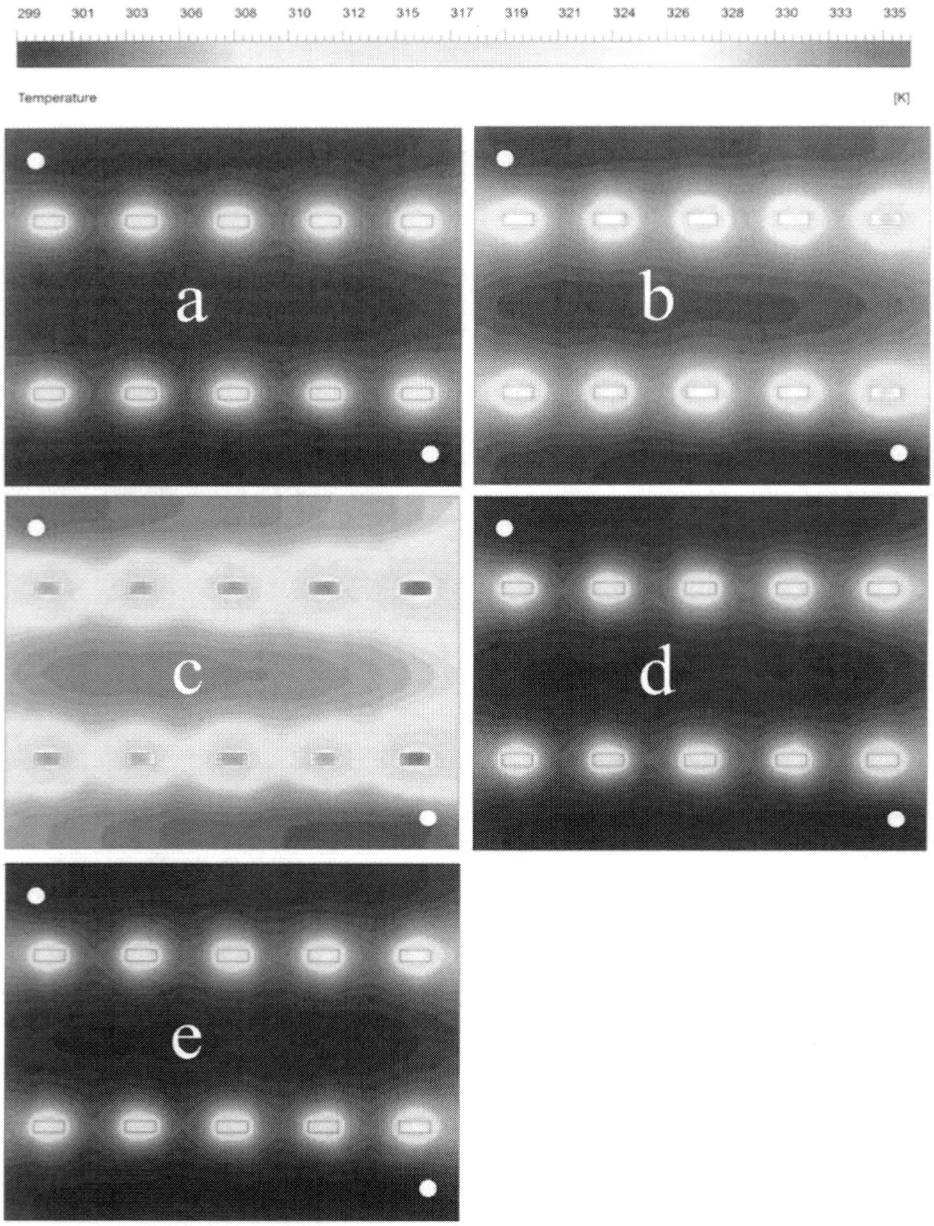

Figure 7. Temperature distribution in the upper surface of the aluminium plate for the five simulations 'a', 'b', 'c', 'd', 'e' of Table 4.

Sharp-edged geometries of the inlets in the manifolds give rise to secondary recirculations reducing the heat transfer.

The position of the electronic components has a significant influence on the performances. The components eccentric to the channels machined in the cold plate show values of temperature higher than the other. However, the position of the components is a parameter that generally can not be changed by the thermal designer and for this reason the hydraulic circuits have to be adapted to each specific layout.

6. Acknowledgments
The authors wish to tank ELENOS Spa, in particular to A Giovannelli, for the interest in the research.

Nomenclature

c	specific heat	J/(kgK)	μ	molecular viscosity	Pa s
h	specific enthalpy	J/(kgK)	ρ	density	kg/m^3
$\overline{n}$	unitary external normal	m			
p'	modified pressure Eq.(5)	Pa	Subscripts		
P	power	W	avg	average	
q	heat flow per unit area	W/m^2	base	base plate	
Q_m	mass flow rate	kg/s	eff	effective	
Re	Reynolds number	-	el	electric	
RTC	thermal contact resistance	m^{2}K/W	electr	electronic component	
T	temperature	°C	in	inlet	
$\bar{u}$	velocity	m/s	out	outlet	
U	heat transfer coefficient	W/(m^{2}K)	t	turbulent	
y^+	dimensionless wall distance	-	tot	total	
			w	wall	
Greek symbols			∞	ambient	
β	pressure drop coefficient	-			
λ	thermal conductivity	W/(m K)			

References

[1] Y A Çengel, *Heat Transfer A Practical Approach*, Mc Graw Hill, 2003.
[2] A D Kraus, A Aziz, J Welty, *Extended Surface Heat Transfer*, John Wiley,. 2001
[3] Y P Zhang, X L Yu, Q K Feng, R T Zhang, Thermal performance study of integrated cold plate with power module, *Applied Thermal Engineering*, 29 (2009) 3568-3573.
[4] F P Incropera, *Liquid Cooling Of Electronic Devices By Single-Phase Convection*, John Wiley, 1999.
[5] T Pesso, S Piva, CFD analysis of liquid cold-plates for applications in electronics, Proc. XXVII UIT Heat Transfer Conference, Reggio Emilia, 22-24 June 2009, 323-328 (in Italian).
[6] ANSYS *CFX-Solver Theory 12.0* (2009).
M J Moran, H N Shapiro, B R Munson, D P DeWitt, *Introduction to Thermal Systems Engineering: Thermodynamics, Fluid Mechanics and Heat Transfer*, John Wiley, 2003

The Influence of Thermistor Location on Temperature Measurement from a Photonic Package

Cormac Eason[1], Marc Rensing[1], Jun Su Lee[1], Peter O'Brien[1]

[1]Photonics Packaging Group,
Tyndall National Institute,
Prospect Row,
Cork City,
IRELAND

cormac.eason@tyndall.ie

This paper begins by describing some commonly used photonic packages. The requirements for optical connections to these packages are then discussed. Photonic packages are different to most electronic packages in that the thermal management requirements usually include maintaining the Photonic Integrated Circuit (PIC) at a fixed, sometimes below ambient, operating temperature rather than with keeping the temperature of a package below an upper limit as with most electronic packages. This means that an active Thermoelectric Module (TEM) based cooling system is required. A thermistor is fitted within the package to provide thermal feedback to the TEM controller. This paper uses finite element modelling to investigate whether there is a good match between the target temperature for the PIC and the temperature registered by the thermistor. The results of the modelling show that the model results are quite stable even with large variations in convection and thermistor thermal properties. The thermistor location influences the temperature measured from the package and its thermal response time, but follows the device temperature well enough to provide the TEM controller with adequate feedback to maintain the PIC at a steady temperature in steady state running conditions.

1. Photonic package types

Photonic packages, similarly to electronic packages, allow the required electrical connections to be made to the Photonic Integrated Circuit (PIC) for transfer of data, powering of the Thermoelectric Module (TEM) and temperature monitoring of the PIC. This is usually accomplished using pins built into the package and insulated from the package case. Wire bonds in gold wire are used to make the connections from the PIC to the package pins.

Combined with the electrical requirements, photonics packages must also provide a means by which light can enter and/or leave the package. Directing the light into or from the package using optical fibres is the most common method. The fibres are either welded or glued in place usually after submicron alignment with the PIC to ensure high coupling efficiency. Where light detectors are required for light sources not coupled to a fibre, a window or lens is integrated within the package case to allow light in and out of the package.

It is for this reason that kovar, an iron alloy with a low thermal expansion coefficient, is used heavily in photonics packaging. A kovar package can be fitted with a glass, quartz or sapphire window/lens with minimal thermal stress between the window and package case material. This allows

Content from this work may be used under the terms of the Creative Commons Attribution 3.0 licence. Any further distribution of this work must maintain attribution to the author(s) and the title of the work, journal citation and DOI.
Published under licence by IOP Publishing Ltd

reliable hermetically sealed photonic packages to be created. Kovar's poor thermal conductivity (≈17 W/mK) does mean there is a thermal cost to using this material.

1.1. 14 Pin butterfly package

One of the most common photonic packages is known as a butterfly package. There are a number of varieties in this package, but the most common is a 14 pin package with a tube through one of the short walls to allow an optical fibre to be fed into the package. A metal lid can be resistance welded over the open top of the package allowing it to be hermetically sealed. The package footprint dimensions are 30 mm × 12.7 mm × 7.8 mm.

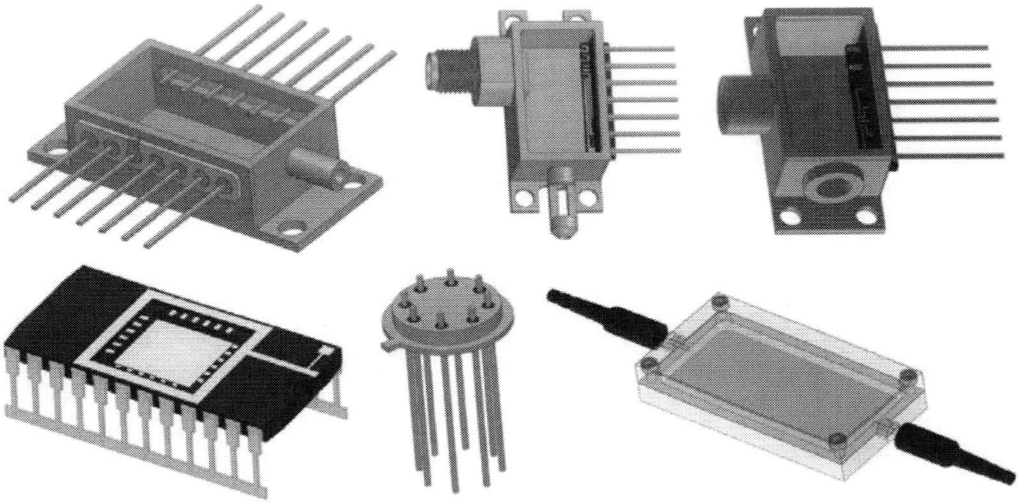

Figure 1. Starting at top left and going clockwise: 14 Pin butterfly Package, 7 Pin SMA package, 7 Pin V-Connector package, custom clamshell package, TO Package, DIL Package. Images are not to scale with each other.

1.2. Butterfly package with high speed connection

This package looks similar to the 14 pin package and matches its dimensions closely, but loses the 7 pins from one side to allow a high speed SMA, V or GPPO (Not shown) connector in the package instead. These connectors can handle signal frequencies in the GHz range. These are required where high speed electrical signals must be converted to or from an optical signal.

1.3. Clamshell package

These packages are generally completely custom made and consist of two nearly identical halves which are bolted together to close the package. These are used when the optical coupling and electrical connection requirements or the size requirements of certain devices prevent them from fitting in off the shelf packages. The dimensions of the package shown are 49 mm × 26 mm × 9.1 mm.

1.4. Transistor Outline (TO) package

These are small mass produced packages, usually used where lower cost or smaller size is important. Lids are available for these packages with various lenses and windows to allow light to access the package. Many mass produced lasers and photodiodes are packaged in TOs. The TO shown in Figure 1 is 4.6 mm in diameter and the pins through it are 15.4 mm long but bigger and smaller TO packages are available.

1.5. Dual In-Line (DIL) package

These are rectangular packages commonly used in the electronics industry with a row of electrical pins along each of the long sides of the package. These can be adapted for photonics use. The DIL shown in Figure 1 is 30.5 mm × 15.6 mm × 9.2 mm in size.

2. Optical coupling to photonic devices

Indium Phosphide (InP) PICs can be used to make lasers and optical detectors. In order to couple light to and from these PICs, the fibre must be aligned with a waveguide in the top surface of the PIC. The fibre axis lies in the same plane as the PIC surface. In most cases the fibre is aligned and held in place using a nickel clip, which is laser welded to a stepped kovar submount that also carries the PIC. This is shown in Figure 2. A photograph of the laser welded fibre and a small laser is shown in Figure 3.

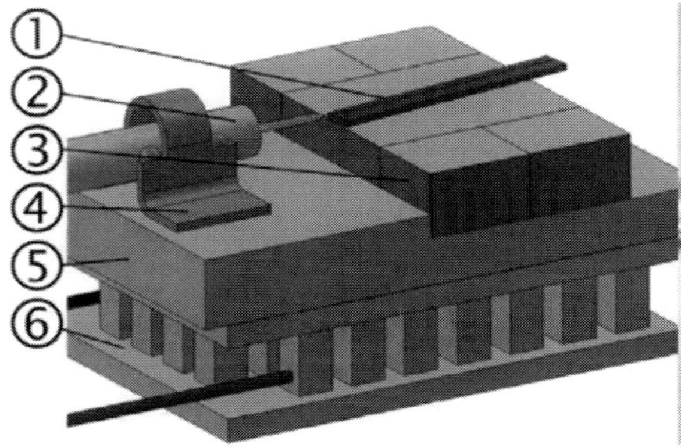

Figure 2. Fibre coupling to edge of PIC. (1) Indium Phosphide (InP) PIC, (2) Optical fibre, (3) AlN Submount, (4) Weld Clip (Nickel) laser welded to fibre and submount, (5) Kovar submount, (6) TEM.

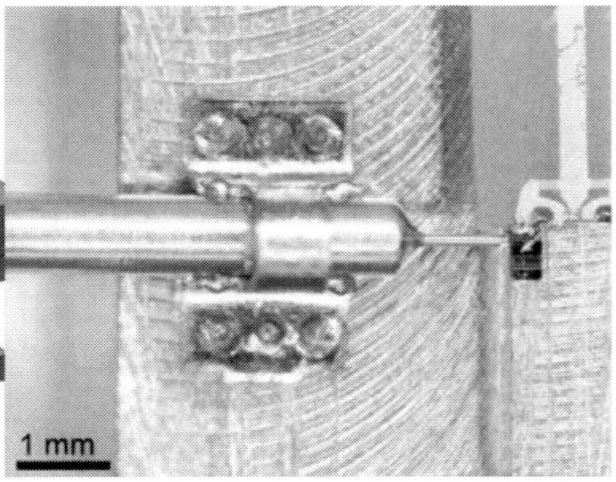

Figure 3. Photograph of laser welded fibre. Laser shown is much smaller than the one modelled in this work. The optical fibre diameter at the weld clip is 0.9 mm

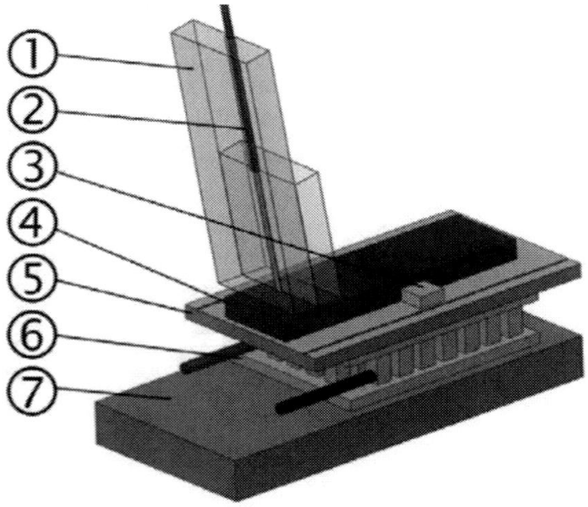

Figure 4. SOI fibre coupling using angled glass block carrying the optical fibre. The block is held in place using UV curing adhesive once it has been aligned. The assembly consists of; (1) Glass Block, (2) Optical fibre, (3) Thermistor, (4) Silicon on Insulator PIC, (5) AlN submount, (6) TEM, (7) Copper heat spreader used to spread the waste heat from the TEM before it reaches the kovar wall of the butterfly package.

Silicon on Insulator (SOI) PICs are a more recent development[1] and allow manufacturing techniques already proven with silicon integrated circuits, to be also be applied to PIC manufacture.

These devices consist of single crystal silicon wafers with an insulating layer of usually silicon dioxide onto which a further much thinner layer of silicon or other materials is deposited. To couple light to these devices, the light must hit a grating on the PIC surface at the correct angle of incidence (Often in the region of 10° from the vertical). To accomplish this, the fibre is usually placed in a glass block polished to the required angle, which is aligned and glued to the grating on the surface of the PIC. This is shown in Figure 4.

Many other materials are used in PICs in small amounts. These materials are engineered to produce or respond to light of various colours based on the band gap of the material. In many cases these materials are used in submicron thick layers on a substrate of silicon or InP. The thermal behaviour of these layers makes very little difference to the bulk temperature of the PIC. Gradients or localised hot spots may measurably alter optical behaviour however.

3. Thermal management

Photonic devices in most cases have low power requirements and as such their thermal management tends to be less demanding than more powerful electronic devices. The difference however is that the optical properties of photonic devices change with temperature, leading to undesirable device behaviour even though the package never overheats.

For this reason most packages are temperature regulated using TEMs rather than just cooled. The operation of these devices adds further waste heat to the package, but this is within ranges that can be dealt with using off the shelf heat sinking even while using mostly kovar packages.

The main uncertainty in most packages is whether the temperature read from the thermistor matches with the temperature experienced by the working parts of the PIC. This will be investigated for the package shown in Figure 5. The package consists of a butterfly package holding a laser. The temperature of the laser is regulated using a TEM and an optical fibre is welded in alignment with the emitting edge of the laser. The laser sits on a metallised Aluminium Nitride (AlN) submount, and wirebonds to connect the laser and thermistor to the package pins. The thermistor is mounted on the top right of the metallised AlN submount.

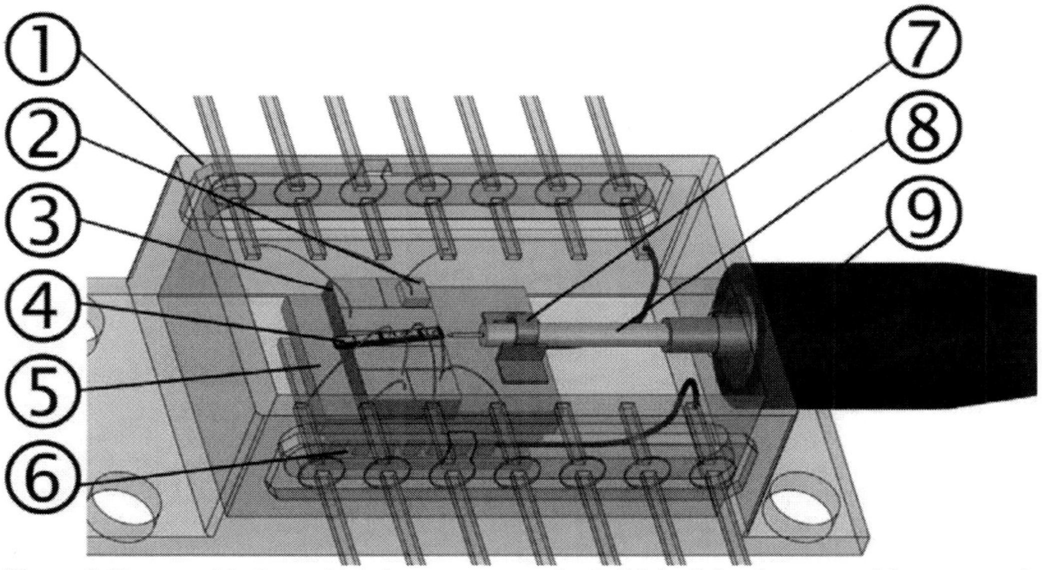

Figure 5. The assembly from which the geometry used to build the finite element model was sourced. Assembly consists of; (1) Kovar butterfly package, (2) Thermistor, (3) AlN submount, (4) Laser, (5) Kovar submount, (6) TEM, (7) Weld clip, (8) Fibre, (9) Fibre strain relief.

The location of the thermistor is clearly less than optimal to measure the PIC temperature but in order to provide a ground electrical connection to the thermistor using one of the metallised islands on the AlN submount, it was necessary to move it away from the laser. This sort of compromise is common in photonic packages, especially in the case of prototypes, as the supporting electrical and optical connections for the PIC are more critical than placing the thermistor in an ideal location.

In terms of relative sizes, the laser in this model is 4.3 mm long, less than 5 μm wide and deep and on a die 0.5 mm wide and 0.1 mm thick while the thermistor is 0.6 mm × 0.6 mm × 0.3 mm in size, which means the thermistor has just under half the total volume of the PIC. Thermistors small enough to match the scale of the features on the PIC are not to the authors' knowledge available. Finding an ohmmeter with a low enough driving current/voltage to prevent the thermistor from self heating and affecting its own resistance and creating a further hot spot within the package is likely to prove difficult at this scale also.

Mounting the thermistor directly to the PIC is avoided as mechanical bonding to the PIC surface can lead to mechanical or thermal strain and the potential malfunction of the waveguides and other components on the PIC[2].

In this package the AlN PIC submount is carried on a further submount, to which the weld clip that holds the fibre is also attached. The submount is made from kovar as this is compatible with the laser welding process. The clip holding the fibre is made from nickel. The thickness of the submount is varied to center the fibre vertically with the surface of the PIC and with the hole in the butterfly package wall through which the fibre enters.

4. Finite element model

A finite element model of the laser package was created in COMSOL Multiphysics to examine the correlation between the temperature measured by the thermistor and the temperature of the laser.

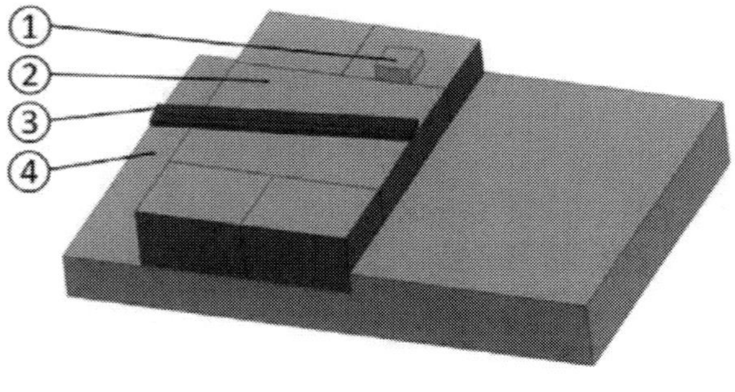

Figure 6. CAD Geometry for Models M1-M7 and M9. (1) Thermistor, (2) Metallised Aluminium Nitride (AlN) submount, (3) Laser (InP), (4) Kovar submount.

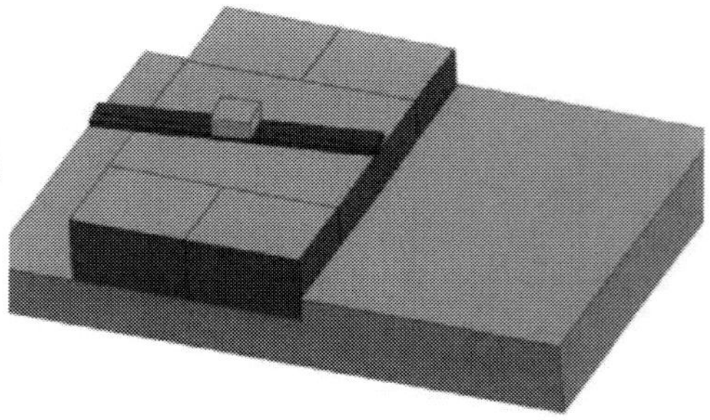

Figure 7. CAD geometry for Models M8 and M10. Thermistor has been placed on top of the laser. This configuration would not be used in a working device.

The stepped submount is made from Kovar, the metallised submount is made from AlN and the laser PIC is made from InP. Neither the thermistor material or its thermal properties are given on the manufacturer's data sheet. For the purposes of the analysis it has been given the properties of Alumina. Models are run to investigate the sensitivity of the results to the thermal conductivity of the thermistor to see whether this measurement is critical.

The laser is modelled as a line heat source of 0.3 W total power. The CAD model of the PIC carrying the laser is split vertically along the path of the laser and the top edge of the split is used as a line heat source representing the laser. Since the cross sections of InP based lasers are typically less than 5 μm × 5 μm and the laser is 4.3 mm long, the laser has an aspect ratio in the region of 1000:1.

Massive numbers of elements would be required to fill this volume with the low aspect ratio elements required for good convergence if a solid model of the laser was used.

All parts of the assembly are glued together using a high silver content thermally and electrically conductive epoxy from Epoxy Technology Inc. (EPO-TEK H20E). The bond line thickness is generally 50 µm. This is modelled as 50 µm thick thermally resistive layer of conductivity 2.5 W/mK at the interface between each component in the assembly.

The cooling provided by the TEM is modelled as a constant temperature of 20°C applied to the underside of the kovar submount while all the other boundary surfaces of the model had a natural convection boundary condition applied based on Equation 1 [3]. It should be noted that the convection coefficient is not the same for all surfaces in reality and that the experimental measurement of natural convection heat transfer from such small areas on a working photonic package presents significant practical difficulties. To account for this, models were run with no convection and with an order of magnitude more convection than the chosen value to investigate the sensitivity of the model to this uncertainty.

$$h = \frac{k}{L}\left(0.68 + \frac{0.67 Ra_L^{1/4}}{\left(1 + (0.492/\mathrm{Pr})^{9/16}\right)^{4/9}}\right)$$

Equation 1

h	= Convection Coefficient (W/m^{2}K)	$h \approx 27$ W/m^{2}K
k	= Thermal Conductivity of air (W/mK)	$k \approx 0.026$ @ 20°C
L	= Characteristic Length of surface of interest (m)	$L \approx 0.001$ m
Pr	= Prandtl number (No unit) = υ/α	$Pr \approx 0.716$
Gr_L	= Grashof number (No unit) = $g\beta(T_s - T_\infty)L^3\upsilon^{-2}$	$Gr_L \approx 0.733$
Ra_L	= Rayleigh number (No unit) = $Gr_L \times Pr$	$Ra_L \approx 0.052$

g = gravitational acceleration (m/s^2), β = volumetric thermal expansion coefficient (1/K),
T_s = temperature of convecting surface (K), T_∞ = temperature of ambient air (K),
υ = kinematic viscosity (m^2/s)

The nominal value of 26 W/m^{2}K found based on this is high for a natural convection coefficient, but arises as a consequence of the small characteristic dimensions of the convecting surfaces. The results from model M2 show that the TEM provides the dominant heat path from the device, making the exact determination of the convection coefficient unimportant in the overall process of predicting the final temperature of the device.

4.1. Grid independence check

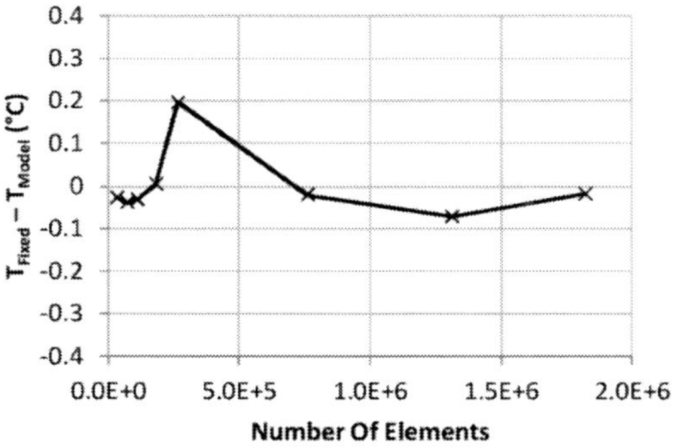

Figure 8. Grid independence check on model. T_{Fixed} is the average of the maximum temperatures measured from each model.

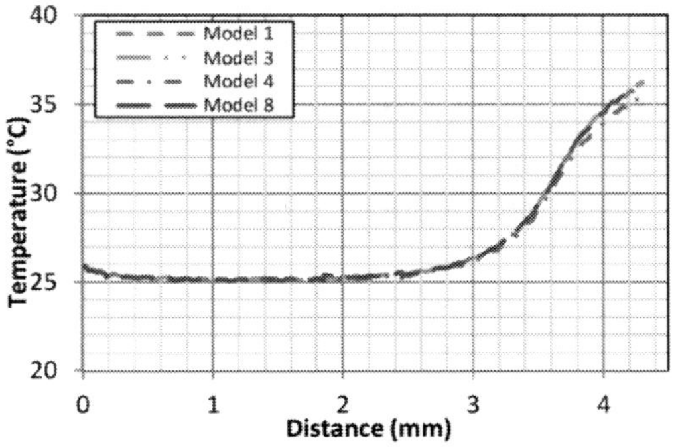

Figure 9. Temperature profile along the laser for various models run. Note high temperature in the laser where it does not contact the submount.

Seven models were run at varying mesh densities from 32 619 to 1 822 326 elements. The largest variation in the temperature results was found to be under ±0.2°C of the average temperature across all the models, with most results being well within 0.1°C of the average as can be seen in Figure 8. This gave good confidence that the 32 619 element model used for the analysis would give acceptably accurate results.

5. Models run and results:

A list of the models run in this study is given in Table 1. The baseline model against which the other models are compared is model M1 (Geometry shown in Figure 10). This uses the best estimates for values of the convection coefficient and the thermistor conductivity. Model M2 was run without modelling the effect of the TEM in maintaining the temperature of the device. It can be seen from this that without the TEM the device runs over 100°C hotter, indicating that the natural convection heat transfer is not the dominant means by which the temperature of the device is maintained.

Table 1. List of finite element models run in this study.

Model	h (W/m^{2}K)	k_{Th} (W/mK)	T_{Max} (°C)	T_{AvgTh} (°C)	T_{AvgL} (°C)	$T_{AvgL} - T_{AvgTh}$
M1	26	27	36.22	20.74	27.12	6.38
M2 (TEM Off)	26	27	144.98	129.88	136.33	6.45
M3	0	27	36.29	20.75	27.14	6.39
M4	260	27	35.55	20.70	26.94	6.24
M5	26	2.7	36.22	20.74	27.12	6.38
M6	26	270	36.22	20.74	27.12	6.38
M7	260	270	35.55	20.70	26.94	6.24
M8 (Thermistor on Laser)	26	27	36.21	25.23	27.12	1.89
M9 (M1 Transient @ 10s)	26	27	36.23	20.75	27.14	6.39
M10 (M8 Transient @ 10s)	26	27	36.21	25.35	27.12	1.77

To further investigate the influence of the convection coefficient on the device temperatures, model M3 was run with no convection heat loss at all. This increased the average temperature of the laser by 0.02°C. Similarly when the convection coefficient was increased by an order of magnitude as in model M4, the laser temperature dropped by 0.18°C. This confirms that the TEM cooling effect is the dominant influence on the laser temperature.

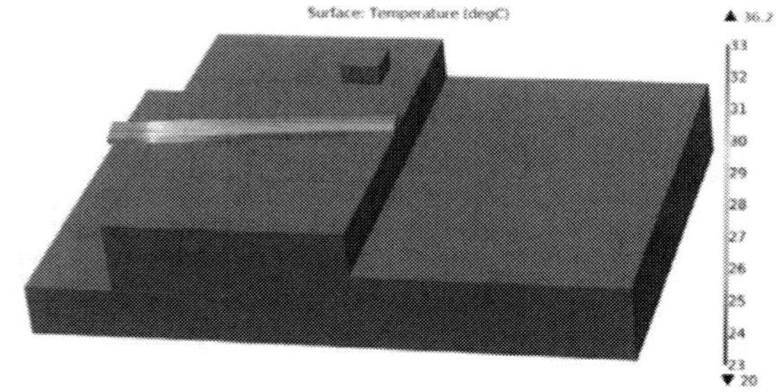

Figure 10. Temperature results from M1.

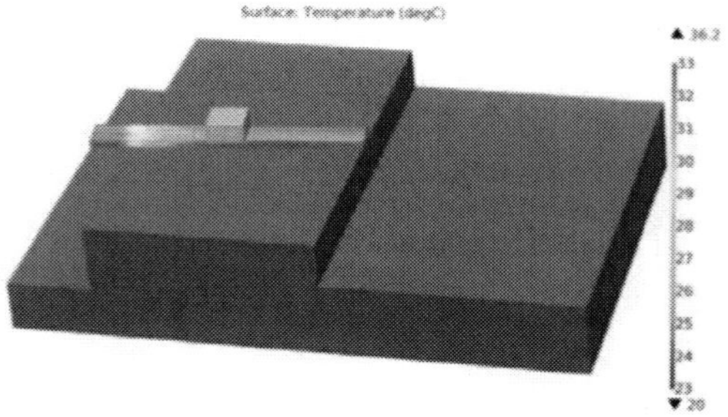

Figure 11. Temperature results from M8.

Models M5 and M6 then looked at the effect of the properties of the thermistor on the temperature it measures. No properties for the thermistor material were available as neither the material nor any of its properties apart from its resistance-temperature characteristics were included on the datasheet. The result from models M5 and M6 shows that a drop or a gain of an order of magnitude in the thermistor's thermal conductivity has no influence on the average temperature of the thermistor to two decimal places compared to the baseline model M1.

Model M7 was then run to see whether the combination of high thermal conductivity and high convection heat transfer would allow the thermistor to act as a heat sink fin and reduce its temperature. The average temperature of the thermistor in model M7 only dropped by 0.04°C from models M1, M5 and M6, indicating that the temperature indicated by the thermistor is a reliable representation of the temperature of the surface it is fitted to even with high uncertainty as to its actual properties and large variation in the convection conditions around it.

The other important result from this is that even though the thermistor in models M1-M7 does not give the correct device temperature, it does give a temperature that follows the device temperature well enough to provide decent temperature feedback to the TEM controller. Even when the model was run without the TEM cooling effect (Model M2), the difference in temperature between the thermistor and the laser shifted from 6.38°C to 6.45°C, a change of just over 1%.

Model M8 was run with the thermistor placed on the laser instead of to the side of it as in the other models. The thermistor conductivity and convection coefficient were left at the same values as model M1. The results from this model show that, as expected, having the thermistor closer to the laser gives a better temperature measurement (See Figure 11). While the thermistor is still 1.89°C below the average laser temperature it is important to note that the average temperature of the area of the laser the thermistor makes contact with is 25.35°C, or 0.12°C more than the thermistor temperature. This confirms that if it was possible to fit the thermistor directly on the laser surface without potential damage to the laser it would represent the temperature of the laser well despite the volume mismatch.

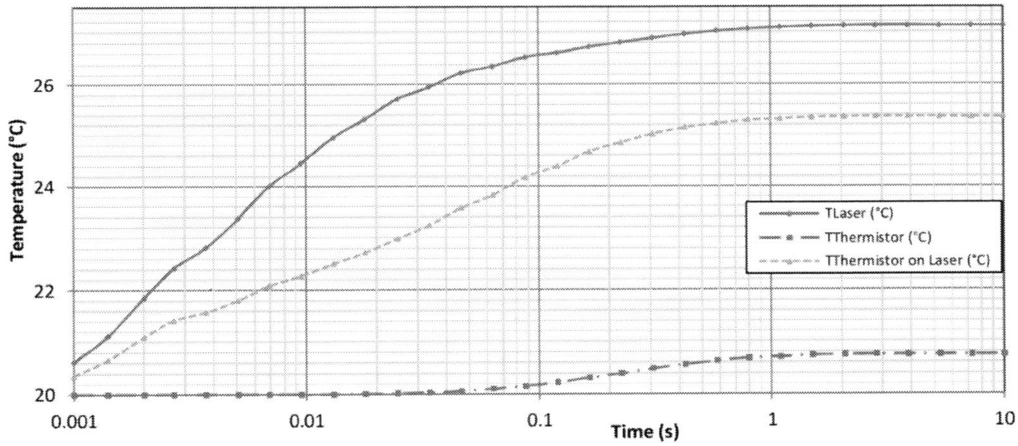

Figure 12. Transient temperature results M9 and M10. T_{Laser} is the average temperature along the line representing the laser in both models, $T_{Thermistor}$ is the average temperature of the volume representing the thermistor and $T_{Thermistor\ on\ Laser}$ is the average temperature of the thermistor for the case where the thermistor is placed on top of the laser.

Model M9 was run at the same settings as model M1 but as a transient model to see how well the thermistor follows the temperature of the laser as the laser is turned on. Thirty logarithmically spaced time steps from 0.001 s to 10 s were used in order to capture the very quick initial thermal response of the laser as well as the steady state final temperatures of the system.

While there is a noticeable delay of the order of 0.02 seconds before the thermistor responds to the laser turning on, both the laser and the thermistor reach a steady state in approximately 2 seconds. This confirms that the thermistor can be used as a representative temperature for the laser for the feedback loop to the TEM. In the case of a laser that is switching on and off however, a much quicker temperature response would be needed.

Model M10 was run with the same time steps as model M9 but with the thermistor on the laser geometry from model M8. The time-temperature curves from models M9 and M10 are shown in Figure 12.

5.1. Further assembly issues

The section of the laser which overhangs the submount can be seen to run significantly hotter than the laser in contact with the submount in Figure 10 and Figure 11. This gave acceptable performance for the prototyping of this particular package but would not be acceptable in many cases. A plot of the temperature profile along the laser is shown in Figure 9. This is detrimentally affecting the average laser temperature and does reduce the correlation between the thermistor temperature and the laser temperature.

It can also be seen in Figure 9 that as a result of the higher convection coefficient used in model M4, the temperature of the unsupported end of the laser can be seen to be lower than in the other models as the unsupported end of the laser is the only part of the model where the cooling isn't completely dominated by the TEM.

6. Conclusions

The results show that the finite element model results are not very sensitive to variation in convection and thermistor properties. As this information is not easily measured or freely available, this is good news for the extraction of meaningful results from finite element models of these assemblies. The important conclusions are summarised in the following points:

- The convection heat transfer from the package components does not play significant part in the temperature regulation of TEM regulated photonics packages of the type analysed.
- The only location where convection was seen to influence the TEM controlled model was where the laser extended past the edge of the submount it was fitted to. A tenfold increase in the convection coefficient altered the temperature of the overhanging end of the laser by almost 1°C.
- Even though properties for the thermistor are not available, the results from the model remain almost completely unaffected by a variation of an order of magnitude each way in the thermistor's estimated thermal conductivity.
- Using a line heat source to model the laser rather than trying to mesh it as a volume massively reduces the computing requirements and solving time for the model while still giving excellent data on the overall behaviour of the package.
- The results show good correlation between the temperature of the surface the thermistor is fitted to and the average temperature of the thermistor volume.
- The thermistor does not return temperature data that matches exactly with the temperature of the photonic components in the package due to unavoidable constraints in the assembly.
- While the thermistor mounted away from the PIC does not give an exact device temperature, it does provide a suitable value to feed to the TEM controller to maintain the packaged device at a constant temperature for steady state running conditions.
- There is a significant delay in the initial thermal response of the thermistor when the laser is turned on. This delay is reduced by mounting the thermistor directly on the component to be measured but creates an unacceptable risk of damage to the optical components due to interfacial stresses between the thermistor and the PIC making this an unacceptable option for real world assembly.
- For more precise temperature control integrating a temperature sensing element directly on the PIC itself during PIC manufacture would eliminate considerable uncertainty in this regard.

References

[1] Kopp, Christophe; Stéphane Bernabé, Badhise Ben Bakir, Jean-Marc Fedeli, Regis Orobtchouk, Franz Schrank, Henri Porte, Lars Zimmermann, and Tolga Tekin (2011). Silicon Photonic Circuits: On-CMOS Integration, Fiber Optical Coupling, and Packaging, IEEE Journal of Selected Topics in Quantum Electronics, vol. 17, No. 3, May/June 2011

[2] Huang, M. and X. Yan (2003). Thermal-stress Effects on the Temperature Sensitivity of Optical Waveguides, J. Opt. Soc. Am. B/Vol. 20, No. 6/June 2003.

[3] Churchill, S.W., and Chu, H. H. S. (1975). Correlating Equations for Laminar and Turbulent Free Convection from a Vertical Plate, Int. J. Heat Mass Transfer, 18, 1323–1329.

Investigation of Vapour Chamber Performance with a Concentrated Heat Source

[2]E. Ravache, [1]S. Siedel, [1,3]R. Kempers, [1]A.J. Robinson
[1]Department of Mechanical and Manufacturing Engineering, Parsons Building, Trinity College Dublin
[2] Université de Lyon, CNRS
INSA-Lyon, CETHIL, UMR5008, F-69621, Villeurbanne, France
Université Lyon 1, F-69622, France
[3]Alcatel-Lucent Ireland, Blanchardstown Business & Technology Park, Dublin 15, Ireland

Email: arobins@tcd.ie

Abstract. This work aims to characterize the performance of a commercially available solid heat sink (SHS) and a vapour chamber heat sink (VCHS) with a small localized heat source. The heat sinks were tested under forced convection conditions in a dedicated wind tunnel. Heat transfer and temperature measurements facilitated the estimation of the source-to-sink thermal resistance whilst thermal imaging on the air side of the heat sink was used to gauge the level of heat spreading. The results indicate that the VCHS was capable of spreading the heat from the localized source over a greater surface area of the heat sink compared with the SHS. However, the improved spreading resistance of the VCHS was offset by the additional contact resistance and/or the thermal resistance of the internal wick structure resulting in a source-to-sink thermal resistance and heater temperature which was commensurate with the SHS. As a result there was no thermal benefit of the VCHS.

1. Introduction

Heat sinks are used as surface area extensions from small electronic components in order to lower the air-side thermal resistance, thus the net source-to-sink thermal resistance. However, there are limits with regard to the size of the heat sink since its effectiveness decreases as it gets larger due to spreading resistance i.e. the temperature of the outer fins decrease as heat is not effectively conducted to them from the source.

One solution to this is the use of a vapor chamber heat sink (Fig. 1) to aid in the lateral spreading of the heat from the localized source to the outer periphery of the heat sink.

As it is illustrated, a VCHS operates as a heat pipe where a working fluid is vaporized within a wick structure at the location of the heat source and condensed in a wick structure in thermal contact with the fin base area. The heat is spread by virtue of the vapor filling the void between the base of the heat sink and the base of the fins. The condensed liquid is returned to the evaporator by means of the wick structure in a manner which is the same as a conventional heat pipe.

Content from this work may be used under the terms of the Creative Commons Attribution 3.0 licence. Any further distribution of this work must maintain attribution to the author(s) and the title of the work, journal citation and DOI.
Published under licence by IOP Publishing Ltd

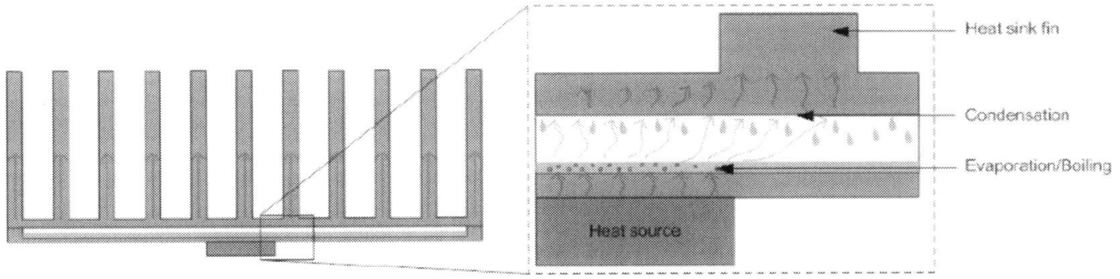

Figure 1. Schematic of a VCHS.

Li et al. [1] compared an aluminum heat sink and a vapor chamber heat sink using infrared thermography. Their results showed that the latter one was more effective with a lower maximum temperature for the same power applied and a lower overall thermal resistance.

Chen et al. [2] compared an aluminum heat sink, a copper heat sink and a vapor chamber heat sink. Once again, the vapor chamber heat sink outperformed the other heat sinks due to its spreading performance. Indeed, they expressed the temperature distribution in terms of the standard deviation and showed that the values associated with a VCHS were lower than for the standard heat sinks. This standard deviation was constant for VCHSs and increased with increasing power for the standard heat sink. Moreover, they showed that the thermal resistance of the VCHS decreases when the power input increases. Conversely, the thermal resistances of the standard heat sinks were constant with respect to the power input.

Boukhanouf et al [3] found similar results and used thermal imaging to show that the thermal spreading resistance of their VCHS was about 40 times lower than that of a similar copper heat sink. Likewise, they showed that the operating temperature was up to 15°C lower than the operating temperature of the copper heat sink for the same heat flux.

Several papers showed that the overall thermal resistance of a VCHS decreases when the power input increases [2, 3, 4, 5]. Conversely, Koito et al. [6] found that the VCHS thermal resistance was constant with respect to the power input and that it decreased as the heat source size became closer to the size of the heat sink. In other studies [5, 6] no comparison was made with solid heat sinks so it is difficult to gauge the benefit of the VCHS.

The results above indicate that VCHSs exhibit lower thermal resistances resulting in lower maximum surface temperatures and that their spreading resistance is smaller than that of a standard heat sink. Even still, as shown by Sauciua et al. [7], it is not always more effective than simply increasing the thickness of the heat sink base plate. In this regard it may be the weight advantage that makes the VCHS more appropriate though this would depend on the particular situation as well as cost considerations.

The design of a VCHS requires careful attention to the design of the wick structure since any gains associated with the heat spreading can be more than offset by the relatively poor effective thermal conductivity of the liquid saturated wick at the evaporator and condenser. This is particularly crucial for concentrated heat sources where the poor effective thermal conductivity is accentuated by a small relative area, which also negatively influences the overall thermal resistance. Further to this, some companies are commercializing VCHSs as 'bolt-on' devices. Here special attention must also be paid to the additional thermal contact resistance associated with the vapor chamber and solid heat sink.

In this investigation, commercially available vapor chambers attached to an aluminum extruded heat sink as well as a solid heat sink are investigated under forced convection conditions. Importantly a discrete heat source is tested with variable heat loading so as to gauge the thermal performance of the VCHS against the solid heat sink.

Journal of Physics: Conference Series **525** (2014) 012005 doi:10.1088/1742-6596/525/1/012005

2. Experimental Apparatus

Three heat sinks were tested under identical thermal loading and flow conditions. The SHS was a solid aluminum extrusion of base dimension of 244 x 300 mm^2 with a total of twenty one 4 mm thick and 45mm high fins. The two VCHS's tested were comprised of the same aluminum extrusion with a commercially sourced 'bolt-on' flat vapor chamber fastened to the base. The geometric parameters associated with the VCHS and SHS are outlined in Table 1.

Table 1. Heat sink dimensions.

	Solid heat sink	VCHS
Length (mm)	300	
Width (mm)	244	
Total height (mm)	48	50
Number of fins	21	
Fin Height (mm)	45	
Fin thickness (mm)	4	
Gap between fins (mm)	9	

The geometry of these heat sinks is more optimal for natural convection operation as they are based on the natural convection heat sinks used in Remote Radio Heads (RRH) used in the telecommunications industry. However, for the purposes of these tests, they were tested under forced convection conditions in order to minimize the convective thermal resistance and increase the relative contribution of the heat sink itself to the measured thermal resistance.

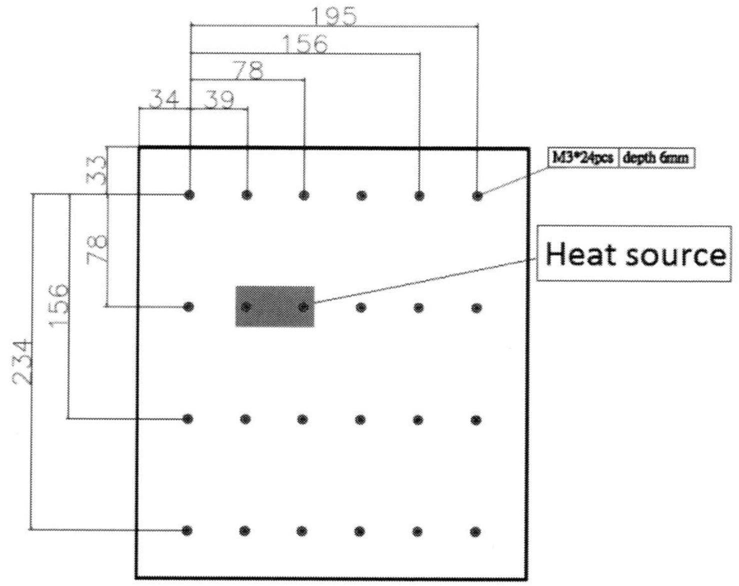

Figure 2. Heat sink base showing heater location and dimensions.

Fig. 2 depicts the base of the heat sink showing the bolt hole locations for affixing the vapor chamber as well as the relative size and location of the heater. Fig. 3 shows the assembled VCHS along with a cut section of the vapor chamber showing the sintered wick structure on both the evaporator and condenser regions.

Figure 3. Image of the VCHS cross section and overall composition.

Figure 4. Heat source attached to the heat sink base.

The heat source was designed to have a similar size and power to the heat sources that are cooled by the heat sink in real operation, such as the RF power amplifiers used in Remote Radio Head applications. To this end, heat is applied to a Ø6.35 mm x 25.4 mm cartridge heater. It is inserted into a copper sheath bolted to the heat source side (Fig. 4). Identical assembly torque settings were used for these tests. Power is supplied to the heaters using a variable voltage transformer that can deliver up to 240 V. Although the heat sink has the capacity to accommodate several heat sources, only one was tested in this investigation. Unwanted heat losses were minimized with a thick layer of insulation on the underside of the assembly.

A schematic of the experimental rig is illustrated in Fig. 5. The rig is composed of a centrifugal fan and a purpose built wind tunnel which contains the heat sink under test. Several measurement devices were used such as an infrared camera, thermocouples, Pitot probe and voltage and current meters. A dedicated computer with DAQ hardware (National Instruments) and software (LabView, ThermaCAM Researcher Professional 2.9) recorded relevant data required for the calculation of the overall source-to-sink thermal resistance.

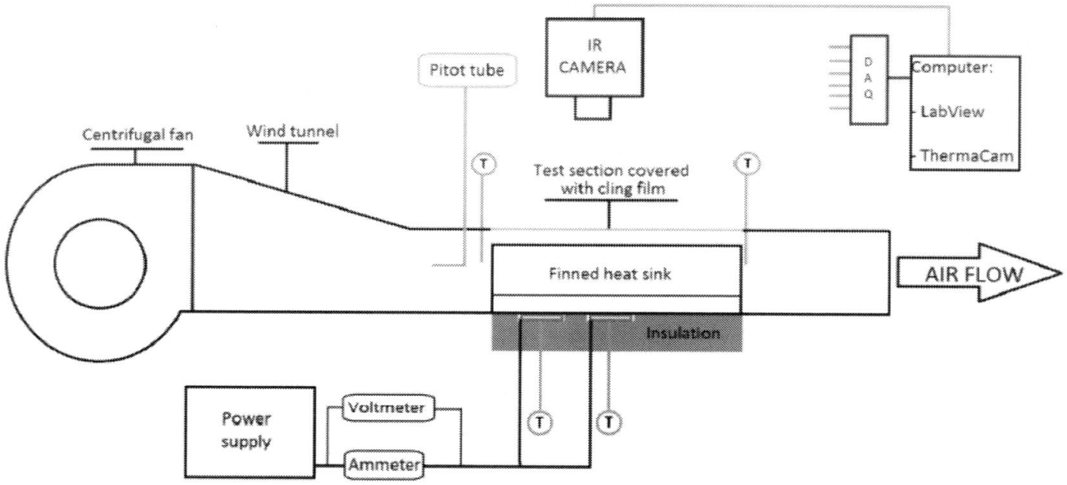

Figure 5. Schematic of the experimental apparatus.

3. Data Reduction

The supplied power Q_{in} was determined from current and voltage measurements taken with digital multimeters. Since heat losses were minimal, a sufficiently accurate estimation of the heat input can be calculated simply as,

$$Q_{in} = IV \tag{1}$$

Using expressions of thermal resistances proposed by Yovanovich and Marotta [8] and Webb [9], the overall thermal resistance of a heat sink can be expressed as a function of the power input and the log-mean temperature difference,

$$R = \frac{\Delta T_{lm}}{Q_{in}} \tag{2}$$

where the log-mean temperature difference is given by,

$$\Delta T_{lm} = \frac{(T_s - T_{air,in}) - (T_s - T_{air,out})}{\ln\left[\frac{(T_s - T_{air,in})}{(T_s - T_{air,out})}\right]} \tag{3}$$

Here, T_s is the source temperature and $T_{air,in}$ and $T_{air,out}$ are the inlet and outlet air temperatures respectively. The temperatures are time averages of 500 measurements once steady state conditions were achieved.

4. Results and Discussion

The performance of the heat sinks was characterized by comparing the various performance metrics such as the overall source-to-sink thermal resistance, heat spreading and source temperature.

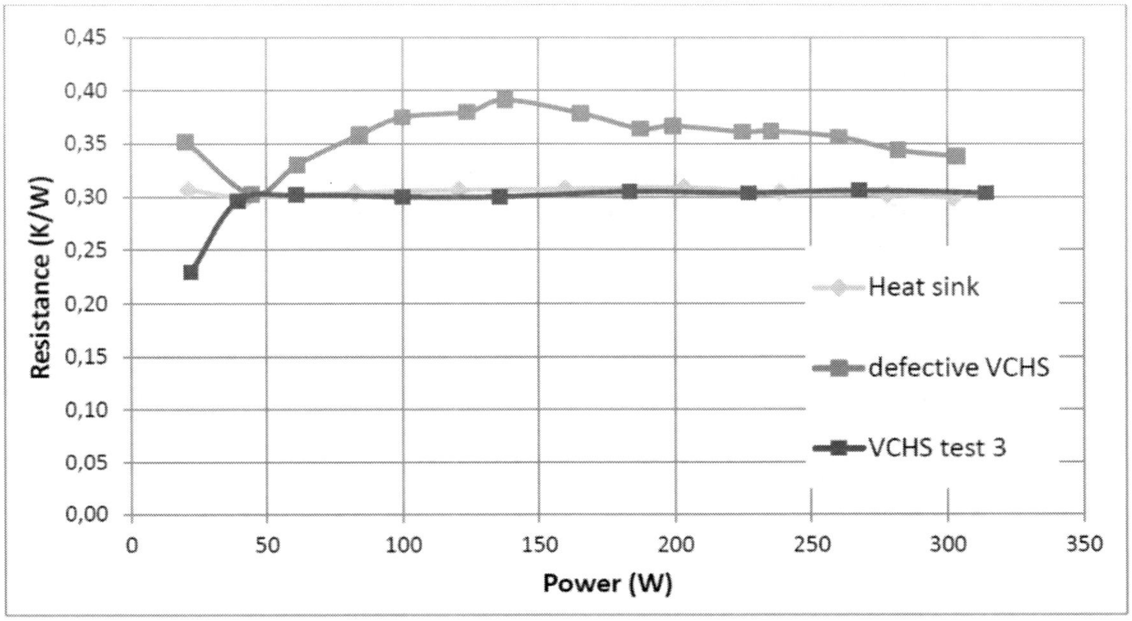

Figure 6. Plot of thermal resistance versus heat load.

Fig. 6 shows the overall source-to-sink thermal resistance for the SHS and two VCHS's. It is first noticed that the thermal resistance of the SHS is independent of the imposed power. This is not entirely surprising since the airflow was kept constant for each experiment and the aluminum thermal conductivity varies very little over the temperature range of the experiments. It does, however, indicate that the spreading resistance within the SHS is not a non-linear function of the imposed power or driving temperature differential.

Fig. 6 also shows the thermal resistance for each of the two VCHSs. It is first noticed that there is a notable discrepancy between the two units, and in fact indicates that one of the VCHSs was faulty as it had a notably higher thermal resistance than both the operational VCHS and the SHS. The operational VCHS shows a thermal resistance which, apart from the lowest power load, was not distinguishably different than the SHS.

Fig. 7 shows thermal images taken of the top-side of the heat sinks for heat loads of 40W, 140W and 240W. The left images are of the SHS and the ones on the right are of the operational VCHS. Here, the temperature scale for the false color representation of temperature is allowed to float so as to accentuate the visualization of the heat spreading. The thermal images clearly show that the VCHS has the ability to spread the heat over a significant surface area compared with the more localized footprint of the SHS. However, considering Fig. 8 where the false color temperature scale is the same for each the SHS and VCHS, it is evident that the VCHS is much cooler indicating that the contact resistance between the vapor chamber and the heat sink and/or the resistance in the wick structures are considerable.

The additional thermal resistances associated with the interface between the vapour chamber and the heat sink and/or the low effective conductivity of the wick structures has a net effect that that the improved heat spreading on the air-side is entirely offset by these additional resistances (Fig. 6).

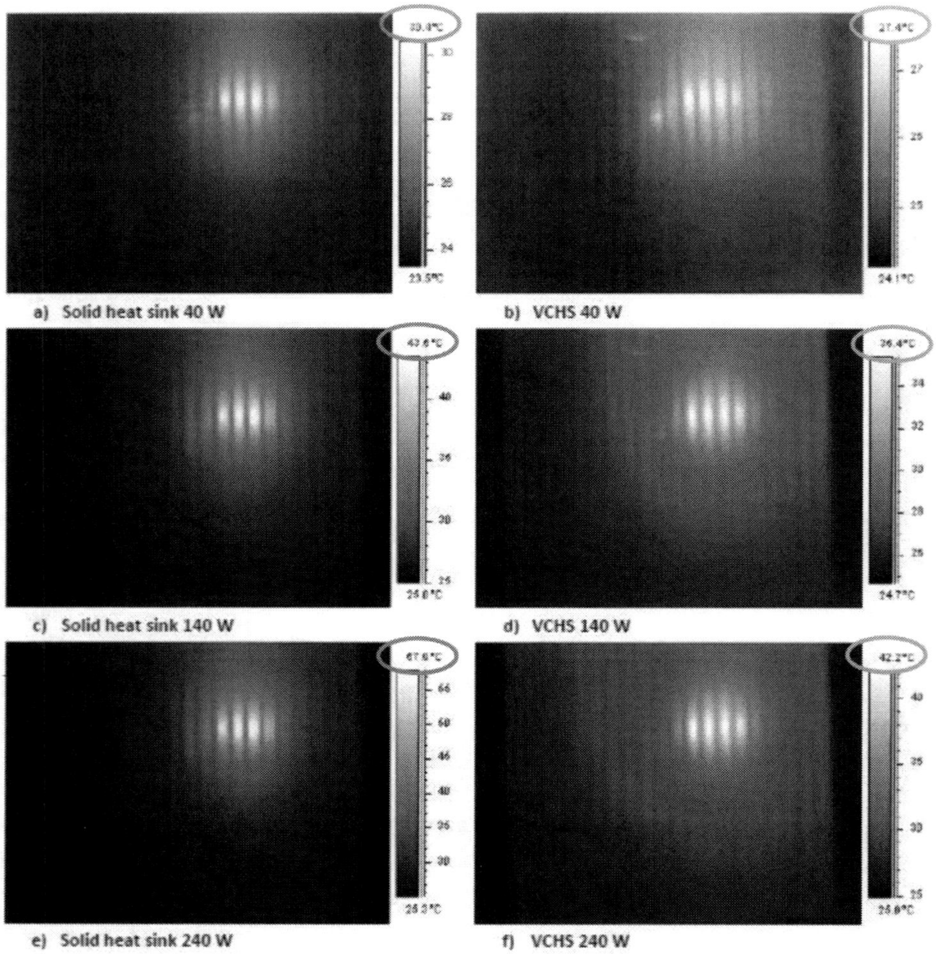

Figure 7. Thermal images showing degree of heat spreading. Floating temperature scales.

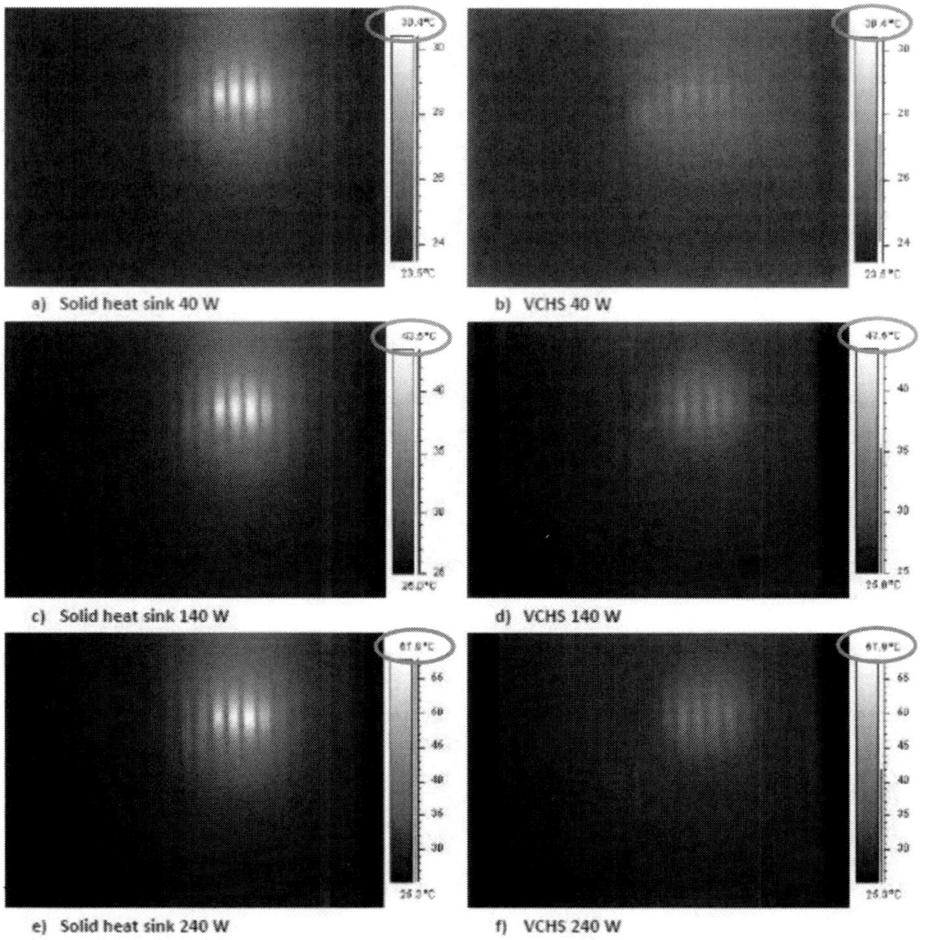

Figure 8. Thermal images showing degree of heat spreading. Fixed temperature scales.

The main purpose of the thermal management system is to keep the source temperature low enough to avoid thermal failure of the electronic component. Even still, other issues such as cost, reliability, complexity and weight are also important in the design of appropriate thermal management systems. Referring to Fig. 9 it is clear that considering all of the aspects mentioned above, this VCHS is a low performing technology. As the figure shows, it does not improve the temperature of the heat source over that of a SHS and in fact worsens the case for one of the two VCHSs tested.

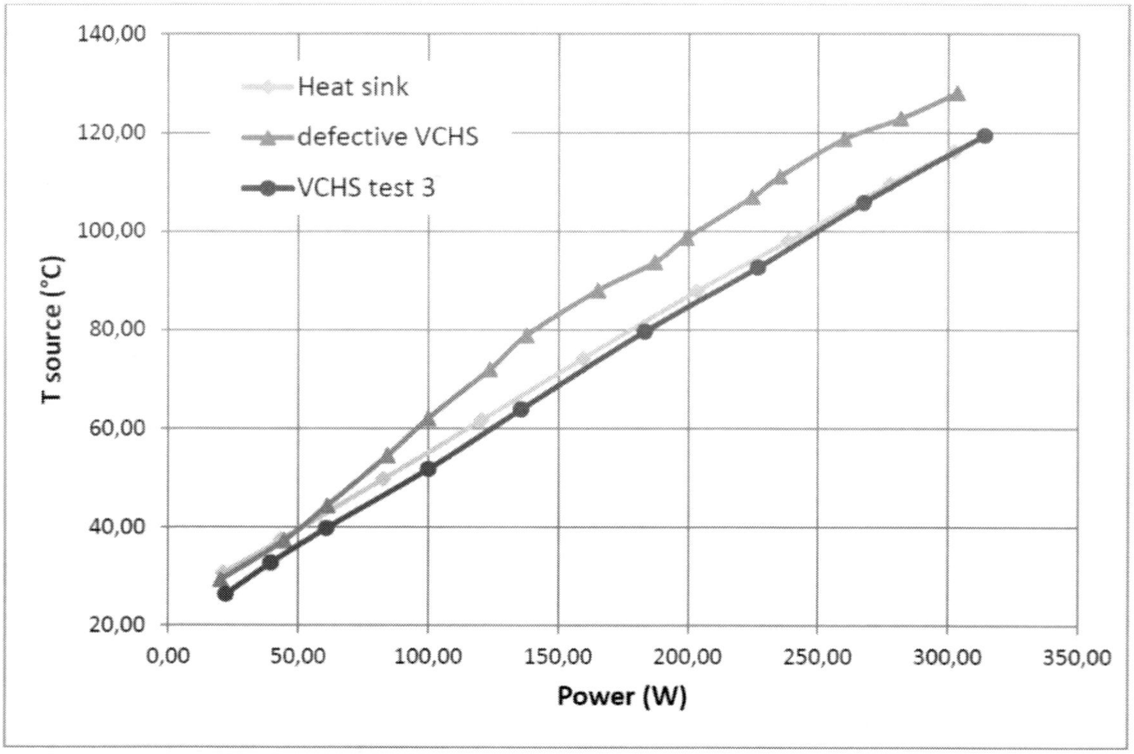

Figure 9. Plot of source temperature versus heat load.

5. Conclusions

The results of this investigation show that 'bolt-on' VCHS's must be used with caution if used with localized heat sources. Although they are effective heat spreaders, they do not necessarily improve the source-to-sink thermal resistance for this scenario. The additional thermal resistances associated with the interface between the vapor chamber and the heat sink together with the wick structures can more than offset any gains associated with the heat spreading. Full integration of the vapor chamber into the heat sink, opposed to the 'bolt-on' system, would of course mitigate the additional contact resistance.

Further, for highly populated boards with a common heat sink the heat spreading may in fact be disadvantageous since the spreading can be considered thermal pollution from high powered components to low powered components on the same heat sink which can compromise device reliability. Another factor to consider is that there are performance limitations, such as the capillary limit, associated with heat pipe types of heat exchangers. In particular, orientation could become a factor and this is not the case with solid heat sinks.

Although VCHSs show promise in terms of becoming another technology in the growing arsenal of advanced thermal hardware for electronics thermal management, more research is still required.

References

[1] Li HY, Chiang M H, Lee C I, Yang W J 2010 Thermal performance of plate-fin vapor chamber heat sinks *Int.Comm.Heat Mass Transfer* 37(7) 731–738

[2] Chen Y S, Chien K H, Wang C C, Hung T C, Ferng Y M, Pei B S 2007 Investigations of the thermal spreading effects of rectangular conduction plates and vapor chamber *J. Electron. Pack.* 129 348–355

[3] Boukhanouf R, Haddad A, North M T, Buffone C 2006 Experimental investigation of a flat plate heat pipe performance using IR thermal imaging camera *Appl. Therm. Eng.* 26 2148–

2156
[4] Tsai M C, Kang S W, Vieira de Paiva K 2013 Experimental studies of thermal resistance in a vapor chamber heat spreader *Appl. Therm. Eng.* 56(1–2) 38–44
[5] Hsieh S S, Lee R Y, Shyu J C, Chen S W 2008 Thermal performance of flat vapor chamber heat spreader *Energy Conversion and Management* 49(6) 1774–1784
[6] Koito Y, Imura H, Mochizuki M, Saito Y, Torii S 2006 Fundamental experiments and numerical analyses on heat transfer characteristics of a vapor chamber (effect of heat source size) *JSME Int. J. Series B* 49 1233–1240
[7] Sauciua I, Chrysler G, Mahajon R 2002 Spreading in the heat sink base: phase change systems or solid metals? *IEEE Trans Compon Pack Technol* 25 621–628
[8] Yovanovich M M, Marotta E E 2003 Thermal spreading and contact resistances, in: A. Bejan, A.D. Kraus (Eds.) Heat Transfer Handbook John Wiley & Sons 261–393
[9] Webb R L Heat Exchanger Design Methodology for Electronic Heat Sinks *J. Heat Transfer* 129(7) 899-901

Integrated smart two-phase spreader to enhance reliability of power electronic modules in aeronautical applications

Mikael Mohaupt[1], Laurent Barremaecker[1], Ian Wilson[2]

[1]Euro Heat Pipes (EHP), Rue de l'Industrie 24, 1400 Nivelles, Belgium
[2]Microsemi Corporation, 14930 East Alondra Blvd, La Mirada, CA 90638-5752

Corresponding author : mikael.mohaupt@ehp.be

Abstract. Reliability of power electronic modules is a key characteristic of existing and innovative modules. An increasing quantity of these modules is used in a large range of applications and addresses from annex up to vital functions, especially with the more electronic aircraft and recent developments in transport applications. To propose a solution to this lifetime issue, Microsemi and EHP have designed, manufactured and tested an improved flat heat pipe to fulfil aeronautic requirements. The frame of this development is based on the existing SP3 power module of Microsemi and significant decrease of thermal resistance is demonstrated by thermal tests.

1. Introduction

Over the last years, significant improvements in several domains such as materials, manufacturing tools and processes, miniaturization and packaging have increased electrical component reliability and power level. Parts of them, power electronic modules, have to deal with power conversion and electronic functions from basic operations up to more complex applications. Among others, transport related applications for railway, aeronautical or space business are using high quantities of power modules in complex and critical products like security systems and primary flight commands for airplanes.

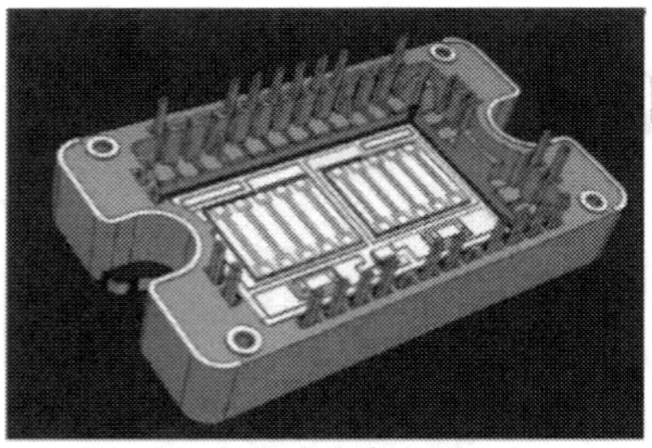

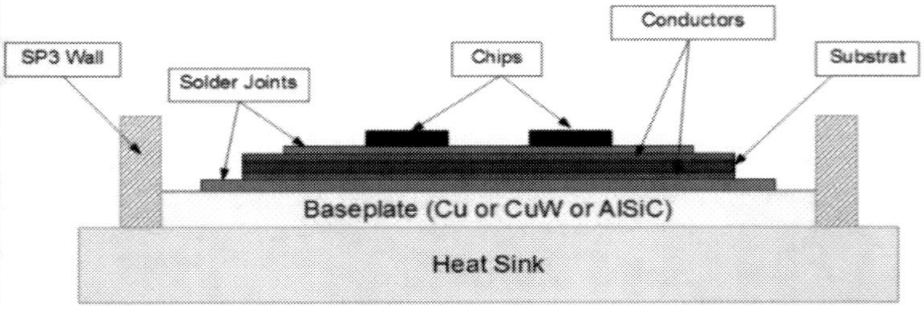

Figure 1 SP3 Microsemi power electronic module (Left side) – Layer implementation (Right side)

Reliability should not be limitative while power levels increase constantly; thus, several developments to raise module lifetimes are in constant development. Difficulties lie in the fact that for almost all transport applications, security, performances and weight aspects are drivers in the maturation of a new technology. Most power electronic modules are built by layer implementation, from chips passing by substrate and baseplate to the heat sink (Figure 1). At the chip level, compared

Content from this work may be used under the terms of the Creative Commons Attribution 3.0 licence. Any further distribution of this work must maintain attribution to the author(s) and the title of the work, journal citation and DOI.
Published under licence by IOP Publishing Ltd

Journal of Physics: Conference Series **525** (2014) 012006 doi:10.1088/1742-6596/525/1/012006

to current technology in Silicon, the step passed by Silicon Carbide technology improve the intrinsic reliability of chips by allowing higher operating temperature thanks to relevant thermal and mechanical properties. Then if substrate parts are designed, at best, to fulfill electronic insulation and then thermal capabilities, their presence induces a thermal gradient. Indeed, the low thermal expansion of ceramics compare to metals of sinks reduces considerably the stresses acting on the chips during high range thermal cycling. Connected to the heat sink, which is most often aluminum and provided by the end-user, the baseplate is the simplest part but also the most problematic in terms of performances. The baseplate thermal and mechanical effects on the module and its mass are not negligible. One of its main functions is to spread the heat on the heat sink to enhance the heat dissipating surface, so it must have high thermal conduction coefficient and the lowest possible mass. To spread heat evenly, its thickness must be large enough but on the opposite, the larger the thickness of the baseplate, the greater the thermal gradient and the mass. These two contradictions are the major issue to design this part and the best compromise must be found.

The purpose of the present study is to expose the implementation of two-phase technologies in power electronic modules to reduce thermal resistance. The designed two-phase spreader fulfils the thermal performance and is designed to sustain severe environment solicitations required by aeronautical applications. After a review of possible implementations of the two-phase spreader in the module, a description of the heat transfer principle and of the design rules is provided and the description of the manufactured device will be done. Then the performed test campaign and test capabilities of the device are presented and analyzed, to give quantitative and qualitative gains, in terms of performances and lifetime, on using such a device in the cooling of power modules. Finally, complementary test campaign to complete aircraft qualification process and foreseen applications and further developments are discussed.

2. Technical overview

2.1. Heat pipes theory

By offering a passive and an efficient way to cool down electronics, two-phase solutions like heat pipes use the latent heat of a fluid to transport the dissipated heat from hot points to remote cold sources. The hot interface converts liquid to vapor which store the heat energy to be transported to the cold side. Condensation occurs due to the connected cold source (through the heat sink). Liquid is then brought back from cold interface to the evaporation area thanks to a passive pump device provided by the capillary structure, called the wick. This passive pump does not need any action or energy: the pressure difference between liquid and vapour phase induces in each capillary a meniscus which cause pressure jump, quantified by the Laplace law, given hereafter and illustrated on Figure 3.

$$\Delta Pcap = \frac{2\sigma\cos(\theta)}{r_p}$$

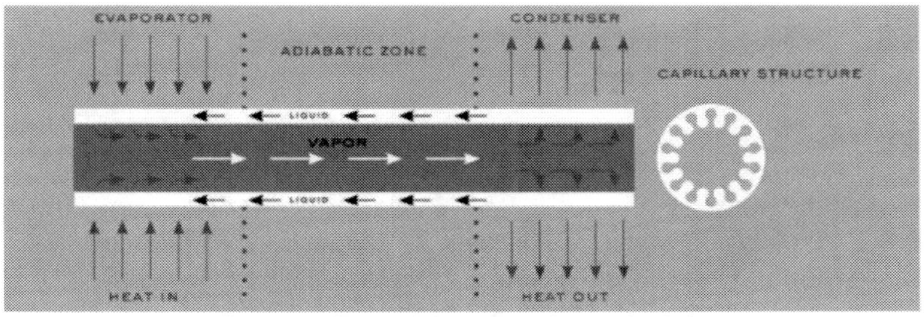

Figure 2 Heat pipes operating principle

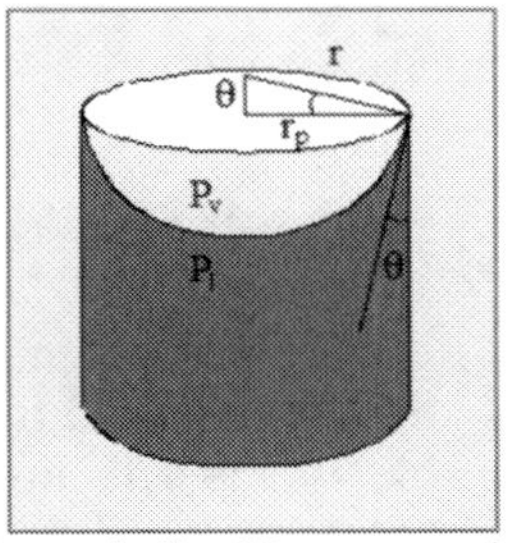

Figure 3 Capillary pumping

2.2. Design considerations of the aeronautic two-phase spreader

The design of a heat pipe needs to comply with hydraulic and thermal physical behaviours. In order to be able to extract the heat, both of these physical properties must allow the fluid loop in the device. Due to its complexity, designing a heat pipe to a given application imply to consider several aspect limits among the viscous, the freezing or the sonic limit if the working temperature is too small or capillary and shear limits for example. These are largely discussed in [1]. Most important for the purpose of this paper, the *boiling limit* appears when the boiling at the evaporation interface is so intense that the vapour flow cannot be extracted and a vapour film is created. Thus, the heat from the isolated source is not extracted at the evaporation interface and its temperature rises drastically and the pressure inside the heat pipe rises accordingly. This limiting consideration represents the major risk when designing a heat pipe and must be defined properly in the specification, especially by specifying the maximum heat flux surface density at the evaporator interface. Also called the critical heat flux (CHF) of two-phase device, the boiling limit is completely dependent on the internal design of the evaporation interface, the wick structure and efficiency and obviously the properties of the working fluid. An exhaustive review of the phenomenon and applicable laws in several configurations has been proposed in [2]. As an example, classical ammonia grooved heat pipes (Figure 4) typical CHF is estimated around 5-10 W/cm². It represents the strongest limitative aspect in the use of cooling high density power electronics thanks to heat pipes.

Most sophisticated two-phase devices like loop heat pipes (LHP) or capillary pumped loops (CPL) (see Figure 5), first addressed respectively in [3] and [4], have a higher boiling limit and could manage up to 100 W/cm² depending on the internal design of the evaporator. The capillary pump is concentrated in the evaporator and thus the evaporation interface could be structured to deal with higher surface heat flux densities. Integration of this kind of device as well as their costs is a real challenge in aeronautic environment and their direct use inside a power electronic module imply significant structural changes.

Figure 4 EHP grooved heat pipe

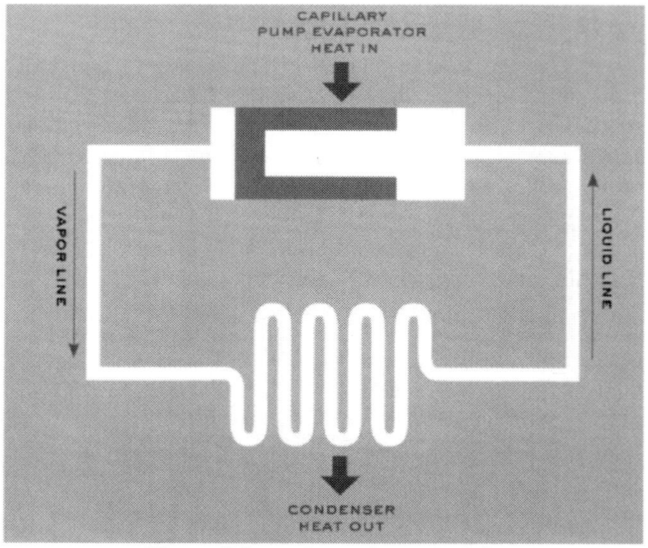

Figure 5 Loop heat pipe principle

Recent developments performed the past few years at EHP and validated on railway application put forward an enhanced internal design of evaporator interface and its connection to the wick structure capable to manage up to 100 W/cm². The porous media is located inside the sealed chamber, connecting the condenser side to the evaporator side (Figure 6). At the evaporator side, the grooves are designed at the best compromise to maximize the contact surface and to give enough space to the created vapour to be extracted. The condenser grooves are manufactured in the body to maximise heat transfer coefficient with the cold source and enhance condensation of the working fluid.

The innovative concept lies in the evaporation area extension thanks to a porous part which bring liquid fluid onto the upper side (according to Figure 6) of the vapour channels. Thus, the connection between the porous and the evaporation area is enhanced in terms of hydraulic connectivity, thermal contact and mechanical strength of this interface.

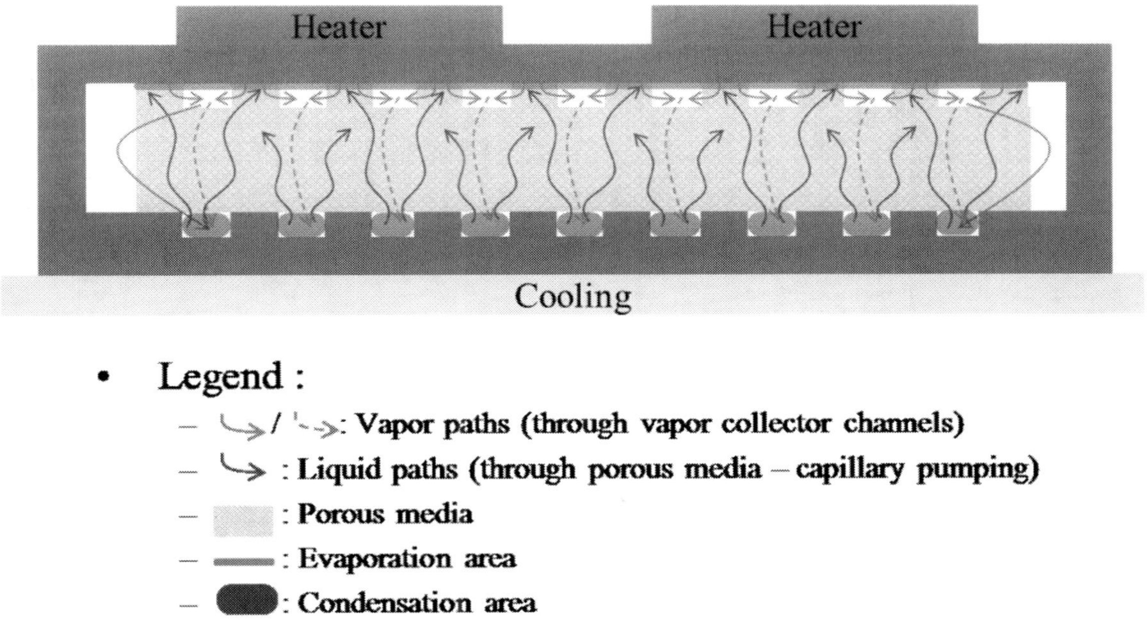

Figure 6 Two-phase spreader internal design

Implementation of this particularly effective capillary structure inside a flat heat pipe, the two-phase spreader, has been developed, manufactured and tested. But before, a strong reflexion about possible ways to implement this smart two-phase spreader in power module application to enhance best thermal properties has been performed and is exposed in the next chapter.

3. Two-phase spreader integration in power electronic module

Several implementations in power electronic modules are possible considering used materials, availability of cold source and possible connections to it and obviously the maturity of the existing or new develop module. Three of them have been studied and are described hereafter with succinct description of maturity and benefits on performances.

3.1. Outside one or several modules

The first and simplest implementation implies no change in the module design. The two-phase spreader is placed between the module baseplate and the same heat sink is used to cool down one or several modules (Figure 7).

Journal of Physics: Conference Series **525** (2014) 012006 doi:10.1088/1742-6596/525/1/012006

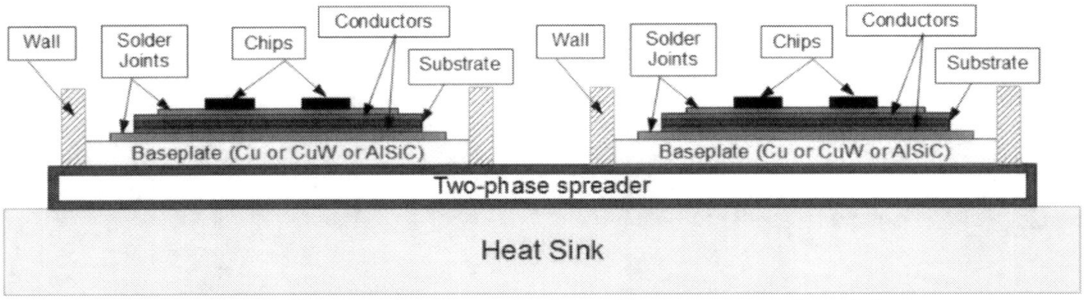

Figure 7 The two-phase spreader is outside one or several modules

If this kind of implementation does not reduce the thermal gradient inside the power module, it allows to share a single cold source between several modules and thus the effective cooling surface (connected to the heat sink) could be largely enhanced, especially in the case of small sized modules (between ~50*50 mm^2 and ~100*100 mm^2).

3.2. Inside the module, replacing substrate and baseplate layers

Another possible implementation consists in removing substrate and baseplate layers replacing them by the two-phase spreader (Figure 8).

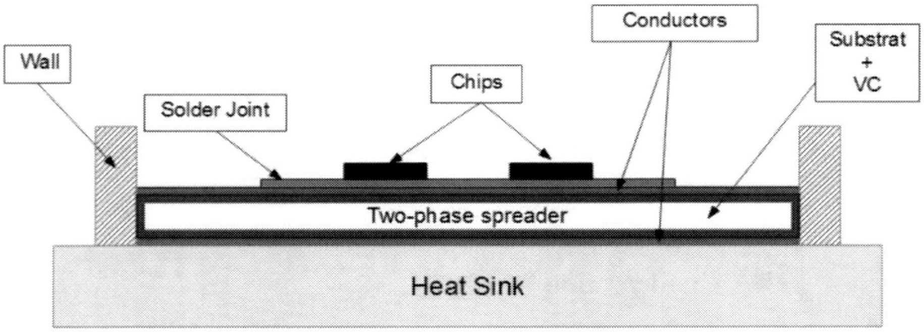

Figure 8 The two-phase spreader replaces the substrate and the baseplate layers

This solution is the most effective considering thermal issues. Indeed, several layers inside the module are removed and the path of heat to dissipate from chips is reduced almost at maximum. Thus, the two-phase spreader ensures two vital functions: the electrical insulation and the heat transfer. The electrical insulation is provided by ceramic materials and the enhanced heat transfer by the working fluid. In addition, low thermal expansion of ceramics is more convenient to reduce the stress on chips during large-range thermal cycling.

Nevertheless, this implementation requires the largest development because the spreader assembly inside the module needs new processes, even if it is possible to keep the original footprint of power modules. In addition, manufacturing ceramics is more complicated and the sealing process of the two-phase spreader needs innovative components. Finally it should be noted that raw ceramics and precise machining of them are expensive compared to metal. Despite these issues, several breadboards have already been realised to validate the feasibility of such a device.

3.3. Inside the module, replacing baseplate layer

This presentation is a good compromise between development progression, thermal performances, manufacturing possibilities and costs. It can be seen as a development step to the ceramic two-phase spreader (§3.2.) as well as a full solution.

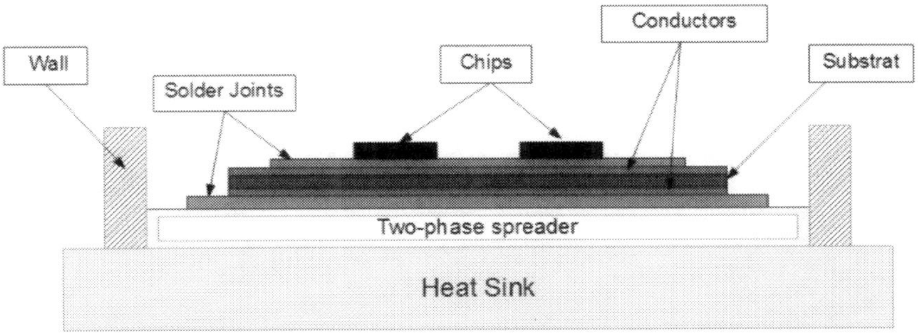

Figure 9 The two-phase spreader replaces the baseplate layer

Implementation inside an existing module is easier and necessitates a reduced number of process developments because the two-phase device is provided in metal like classical baseplates. Thus, the whole module except the baseplate remains unchanged. A significant reduction of the thermal gradient is obtained with only minor changes in power electronic module design and manufacturing.

This is the chosen integration tested by EHP on a typical aeronautic application of the Microsemi SP3 power module.

4. Thermal characterisation tests

A dedicated test campaign has been performed to demonstrate thermal performances of the two-phase spreader design under a general aeronautic specification. Tests have been performed under laboratory conditions but their design has taken into account margin to fulfil aeronautic requirements.

4.1. Specification

The specification set corresponds to a typical aeronautic application requirement and is resumed in **Table 1**.

Table 1 Present SP3 specification set

Maximum power dissipation	P (Tc=25°C)	250 W
Maximum surface heat flux	q	50 W/cm²
Baseplate SP3 junction to case Thermal Resistance	Rthjc	0.26 °C/W
Operating junction T° range	Tj	[-40°C; 150°C]
Storage T° range	Tstg	[-40°C; 125°C]
Operating case T° range	Tc	[-40°C; 100°C]
Operating sink T° range	Tsink	[-40°C; 70°C]
Mounting torque to heatsink (M4)	Torque	[2.5 N.m; 4.7 N.m]
Max thickness	e	8 mm
Gravity constraint	g * G	9.81 * 2.6
Lifetime	/	150 000 h
Spreader inside the module – Same footprint as present SP3	/	/

Data to challenge is the Rthjc, which represents the thermal resistance between the *junction* of the chip which is connected to the module and the *case* which is the baseplate to heat sink interface.

A trade-off study has been performed considering the goal of TRL-3/4. This trade-off takes into account parameters like thermal efficiency, mechanical properties, fluid characteristics as well as costs of development and maturity of processes. The conclusion figures a nickel body filled with methanol is a good first choice to test thermal performances and manufacturing validity. Nevertheless, other fluids such as Acetone, Water or Refrigerants are possible to be used depending on the wanted performances. The wick is composed of nickel sintered porous media of about 5 µm pore radius.

4.2. Test set-up and monitoring

A complete description of the test set-up is given on Figure 10. Dedicated tooling has been manufactured to represent at best the chip distribution and dissipation. Two ceramic heaters are clamped on a heating tool which is dedicated to concentrate the heat flux on the equivalent surface of the most dissipative chips in the module, the IGBTs. It is assumed that all the heat power to dissipate from the module comes from these chips. The cold plate temperature is controlled thanks to a cryostat connected to it with thermal grease. Rains are manufactured in it to measure temperatures, homogeneity and so spreading effects.

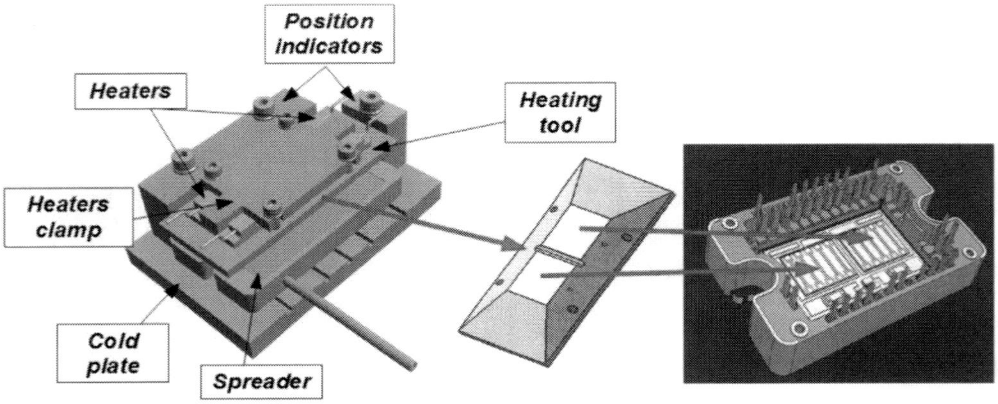

Figure 10 Representative SP3 tooling and test assy of the two-phase spreader

The two-phase spreader, its tooling and ambient is fully instrumented to measure temperatures, as presented on Figure 11. The whole device is isolated in polystyrene insulation to minimize the impact of laboratory ambient.

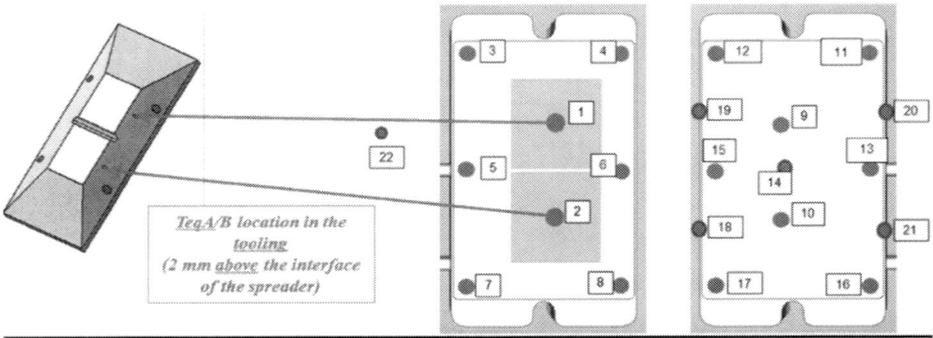

N°	Temperature Sensor	Localisation
1	Teq A	Inside the heating tool, under the heater A
2	Teq B	Inside the heating tool, under the heater B
3 to 8	Tevap	On the evaporator face
9	Tcond A	On the condenser face under the heater 1
10	Tcond B	On the condenser face under the heater 2
11 to 17	Tcond	On the condenser face
18 to 21	Tside	On both longitudinal face of the spreader
22	Tamb	Ambient temperature

Figure 11 Thermocouple location and definition

Even with insulation placed all around the device, to characterize the influence of the laboratory ambient, the heat loss coefficient needs to be estimated properly. Thus, the effective power from the heaters transferred to the two-phase device. A small and constant power is applied to the inactive set-up which is empty of fluid. A long time is needed to reach stabilization of temperatures meaning that all the power is dissipate to ambient. The thermal conductance of the heater to ambient can be calculated, as described below and presented in Figure 12.

$$K_{loop} = \frac{Power}{T_{evap} - T_{cond_{mean}}}$$

$$Q_{eff} = Q_{elec} - K(T_{Equip} - T_{Amb})$$

Stabilization
Kloop = 0.36 W/K

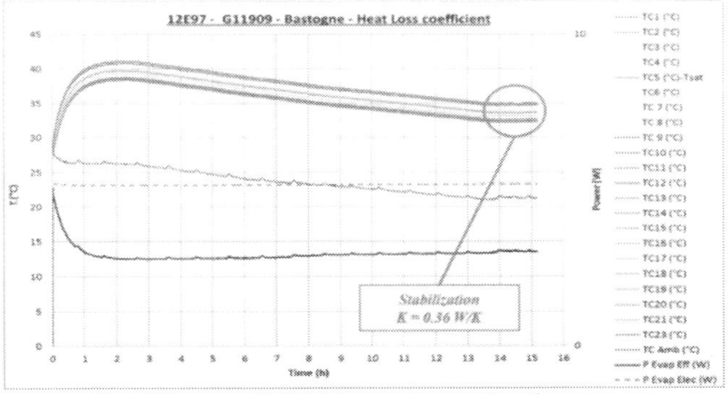

Figure 12 Heat loss determination

On the condenser side, the connection to the heat sink should not be studied in terms of ambient losses because the temperature sensors are brazed directly to the condenser side of the spreader and so additional resistance is assumed to be negligible. Once the test set-up ambient influences are fully characterized, the test sequence shows mechanical and thermal performances.

4.3. Test sequence

The test sequence is divided into two separate parts: the mechanical resistance test and the thermal characterisation. The mechanical test is a proof pressure resistance test success at 40 bars. The thermal test sequence is summarized in **Table 2**.

Table 2 Thermal test sequence

Operation	*Effective Power [W]*	*Cold source Temperature [°C]*
Heat Loss determination	3.0	Ambient T°
Filling	124	74
Test 1	283.1W - Up to Security	25
Test 2	220.3W - Up to Security	50
Test 3	181.9W - Up to Security	70

The security of the setup is the melting point of the polystyrene insulation which is 125°C. According this limiting value, software and hardware security has been implemented respectively at 115°C and 120°C.

4.4. Test results

As with all two-phase devices, the two-phase spreader thermal performance is correlated to the working fluid filling quantity. The exact filling of fluid to obtain the best spreader conductance is a function of boundary conditions that are the cold source temperature and the effective power passing through the two-phase device. It is important to notice that this optimal point should be defined properly considering application needs and could easily be adapted to specific requirements. In present study, the filling point has been defined at condenser side temperature fixed at 80°C and 124W of heat power to dissipate. In that particular case, the single two-phase spreader thermal resistance is calculated around 0.11 K/W. Performances differ slightly on an acceptable range around the best operating point, as shown on following test results performed with different sink temperatures to cover especially high values, from 25°C to 70°C.

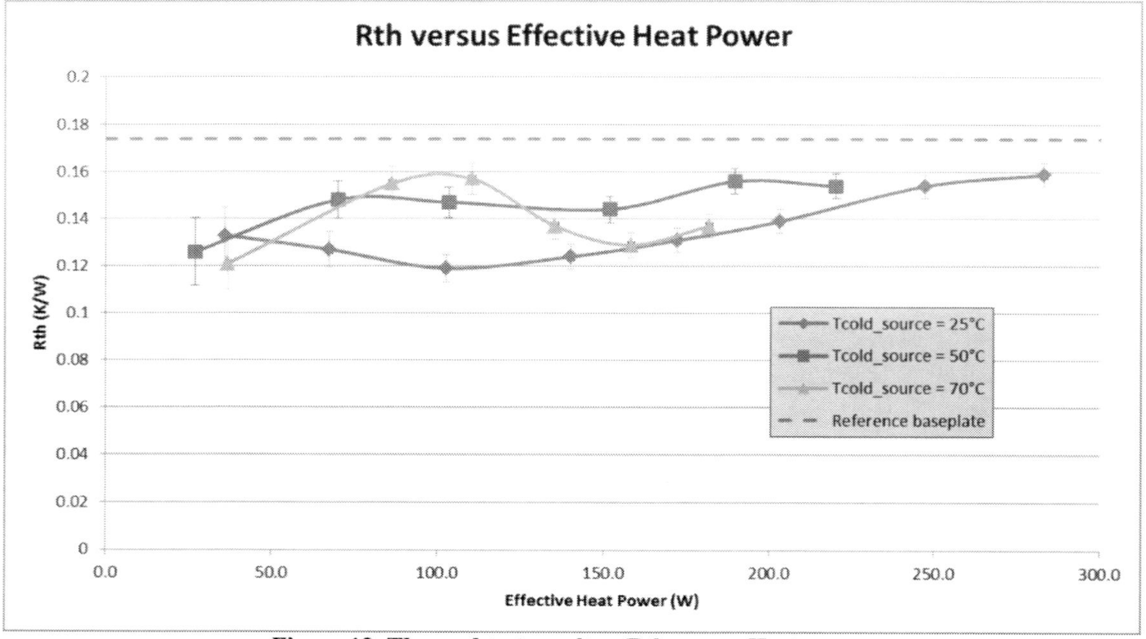

Figure 13 Thermal test results – Rth versus Heat power

The two-phase spreader demonstrates its ability and robustness by offering a thermal resistance competitive with respect to isolated baseplate, estimated by analytical model not lower than 0.174 K/W, part of the 0.26 K/W of the SP3 module (see **Table 1**). *A gain between 10% and 35% is thus put forward, which could represent up to 16°C gain on junction temperature under 250W of dissipative heat power.*

As the size of the interface plays a preponderant role in the estimation of Rth, it should be noticed that the spreading effect rise when the size of the module grows. The spreading, thanks to the working fluid, fixes the case temperature (on the condenser side of the spreader) to the exact same temperature. By eliminating hot spots due to the bad spreading of purely conductive materials, the sink to case interface is used at its best capabilities.

Although the impact on reliability for SiC chips technologies with this thermal junction temperature drop is not well known, such a gain on Si chips would more than double lifetime for the power electronic module. As demonstrated by these representative tests, a two-phase device enhances significantly the cooling solution whatever the sink of application and opens a way to greatly increase power module electronic lifetime. New applications could be foreseen thanks to the proposed development.

5. Application

A 5kW power module for the more electric commercial aircraft under ever increasing reliability and dynamic requirements is in development. Flight critical components on next generation aircraft have a mission life goal equal to 150,000 flight hours. At 95% efficiency, the module requires 250W of power dissipation. Due to the very large power requirements and small size of this 5kW power module, thermal management is a critical driver for system reliability and mission life. Use of the two-phase spreader also allows heat to be directed through specific higher level systems. This direction of heat paths provides an opportunity for differential cooling. Through differential cooling, heat from the power module will not impact control circuitry, lowering junction temperatures for control circuitry, and increasing system reliability.

Reduced internal assembly temperatures resulting from the use of a two-phase thermal spreader (instead of a conductive one) can significantly increase the overall reliability of an assembly. Based upon parameters in **Table 1** the maximum internal power module temperature range is almost 184° (140°C due to the operational limits for the case and 43.5°C due to the thermal resistance of the isolated baseplate estimated by calculus to be at minimum 0.174°C/W). However, with the two-phase thermal spreader, thermal resistance drops to 0.11°C/W and the temperature range falls to only 167.5°C. Assuming a conservative fatigue coefficient of 2, the 16°C reduction in temperature range results in almost 17% more life on solder joints internal to the power module [5]. On the other hand, because of the reduced thermal resistance, the power module output can be increased to 7.5 kW and have the same size assembly, with higher powers, and a lower internal temperature delta than the existing 5 kW assembly.

6. Conclusion and further work

Feasibility and interest of the two-phase spreader solution to replace the baseplate of a typical power electronic module is demonstrated by the present study. Thermal performances cover the specification points and better results could be obtained by choosing the appropriate filling of working fluid for a given application and a more specific test set up.

The proposed next step is a thermal test in laboratory conditions with a two-phase spreader integrated inside a real SP3 module. This test should validate first results exposed in this paper as well as manufacturing processes of integration. In most cases, experience shows manufacturing issues correspond to the development of brazing processes.

Once manufacturing of such an enhanced power electronic module and thermal performances in laboratory conditions are validated, environment tests should be performed for a representative aeronautic application. Constraints like ambient temperatures from -40°C to 70°C, pressure variations due to altitude and effects of gravity, all taken into account nowadays by design rules, must be validated by dedicated tests.

The last development is performed in parallel of the others and consists of converting the solution developed in Nickel material for the body (presented here) with a ceramic material in contact with the rest of the module (see Figure 8). One potential gain of this change is removing several layers in the module and replacing all the substrate and baseplate (and their functions) by the two-phase spreader. Nowadays, costs of materials and developments, as well as coating or sealing manufacturing processes are the key points to consider when designing a competitive and efficient two-phase thermal solution.

References

[1] Faghri A 1995 *Heat Pipe Science And Technology* (Washington, Taylor & Francis)
[2] Katto Y 1994 Critical Heat Flux *Int. J. Multiphase Flow* **20** 53-90
[3] Maidanik J F Vershinin S V Kholodov V F and Dolgirev J E 1985 Heat Transfer Apparatus U.S. Patent No. 4515209
[4] Stenger F J 1966 Experimental Feasibility Study of Water-Filled Capillary-Pumped Heat Transfer Loops *NASA Technical Memorandum* NASA TM-X-1310
[5] Association Connecting Electronics Industries 1992 Guidelines for Accelerated Reliability Testing of Surface Mount Solder Attachments IPC-SM-785

Development of a naturally aspired thermosyphon for power amplifier cooling

[1]**S. Siedel,** [1]**A.J. Robinson,** [1,2]**R. Kempers and** [3]**S. Kerslake**
[1]Department of Mechanical and Manufacturing Engineering, Parsons Building, Trinity College Dublin
[2]Alcatel-Lucent Ireland, Blanchardstown Business & Technology Park, Dublin 15, Ireland
[3]Alcatel-Lucent Telecom, Swindon, United Kingdom

Email: samuel.siedel@tcd.ie

Abstract. This paper details the early development steps of a two-phase thermosyphon thermal management solution for power amplifiers (PA) in the telecommunication industry. These components, attached to a vertical PCB within an enclosure between the RF filter and a natural or forced convection heat sink, dissipate a large amount of heat with a high heat flux density. Currently cooled by direct contact to a shared heat sink, they tend to spread heat towards other components of their board, affecting their reliability. A thermosyphon thus appear as an ideal thermal management solution to transport the heat from the power amplifiers in order to dissipate it to a remote and dedicated natural convection heat sink. In the present study, the performance and the heat spreading of a forced convection unit is measured. A thermosyphon solution is then designed with a flat vertical evaporator and a radial natural convection heat sink and condenser. The performance of the thermosyphon thermal management solution is measured and compared to the initial solution. The limits and improvement needs of the thermosyphon solution are then discussed.

1. Introduction

The reliability and failure of electronic components is highly related to the temperature levels and variations to which they are subject [1]. Appropriate thermal management solution is thus a key design feature for enhancing their reliability. However, although each component has different temperature sensibility characteristics, different components often share the same thermal management solution on the same PCB.

This is the case for telecom equipment such as Remote Radio Head (RRH). This equipment is composed of several parts including the radio-frequency (RF) filter and the main PCB comprising a large number of electronic components. This PCB is often enclosed between the RF filter or a radio shield, and a large heat sink acting as thermal management for the electronic components. The heat sink usually dissipates heat to ambient air by natural convection on outdoor units and by forced convection on indoor units. Among the different electronic components that need to be maintained at adequate temperature, the power amplifiers (PA) dissipate high heat fluxes, up to 140 W. As they share the heat sink with all other components, they tend to thermally pollute them through the heat sink and compromise their reliability.

Content from this work may be used under the terms of the Creative Commons Attribution 3.0 licence. Any further distribution of this work must maintain attribution to the author(s) and the title of the work, journal citation and DOI.
Published under licence by IOP Publishing Ltd

In order to improve the reliability of both the power amplifiers and the other electronic components, a thermosyphon solution is considered in order to transport the heat flux produced by the PA to a remote, dedicated heat sink. The objective of this solution is not only to enhance the thermal management of the PA, but also to decouple thermally the PA from the other components of the same board, unlike other solutions such as vapor chamber heat sinks [2].

Thermosyphons are recognized as an attractive thermal management solution in many applications due to their passive operation, low cost, and ability to transfer heat over relatively large distances with small temperature differences. Unlike heat pipes, thermosyphons operate under the influence of gravity and without the use of a wick to return the condensate to the evaporator. For this reason thermosyphons are orientation specific and most commonly they consist of straight tubes placed vertically to ensure the liquid fully returns to the evaporator.

2. Experimental Apparatus

The first step of the development of a thermosyphon thermal management solution for RRH consists in a proof of concept *in situ* of a thermosyphon prototype, and of the estimation of the performance in terms of thermal resistance and thermal spreading of both the original and the modified version. This section describes the indoor forced-convection Transmit Receive Duplex Unit (TRDU) chosen and the design of the thermosyphon prototype.

2.1. Test telecommunication equipment

The indoor unit on which the prototype thermosyphon is tested consists of a different parts stacked, as depicted in Figure 1. The main part of interest is the PA board. It is on one side protected by a radio shield and placed against the radio filter. On the other side, the board is in direct contact with an aluminium *casted* finned heat sink. Another PCB is located 3 mm apart of the heat sink. The whole system is enclosed into a rack structure. A fan array at the bottom ensures the air flow and subsequent forced convection on the heat sink.

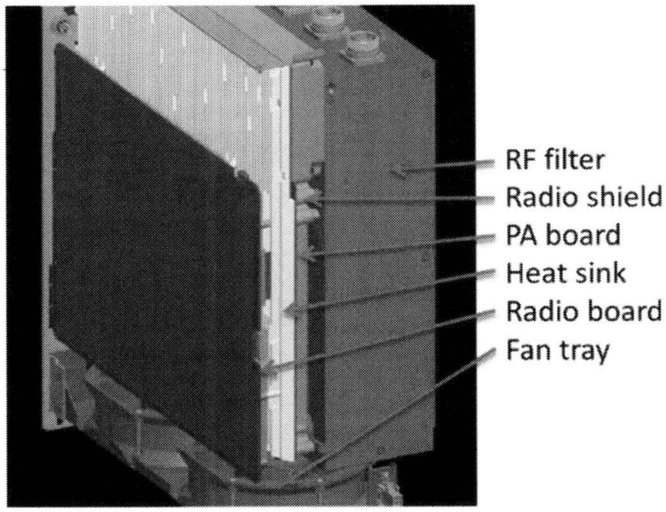

Figure 1: General layout of the system

The PA board comprises many electronic components. The main heat dissipating components are the power amplifiers, which are grouped into 3 pairs (Figure 2). Each pair consists of a main and a peak power amplifier. In normal operation, most of the power, and thus most of the heat generation concerns the main PA. The objective of the current development is to provide an adequate thermal management of the PA without polluting thermally other components of the same PCB which may be more temperature sensitive. The PA board of the test apparatus has been modified in order to directly control the power consumed and the subsequent heat flux generated by the power amplifiers.

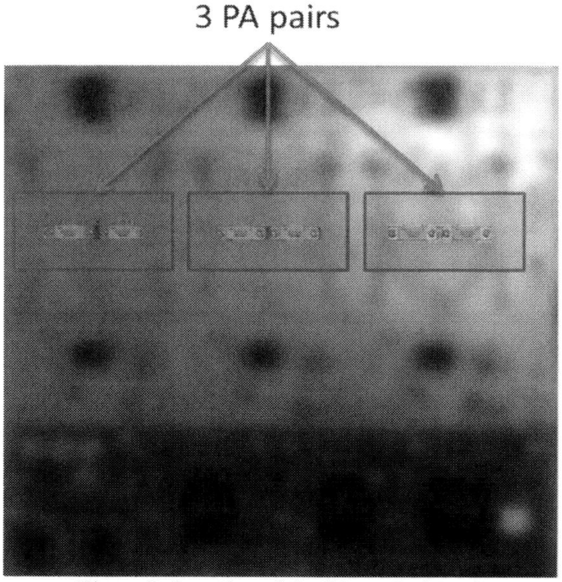

Figure 2: Location of the PA on the board

2.2. *Thermosyphon prototype*

A thermosyphon is a closed system containing a fluid (see Figure 3). It comprises an evaporator in contact to the heat source, an adiabatic section for thermal transport and a condenser which releases the heat transported. In the developed solution, the evaporator is a rectangular copper container. Each pair of power amplifier shares a same evaporator, to which they are in contact using a thermal interface material (TIM). The heat is conducted through the copper wall of the evaporator and transferred to the working fluid, which is water. The water thus boils and the heat is transformed into latent heat in the vapour phase.

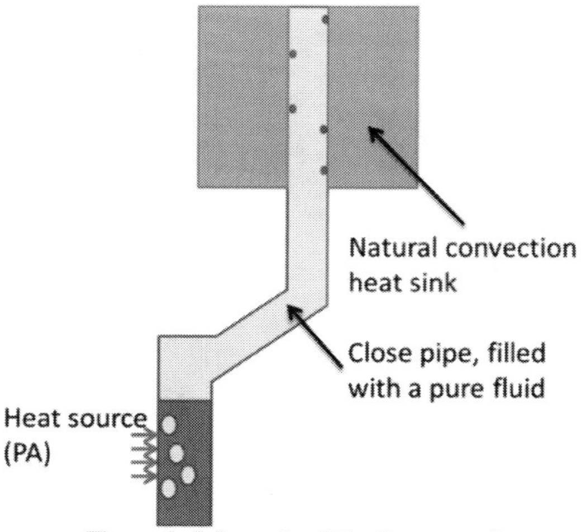

Figure 3: Schematic of the thermosyphon

The vapour is transported through the adiabatic section, which consists in a 10 mm inside diameter, 50 cm long stainless steel pipe. The vapour then enters the condenser, which is in direct contact to a natural convection heat sink. The latent heat is released to heat sink by condensation. The heat is then conducted through the heat sink and released to the ambient air by natural convection. The condensate formed into the condenser flows back to the evaporator due to gravity.

Both advantages of a thermosyphon solution are illustrated in Figure 4. First, the evaporating and condensing temperature are close to each other, at the saturation temperature of the fluid, which allows dissipating efficiently the heat flux to the air with a hot heat sink. Also, the adiabatic transport of the latent heat in the pipe gives the opportunity to separate the heat sink at a remote location from the heat source.

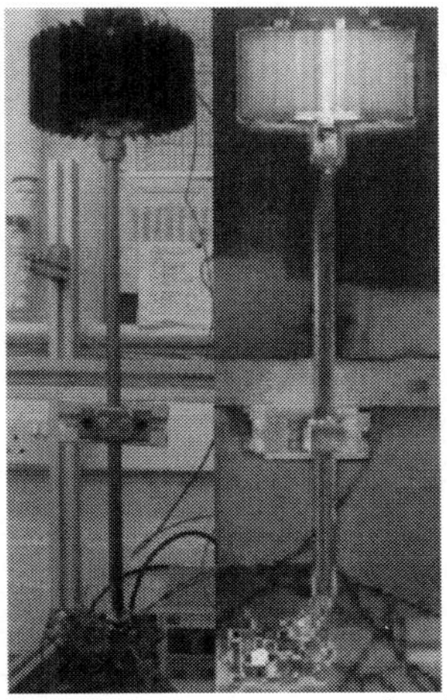

Figure 4: Thermosyphon in operation, normal and infrared visualisation

For the tests using the thermosyphon prototype, the test facility described in Section 2.1. is modified by cutting a hole through the heat sink, the radio board and the case in order to bolt the thermosyphon evaporator at the back of the central PA pair.

3. Thermal characterization

In order to properly characterize the performance of the thermosyphon solution, a methodology is developed with its adequate metrology in order to define the cooling efficiency and the thermal spreading on the board. The performance is measured for both the initial forced convection thermal management solution and with the use of the thermosyphon prototype. The two solutions are then compared, and conclusions are discussed in order to define the improvements needed for the next steps of development.

During the tests, only the main power amplifier of the central PA pair is powered, in order to best estimate the heat dissipation in the board. As the components are cooled by the back of the board (either by the forced convection heat sink or by the thermosyphon), the board temperature is measured by means of IR visualization on the front of the board (see Figure 5). A black paint of homogeneous and known emissivity is applied at four key locations on the board in order to measure accurately the local temperature (see Figure 6). In addition, the ambient air temperature is measured by a thermocouple.

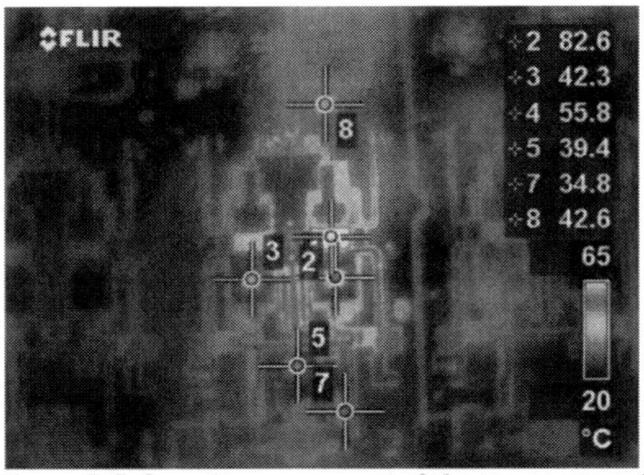

Figure 5: Infrared measurement of the temperature field

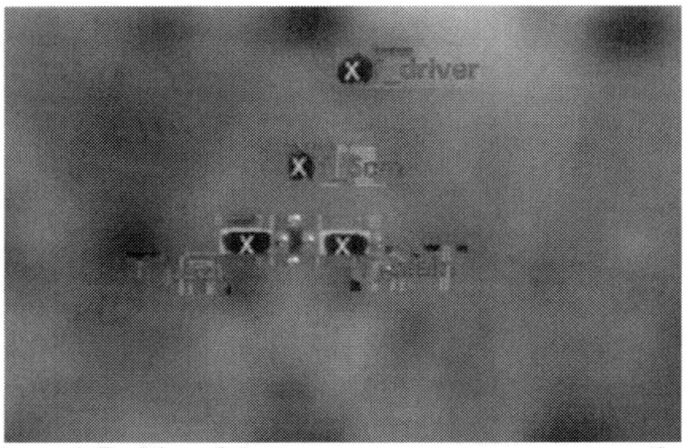

Figure 6: Position of the temperature measurements on the PCB

The temperature measurements are summarized in Table 1. The total thermal resistance R_{th}, or cooling efficiency, is defined in [K/W] as a function of the temperature difference between the main PA and the ambiant air and of the heat flux P dissipated by the transistor, such as:

$$R_{th} = \frac{T_{main} - T_{amb}}{P}$$

Table 1: Description of the temperature measurements

Label	Description
T_{main}	Temperature on the main power amplifier component
T_{peak}	Temperature on the peak power amplifier component
T_{5cm}	Temperature at a central location about 5 cm up from the PA pair (downstream)
T_{driver}	Temperature close to the driver chip, thermally sensitive component, about 10 cm downstream from the main PA
T_{amb}	Ambiant temperature, entering either the forced convection heat sink or the natural convection dedicated heat sink of the thermosyphon

The heat spreading S_i is defined in [%] at each location i on the board such as:

$$S_i = 100 \times \frac{T_i - T_{amb}}{T_{main} - T_{amb}}$$

4. Results and Discussion

Table 2 presents three sets of results, with a power of 100 W dissipated in the main power amplifier. The first set was measured with the original forced convection thermal management solution. The second set was measured with the thermosyphon installed as a cooling solution for the central PA pair, keeping forced convection on the remaining area of the heat sink to cool the rest of the PA board. For the third set, forced convection was avoided on the heat sink by turning of the fans. Table 3 presents similar sets of results for a power of 140 W provided to the main PA.

Table 2: Heat load of 100 W on the main PA

	Total thermal resistance K/W	Spreading Peak %	Spreading 5cm %	Spreading driver %
Forced convection	**1.12**	**18.1**	**25.5**	**13.8**
TS and fan on	**0.94**	**39.8**	**15.3**	**6.7**
	-16%	119%	-40%	-51%
TS and fan off	**1.02**	**50**	**27**	**16.5**
	-9%	176%	5%	20%

Table 3: Heat load of 140 W on the main PA

	Total thermal resistance K/W	Spreading Peak %	Spreading 5cm %	Spreading driver %
Forced convection	**1.05**	**19**	**27.5**	**15.5**
TS and fan on	**0.79**	**40.1**	**14.7**	**5.2**
	-25%	111%	-47%	-66%
TS and fan off	**0.94**	**49.4**	**27.6**	**17.6**
	-10%	160%	0%	14%

The cooling efficiency has been improved using the thermosyphon prototype by decreasing by up to 25 % the total thermal resistance between the main PA and the ambient air. The efficiency increases for both the forced convection and the thermosyphon solutions when increasing the power dissipated, due to enhanced convection on higher temperature heat sinks. When the fan is on, the forced flow contributes to cooling the thermosyphon evaporator, improving its cooling efficiency.

As both main and peak power amplifiers share the same evaporator in the thermosyphon solution, they are thermally linked. Consequently, the thermal spreading between the PA is greatly enhanced. Concerning the heat spreading to other locations on the board, it is significantly decreased using the thermosyphon with the fan on, as the PA are thermally decoupled from the board. However, there is no improvement with the fan off as the stagnant air in the fins of the board heat sink is slowly heated by the thermosyphon evaporator.

5. Conclusions

A thermosyphon prototype equipped with a dedicated natural convection heat sink has been successfully tested as a thermal management solution for power amplifiers on an existing RRH unit. The design of this first prototype was sufficient to slightly improve the cooling efficiency of the PA, while reducing heat spreading from the PA towards other components sharing the same PCB.

References

[1] Ohring M 1998 *Reliability and failure of electronic materials and devices* (London: Academic Press)

[2] Skandakumaran P, Khanikar V, Smalc M, Norley J, Reis B 2008 Passive, lightweight thermal solutions for Remote Radio Head (RRH) electronics *11th Intersociety Conference on Thermal and Thermomechanical Phenomena in Electronic Systems (Orlando, FL, 28-31 May 2008)* pp 25-30

Journal of Physics: Conference Series **525** (2014) 012008 doi:10.1088/1742-6596/525/1/012008

Experimental characterization of novel microdiffuser elements

L Ehrlich[1], J Punch[1], N Jeffers[2] and J Stafford[2]
[1]CTVR, Stokes Institute, University of Limerick, Limerick, Ireland
[2]Bell Labs Ireland, Thermal Management Research Group, Alcatel-Lucent, Dublin, Ireland

E-mail: leicester.ehrlich@ul.ie

Abstract. Micropumps can play a significant role in thermal management applications, as a component of microfluidic cooling systems. For next-generation high density optical communication systems, in particular, heat flux levels are sufficiently high to require a microfluidic circuit for cooling. Valveless piezoelectrically-actuated micropumps are a particularly promising technology to be deployed for this application. These pumps exploit the asymmetric flow behaviour of microdiffusers to achieve net flow. They feature no rotating or contacting parts, which make them intrinsically reliable in comparison to micropumps with active valves. In this paper, two novel microdiffuser elements are reported and characterized. The micropumps were fabricated using a 3D Printer. Each single diffuser had a length of 1800 μm and a depth of 400 μm. An experimental characterization was conducted in which the flow rate and differential pressure were measured as a function of operating frequency. In comparison with standard diffuser, both elements showed an increase in differential pressure in the range of 40 – 280 %, but only one of the elements exhibited an improved flow rate, of about 85 %.

1. Introduction

Next generation high density Photonic Integrated Circuits (PICs) show significant advantages in terms of performance, miniaturisation and energy efficiency. However the thermal management of these devices is extremely challenging. Especially, the integrated laser arrays produce high heat fluxes over 10^2 W/cm^2 and furthermore require tight operating temperature limits of about ±0.1 K. Thermoelectric modules (TEMs) are currently used to achieve thermal control, but these are limited in terms of heat flux to about 10^1 W/cm^2 and are notably inefficient in comparison with conventional vapour phase refrigeration. To overcome this thermal issue for future generation PICs, a microfluidic circuit for cooling individual devices is needed. Micropumps play a significant role for thermal control and are essential for running a microfluidic cooling system. However, these pumps have to fulfil several requirements such as a long lifetime, easy maintenance, low cost and miniature structure [1]. As valveless piezoelectric actuated micropumps feature no rotating or contacting parts (high reliability) and have a simple structure, they are a particularly promising technology to be deployed for this application. These pumps exploit the asymmetric flow behaviour of microdiffusers to achieve net flow. As the flow in artefacts such as microchannels and microgaps require high pressure heads [2], an

Content from this work may be used under the terms of the Creative Commons Attribution 3.0 licence. Any further distribution of this work must maintain attribution to the author(s) and the title of the work, journal citation and DOI.
Published under licence by IOP Publishing Ltd

enhancement of the performance of these pumps is needed to fulfil the requirements needed for the cooling of next generation PICs.

In recent years, different configurations of micropumps have been developed and studied, however piezoelectrically-actuated micropumps are the most reported and deployed. The advantages of piezoelectric actuation are a relatively high displacement, force and fast mechanical response as well as high flow rates per unit area, compared to other actuation methods [3]. The pumping action of piezoelectrically-actuated pumps can be divided into two modes, as illustrated in Fig. 1. During the supply mode, the volume of the pumping chamber increases as the piezoelectric element moves upwards. This causes a decrease of pressure inside the pumping chamber which forces the fluid to flow into the chamber through the inlet. During the pumping mode, the pressure increases inside the chamber due to the downward movement of the piezoelectric element, pushing the fluid through the outlet. For this mechanism to achieve a positive fluid displacement, valves at the inlet and outlet must be employed.

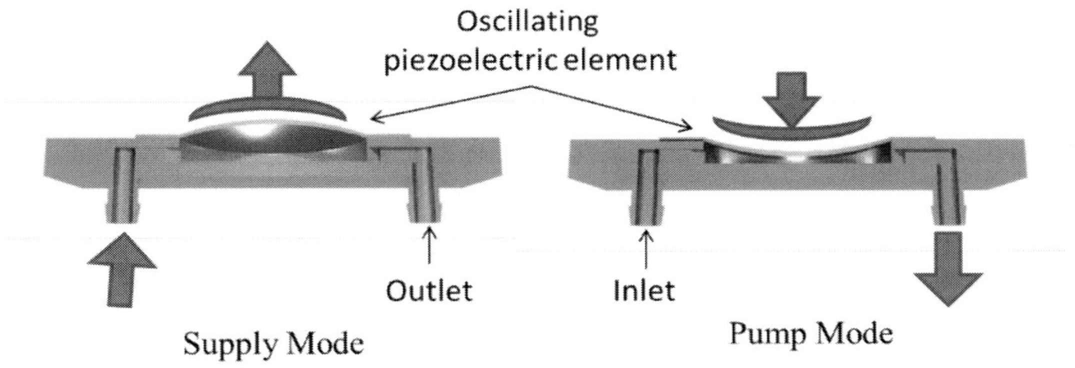

Figure 1. Operation of piezoelectrically-actuated micropump

Commercially available valveless micropumps use standard diffuser/nozzle elements to bias the flow. Several studies have been undertaken to optimize the standard diffuser/nozzle element [4-5]. However, these studies mainly concentrated on the effect of different design parameters such as diffuser angle and length. In this study, two novel diffuser elements are presented with the intention to improve the performance compared to the standard diffuser/nozzle element. The novel elements employ protrusions inside the diffuser chamber as well as side channels to reduce the backflow and consequently to increase the efficiency of the diffuser element. The micropumps were manufactured using 3D printing methods. A 3D printer uses an additive process, in which successive layers of a specific material are laid down to create the intended shape. Differential pressure and flow rate were measured for a range of operating frequencies to characterize the micropump design and experimentally assess the diffuser elements. For the two novel diffuser elements, a differential pressure improvement was achieved in the range 40 – 280 % compared to the standard diffuser/nozzle element. A net flow rate improvement of about 85 % was achieved with only one of the novel diffuser elements.

2. Novel Microdiffuser Elements

One important component responsible for the performance of piezoelectrically-actuated micropumps are the microdiffuser elements located at the inlet and outlet. As mentioned before, several studies have been undertaken to investigate the effects of geometrical parameters in standard diffuser/nozzle elements. In this study, however, the performance of a piezoelectrically-actuated micropump is intended to be enhanced by adding channels and protrusions inside the diffuser chamber to reduce the backflow during supply mode. The length and height of the investigated microdiffuser were set to 1800 μm and 400 μm, respectively. These dimensions were based on the dimensions published by Sun et al. [6]. The effects of different diffuser lengths and widths were not investigated as the purpose of this study was to characterize the effects of the additional channels and protrusions inside the diffuser chamber. The two novel diffuser elements, along with the standard diffuser/nozzle element, are shown in Fig. 2.

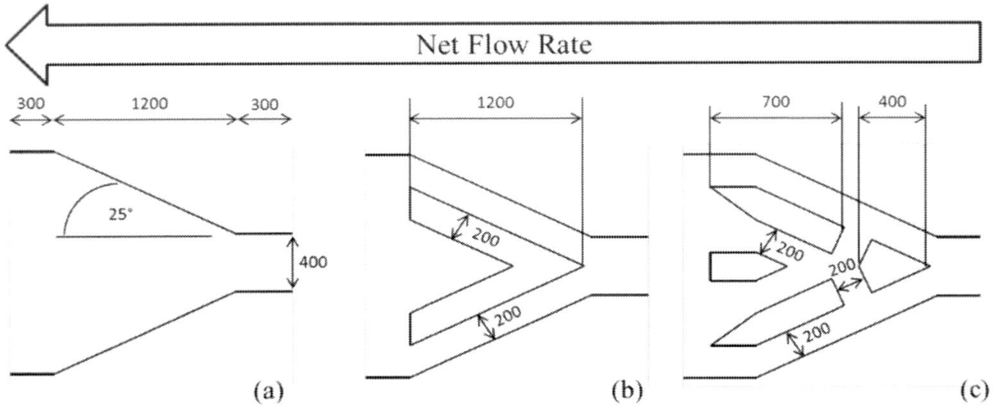

Figure 2. Microdiffuser elements (a) diffuser/nozzle, (b) V-shaped, (c) X-shaped (dimensions in μm)

The inlet width, length and height of the diffuser were set to 400 μm, 1800 μm and 400 μm respectively. All three diffusers had the same outline dimensions (illustrated in Fig.2(a)). The first novel element (V-shaped diffuser Fig.2(b)) featured a V-shaped protrusion in the centre of the diffuser. The protrusion width was set to 200 μm. During supply mode, it is anticipated that the fluid will be trapped in the V-shaped pocket, reducing the backflow and increasing the efficiency of the diffuser. The second novel element (X-shaped diffuser Fig.2(c)) has four protrusions located inside the diffuser. During pumping mode, the fluid is mainly expected to flow along the two channels located at the edge of the diffuser. During supply mode, however, the flow is anticipated to split into four different flows. The two flow paths in the centre are directed so that they join the two flow paths close to the walls perpendicularly. The convergence of the two flows is postulated to create vortices leading to a reduction in velocity and a reduction of the overall backflow.

3. Evaluation of 3D Printed Channels and Protrusions

3D printing is showing promising benefits to be used in the future as a cost-effective alternative to conventional silicon micromachining. Although the accuracies of 3D printers rapidly increased in the last few years, printed parts in the micro scale can still show defects or blockages, especially with the Pro Jet HD 3000 Plus 3D printer used in this study. Consequently, an optical analysis was performed to determine the minimum feature sizes that the 3D printer was capable of printing for a functional diffuser. For printing the material EX 200 from 3D Systems was used. The desired height of the diffusers was fixed to 400 μm. To this end, several single protrusions and channels with different dimensions were printed. Each protrusion and channel to be investigated was printed five times. The protrusion widths were varied from 40 to 480 μm, and the channel widths were varied from 100 to 800 μm. Each protrusion and channel was printed inline and spanwise to see if any differences in accuracy appeared. (The term inline means that the printed layers are laid down along the orientation of the protrusion or channel; spanwise means the layers are laid down tangential to the orientation of the protrusion or channel). The single channels and protrusions were examined under a Keyence VHX

Digital Microscope. From the recorded 2D images, the area, height and width were measured with the included software. The measured heights for different widths protrusions and channels are shown in Fig. 3.

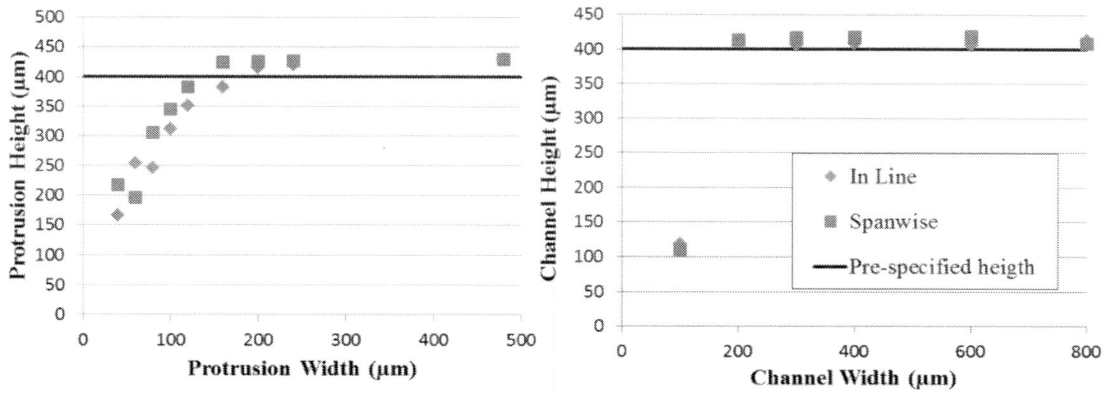

Figure 3. Measured protrusion and channel heights for different widths

It can be seen that the desired height of 400 µm for the protrusions (pre-specified height of the diffusers) is reached at a protrusion width of 200 µm. No significant difference between inline and spanwise printing was observed. The results for the channels show a similar outcome: the desired height of 400 µm was also reached at a channel width of 200 µm for both inline and spanwise printing. Based on these results, the dimensions of all protrusions, openings and distances between the two printed parts were set to be greater than 200 µm. Fig. 4 shows, in extracts, the profiles of channels and protrusions for different widths.

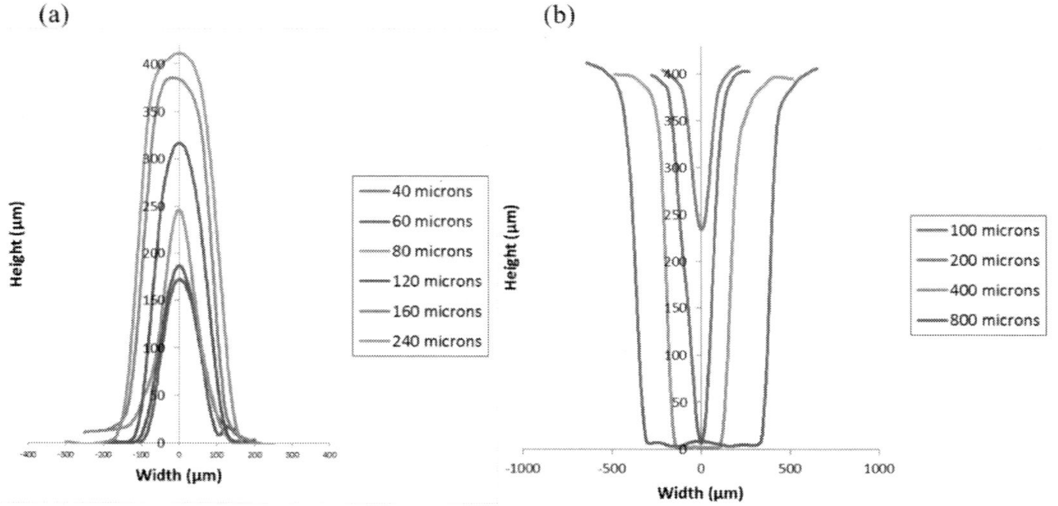

Figure 4. Measured profiles: (a) protrusions, (b) channels

It can be seen the bigger the width, the more rectangular (anticipated shape) the channel and protrusion becomes. For the diffuser elements, an inlet width of 400 µm was chosen (described in section 2) as profiles showed almost complete rectangular structure. The protrusion width was set to 200 µm. Microscope images of the two 3D printed novel diffuser elements are shown in Fig. 5. The images show that the diffusers were printed with neither defects nor blockages, and that their functionality was ensured.

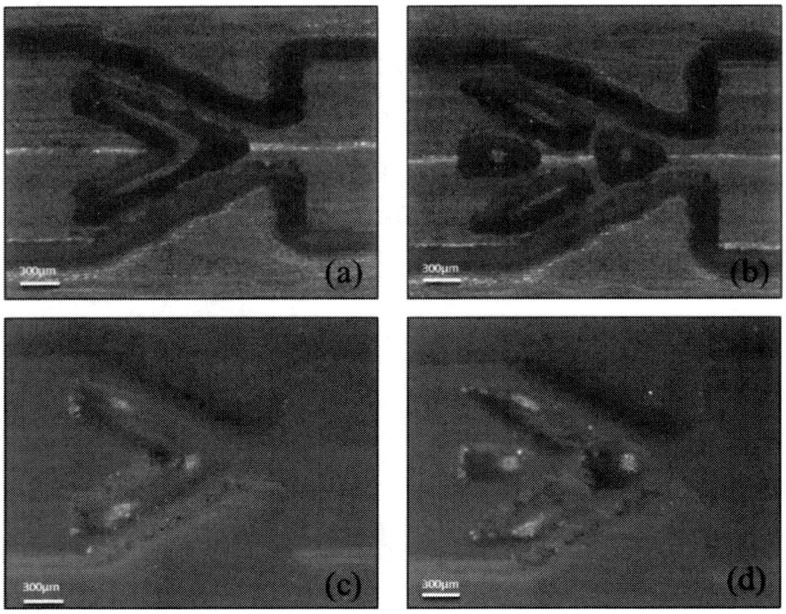

Figure 5. Microscope image: (a) V-shaped, (b) X-shaped; angular microscope image: (c) V-shaped, (d) X-shaped

4. Micropump Design and Experimental Set Up

The design of the piezoeletrically-actuated micropumps and the experimental set up for characterization is discussed in section 4.1 and section 4.2 respectively. Section 4.3 describes the uncertainties in the experimental set up for characterization of the micropumps.

4.1. Micropump Design

The micropump assembly was manufactured using the material EX 200 on the Pro Jet HD 3000 Plus 3D printer (same as described in previous section) and is seen in Fig. 6.

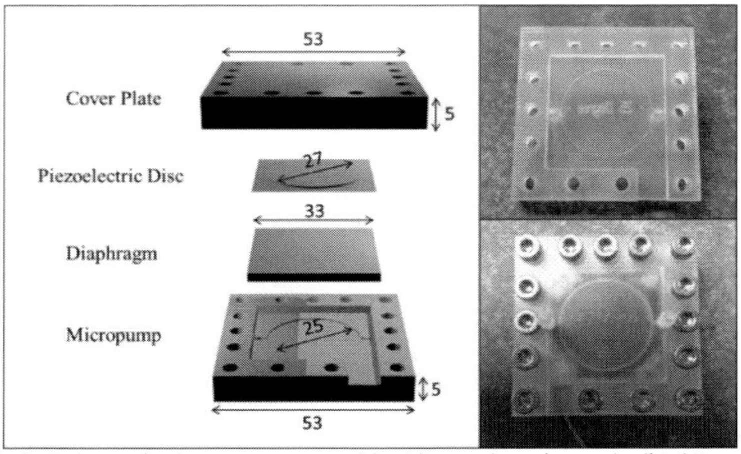

Figure 6. (a) exploded view of micropump assembly (dimensions in mm), (b) 3D printed micropump, (c) fully assembled 3D printed micropump

The pumping chamber had a diameter of 25 mm and a height of 400 μm. A steel sheet with a thickness of 200 μm was used as a diaphragm. For actuation, a piezoelectric element from Radionics (Part Number: 724-3162) was employed. The piezoelectric element had a diameter of 27 mm, thickness of 520 μm and was attached to the diaphragm using double-sided adhesive tape. The diaphragm with the attached piezoelectric element was then clamped on top of the pumping chamber and diffuser elements. Clamping was achieved by using fifteen M4 screws with a force sufficiently high to seal the pumping chamber.

4.2. Experimental Set Up

The experimental set up is illustrated in Fig. 7. For the differential pressure measurements, a PX409-015DWUV pressure transducer from Omega was used. The transducer was pre-calibrated and certified by the manufacturer. The flow rate was determined using the DUAL-NE-1000X syringe pump from New Era Pump Systems Inc. To drive the piezoelectric element, a TTI TG2000-20MHz DDS function generator was used to input a sine wave to a Piezo System Inc. EPA-104 power amplifier. Stainless steel Swagelok T-fittings and straight fittings were utilized and flexible vinyl tubing (ID 3.175 mm) was used to couple the experimental components together.

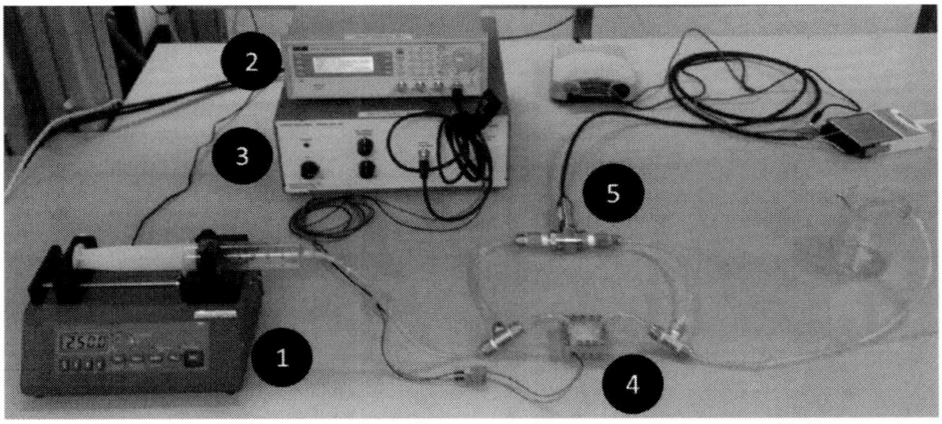

Figure 7. Experimental Set Up: 1) Syringe Pump, 2) Function Generator, 3) Power Amplifier, 4) Micropump, 5) Pressure Transducer

Water was used as the working fluid for the experiments. The voltage for driving the piezoelectric element was fixed to 170 V and the frequency was increased between 0 – 100 Hz in 10 Hz steps. Above 100 Hz, the steps were increased to 25 Hz. For every characterization, the frequency was increased until the differential pressure and flow rate went back to zero. For each investigated frequency, the differential pressure was recorded when the system reached a steady state condition. During the differential pressure measurements, the syringe pump was turned off in order to ensure zero flow rate in the system. When the micropump was powered with zero flow in the system, the maximum achievable differential pressure for the given frequency was measured directly at the ports of the pressure transducer.

After a consistent differential pressure was recorded, the syringe pump was turned on with a flow rate of 10 μl/min. The flow rate was then slowly increased in 10 μl/min steps. After every increasing step in flow rate, the system was given sufficient time to stabilize. With increasing flow rate of the syringe pump, the differential pressure reduced at the pressure ports as the system gets gradually opened. At a specific flow rate of the syringe pump (system completely open), the differential pressure returned to zero. The flow rate at which the differential pressure was zero corresponds to the maximum flow rate of the micropump at the specified operating frequency.

This differential pressure and net flow rate measurement procedure was conducted for every investigated frequency.

4.3. Experimental Uncertainty

The pressure transducer used in the experimental testing had a range of 100 kPa. This range was chosen as maximum differential pressures of 30 – 50 kPa were anticipated. The total uncertainty of the pressure transducer was given as ± 0.08 % over the full scale. However, lower differential pressures were recorded during testing which lead to higher percentage uncertainties than expected. The maximum recorded differential pressures of the three diffuser elements varied between 0.169 – 0.658 % over the full scale of the transducer. As the differences in the measured differential pressures of the diffuser elements were quite high (~0.489 % between diffuser/nozzle and X-shaped), the transducer was assumed to be accurate enough to show if a performance improvement was achieved with the novel elements.

The syringe pump used for the net flow rate measurement had an uncertainty of ± 1 % over the dispensing range, which was sufficiently accurate to record the maximum flow rates (in the range of 6 – 16 % of the dispensing range).

5. Results and Discussion

The flow rate measurements and differential pressure measurements are discussed in section 5.1 and 5.2 respectively

5.1. Differential Pressure Measurement

Fig. 8 shows the results of the differential pressure measurements for the three diffuser elements as a function of frequency.

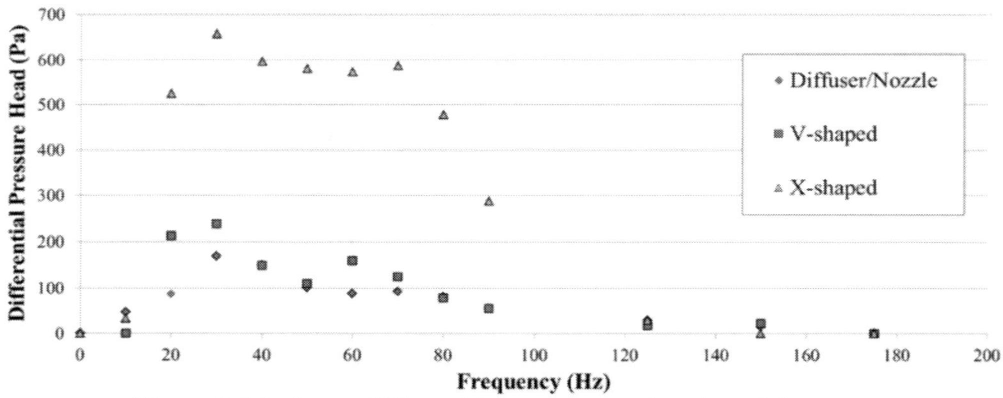

Figure 8. Maximum differential pressure as a function of frequency

All three diffuser elements achieve their maximum differential pressure at 30 Hz and, at the highest frequency, of 175 Hz, the differential pressure reverts to zero for all diffuser elements. It was anticipated, however, that the different diffuser elements would reach their maxima at different frequencies. This indicates that the assembly of the piezoelectric element and diaphragm shows inferior peak differential pressure. The lower the displacement of the piezoelectric element, the lower the pressure produced inside the pumping chamber, leading automatically to a reduction of the differential pressure. Higher differential pressures for all three diffuser elements would most likely be achieved with an improved piezoelectric element and a diaphragm with lower damping effects. Most importantly, however, both novel diffuser elements show an increase in their maximum differential pressure in comparison with the standard diffuser/nozzle element. The V-shaped diffuser shows an increase of the maximum differential pressure of 70 Pa. Although this is within the range of the uncertainty of the pressure transducer, the V-shaped element is most likely to achieve superior differential pressures. Higher values are also obtained between 20 and 40 Hz, as well as between 50 and 80 Hz. Between 40 and 50 Hz and above 80 Hz, however, the V-shaped diffuser element shows similar values to the diffuser/nozzle element. To verify the improved performance of the V-shaped element, measurements with a more accurate pressure transducer are required. The X-shaped diffuser shows an increase of the maximum differential pressure of 489 Pa compared to the diffuser/nozzle element, which is an increase of approximately 280 %. In addition, the X-shaped diffuser shows almost four times the differential pressure values of the diffuser/nozzle element over the whole frequency range of 175 Hz. Even allowing for the uncertainty in the pressure transducer, a significant improvement in differential pressure is achieved with the X-shaped element.

5.2. Flow Rate Measurement

Fig. 9 shows the results for the flow rate measurement of the three diffusers for different frequencies at zero differential pressure.

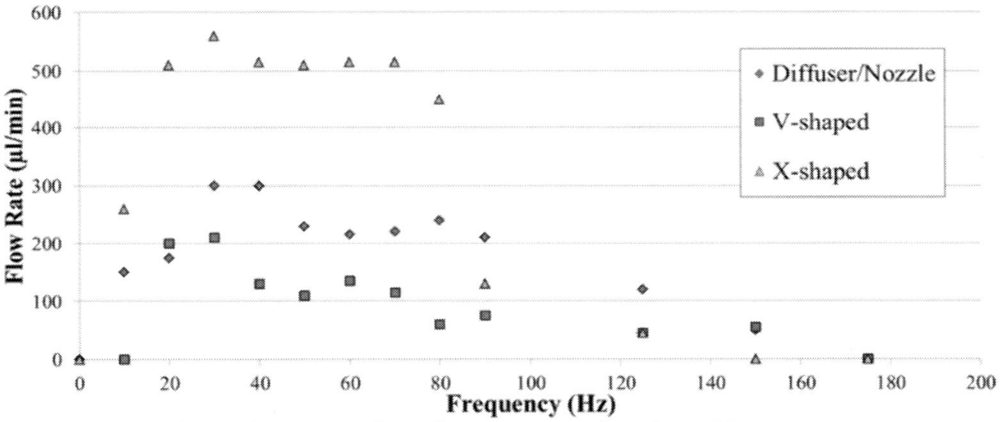

Figure 9. Measured net flow rates as a function of frequency

The results of the flow measurement reflect the results for the differential pressure characterization. As all discussed phenomena in the differential pressure measurement section are relevant for the flow rate measurements, only distinct features will be discussed in this sub-section. The diffuser/nozzle element reaches its maximum flow rate of 300 µl/min at 30 Hz. This trend in the flow rate is as expected from the differential pressure measurements, and it is similar to the behaviour of diffuser/nozzle elements investigated by other authors [7-9]. Although the V-shaped diffuser element shows an increase in the differential pressure, the achieved flow rates are lower over the frequency range than those of the diffuser/nozzle element. This reduction in flow rate occurs as the V-shaped diffuser element has a high flow resistance in the diffuser direction, leading to smaller flow rates during pumping mode than for the diffuser/nozzle element. In the nozzle direction and during supply mode, however, the flow resistance of the V-shaped protrusion is even higher, which reduces the backflow significantly. This results in a low flow rate – high pressure rise pump characteristic. The maximum net flow rate is measured to be 210 µl/min at 30 Hz. This is 90 µl/min lower than the maximum value for the diffuser/nozzle element. Similar to the differential pressure measurements, the X-shaped diffuser shows a large increase in the flow rate compared to the diffuse/nozzle element. It reaches its maximum of 560 µl/min at 30 Hz, about twice the value of the standard diffuser/nozzle element. The flow rate increase indicates that the backflow gets reduced significantly by this element. However, the measured results do not prove if this is caused by the anticipated vortex areas, or by a much higher flow resistance in the nozzle direction.

6. Conclusion

This paper presents the characterization of 3D printed valveless piezoelectrically-actuated micropumps with novel microdiffuser elements. As microchannel cooling requires high pressure heads and high flow rates, the novel X-shaped microdiffuser shows promising results to be used for the application of next generation PICs cooling. With the presented design, the backflow during supply mode was successfully reduced leading to a maximum reached differential pressure of 658 Pa and a maximum flow rate of 560 µl/min, both at an operating frequency of 30 Hz. This is an improvement of 280% in differential pressure and 85% in net flow rate compared to the standard diffuser/nozzle element. A secondary outcome of this work is the successful demonstration of 3D printing as a fabrication technique for a microfluidic component. Although 3D printing still shows limitations concerning manufacturing in the micro scale, with the rapid increase in technological advances this quite young manufacturing technique will very likely play an important part in the future of microfluidics. For further enhancement of the introduced microdiffuser elements, a numerical optimization of the protrusions and channels will be conducted. A re-characterization of the micropumps with a pressure transducer with higher accuracies is expected to confirm the performance improvement, especially for the V-shaped element.

Acknowledgement

The authors acknowledge the financial support of Science Foundation Ireland under Grant No. 10/CE/11853. Bell Labs Ireland would like to thank the Industrial Development Authority (IDA) for the continued support.

References

[1] Singhal V., Garimella S. V., Raman A., "Microscale pumping technologies for microchannel cooling systems", Appl Mech Rev vol 57, no 3, pp 191-248, (2004).More references
[2] Garimella S.V. and Sobhan C.B., "Transport in microchannels: A critical review", Annu. Rev. Heat Transfer 15, pp. 1–50, (2003).
[3] Iverson B, Garimella S, "Recent advances in microscalepumping technologies: a review and evaluation", Microfluid Nanofluid, pp 145-174, (2008).
[4] Sun C, Yang Z, "Effects of the half angle on the flow rectification of a microdiffuser", J Micromech Microeng 17:2031–2038, (2007).
[5] Nabavi M, Mongeau L, "Numerical analysis of high frequency pulsating flows through a diffuser-nozzle element in valveless acoustic micropumps". Microfluid Nanofluid, pp 659-681, (2009).
[6] Sun C. L., Huang K. H., "Numerical characterization of the flow rectification of dynamic microdiffusers", Journal of Micromechanics and Microengineering, pp 1331-1339, (2006).
[7] IzzoI., Accoto D., Menciassi A., Schmitt L., Dario P., "Modeling and experimental validation of a piezoelectric micropump with novel no moving part valves", Sensors and Actuators A 133, pp 128-140, (2007).
[8] Chandika S., Asokan R., Vijayakuma K.C.K., "Flow characteristics of the diffuser/nozzle micropump – A state space approach", Flow Measurement and Instrumentation, pp 28-34, (2012).
[9] Choi A., Vatanabe S. L., Lima C., Silva E., "Computational and experimental characterization of a low-cost piezoelectric valveless diaphragm pump", Journal of Intelligent Materials and Structures 23(1), pp 53-63, (2011).

Eurotherm Seminar 102: Thermal Management of Electronic Systems IOP Publishing
Journal of Physics: Conference Series **525** (2014) 012009 doi:10.1088/1742-6596/525/1/012009

A dimensional comparison between embedded 3D-printed and silicon microchannels

J O'Connor[1], J Punch[1], N Jeffers[2] and J Stafford[2]

[1] CTVR, Stokes Institute, University of Limerick, Limerick, Ireland

[2] Thermal Management Research Group, Bell Labs Ireland, Alcatel-Lucent, Blanchardstown, Co. Dublin, Ireland

E-mail: `john.patrick.oconnor@ul.ie`

Abstract. The subject of this paper is the dimensional characterization of embedded microchannel arrays created using contemporary 3D-printing fabrication techniques. Conventional microchannel arrays, fabricated using deep reactive ion etching techniques (DRIE) and wet-etching (KOH), are used as a benchmark for comparison. Rectangular and trapezoidal cross-sectional shapes were investigated. The channel arrays were 3D-printed in vertical and horizontal directions, to examine the influence of print orientation on channel characteristics. The 3D-printed channels were benchmarked against Silicon channels in terms of the following dimensional characteristics: cross-sectional area (CSA), perimeter, and surface profiles. The 3D-printed microchannel arrays demonstrated variances in CSA of 6.6-20 % with the vertical printing approach yielding greater dimensional conformity than the horizontal approach. The measured CSA and perimeter of the vertical channels were smaller than the nominal dimensions, while the horizontal channels were larger in both CSA and perimeter due to additional side-wall roughness present throughout the channel length. This side-wall roughness caused significant shape distortion. Surface profile measurements revealed that the base wall roughness was approximately the resolution of current 3D-printers. A spatial periodicity was found along the channel length which appeared at different frequencies for each channel array. This paper concludes that vertical 3D-printing is superior to the horizontal printing approach, in terms of both dimensional fidelity and shape conformity and can be applied in microfluidic device applications.

1. Introduction

Industry has benefited greatly from the introduction of microfluidic technologies including thermal management, biotechnology, and health care. As an example, thermal management has seen improved heat flux dissipation from increasingly smaller areas which serves in maintaining chip temperature and reducing localised hot spots in integrated circuit devices [1]. When developing microfluidic devices or structures, a number of different channel manufacturing techniques exist for both thermal management and Lab-on-a-chip(LOC) applications. For thermal management applications, chemical etching of Silicon wafers using deep reactive ion etching (DRIE) or wet-etching techniques (KOH) is the conventional method of channel fabrication. Anodic bonding is then used to create a permanent seal between a glass slide and the Silicon wafer [2]. In LOC applications, Polydimethlysiloaxane (PDMS) or Poly(methyl methacrylate) (PMMA), casted from a Silicon mold, is the conventional method of microchannel fabrication and was initially demonstrated by Martynova *et al.*[3]. The channels are sealed by bonding the microchannel to a glass slide using PDMS or some other Silicone based polymer. Although it can be important to have dimensionally accurate and uniform channels, there are

Content from this work may be used under the terms of the Creative Commons Attribution 3.0 licence. Any further distribution of this work must maintain attribution to the author(s) and the title of the work, journal citation and DOI.
Published under licence by IOP Publishing Ltd

Journal of Physics: Conference Series **525** (2014) 012009 doi:10.1088/1742-6596/525/1/012009

limitations associated with these methods of manufacture: primarily cost, flexibility of design, and fabrication time. Rapid-prototyping, in particular 3D-printing, is a cost-effective and swift alternative.

Rapid-prototyping is the fabrication of a structure or assembly using computer aided design files in a stereo-lithography file format. Specifically, the type of rapid-prototyping under investigation in this paper is 3D-printing. Modern commercially-available 3D-printers are capable of printing to a resolution of 16 μm with a range of polymeric materials. Typically, the 3D-printing process involves depositing a sacrificial base layer followed by the print material and in areas where there is an embedded or overhanging feature, sacrificial material is deposited to maintain the parts' structure. High layer resolution and flexible design offered by commercial 3D-printers facilitates the creation of embedded microchannels within structures, provided there is a method to expel the sacrificial material from the embedded feature. Microfluidic studies using 3D-printing as an alternative to conventional techniques are becoming more commonplace in the literature and the following paragraph details studies which use 3D-printing in microfluidic research.

Payder *et al.* [4] designed and characterized 3D-printed microfluidic interconnects integrated with an O-ring. This device provided an effective interconnection, as it could sustain pressures exceeding 400kPa with no leaks. McDonald *et al.* [5] investigated 3D-printed molds as an alternative to Silicon molds for casting PDMS microchannels. It was found that channel quality increased at higher resolutions, and that surface roughness was measured to be 10-20 μm with no periodic patterning evident along the channels. Moore *et al.* [6] detailed the cost effectiveness and the fast fabrication times of 3D-printing, and used this method to evaluate a passive valve design for a centrifugal LOC disk. This work characterized the surface profile of the 3D-printed part and revealed a high degree of spatial periodicity across the top surface, while the microchannel structure printed had a distorted base with clear passive valve features present at the side-wall. Bonyár *et al.* [7] reported similar findings, whereby molds were 3D-printed for PDMS casting. This work also investigated 3D-printed LOC structures which were sealed using an adhesive film. Krejcova *et al.* [8] used a similar methodology to fabricate a chip which was developed as a virus detection system. Finally, McCullough and Yadavalli [9] investigated the suitability of poly(acrylonitrile butadiene styrene) for 3D-printed biomedical device fabrication using fusion deposition modelling; this involved surface treatment and surface characterization of the microchannel. The surface modifications created bio-compatible, hydrophilic, and water impermeable microchannels.

The studies described here use 3D-printing techniques which created troughs on a single plane of material. It is also important to note that no studies exist that characterize embedded 3D-printed channels. The aim of this paper is the characterization of these 3D-printed embedded microchannel arrays. This paper will discuss the objectives, methodologies used to characterize the channels, and measured dimensional characteristics of the microchannels. It paper concludes with the findings of the dimensional characterization and a summary of the implication for microchannel fabrication via 3D-printing.

2. Objectives

The overall objective of this paper is the dimensional characterization of embedded microchannel arrays. To this end, the specific objectives are as follows:

- To compare the twenty-two channel DRIE and KOH Silicon microchannel arrays characterized by Eason *et al.* [2] with an equivalent 3D-printed design using vertical and horizontal 3D-printed techniques. The dimensions of the channel geometries are illustrated in figure 1.
- To evaluate the performance of both vertical and horizontal 3D-printing techniques with regard to fabricating embedded microchannel arrays. A total of twelve microchannel arrays

Eurotherm Seminar 102: Thermal Management of Electronic Systems IOP Publishing
Journal of Physics: Conference Series **525** (2014) 012009 doi:10.1088/1742-6596/525/1/012009

were investigated; a statistical sample of three was taken for each shape and these shapes were printed in both orientations.

- To characterize 3D-printed microchannel arrays geometrically by measuring cross-sectional area (CSA), perimeter, aspect ratio and surface profiles in order to quantify the dimensional fidelity and precision of 3D-printed microchannels.

3. Methodology

This section describes the fabrication process for both the vertical and horizontal 3D-printed microchannel arrays and details how these arrays were prepared after printing. The dimensional characterization procedures for both cross-sectioning and surface profiling are described.

3.1. Manufacture

The 3D-printed microchannel arrays characterized in this paper were created using a Projet HD 3500 3D-printer at its highest resolution setting. The highest achievable layer resolution, as stated by the manufacturer, was 16 μm in the x, y, and z directions. The materials used to create the microchannel arrays were Acrylonitrile as the 3D-printing material and wax as the sacrificial material. Two different printing orientations were explored to examine the influence of print direction on channel characteristics: vertical and horizontal 3D-printing approaches, as illustrated in figure 2.

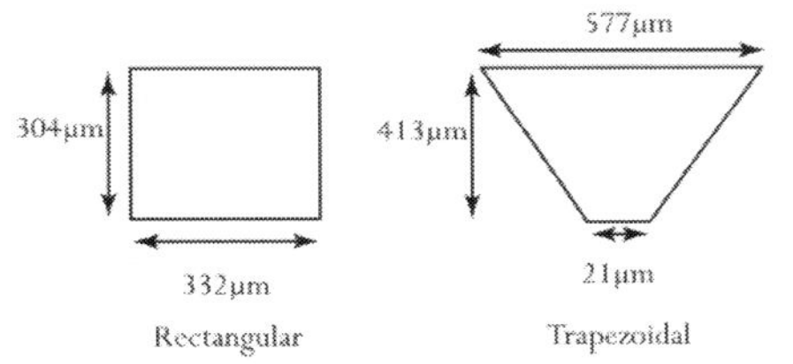

Figure 1. Nominal channel shapes and dimensions.

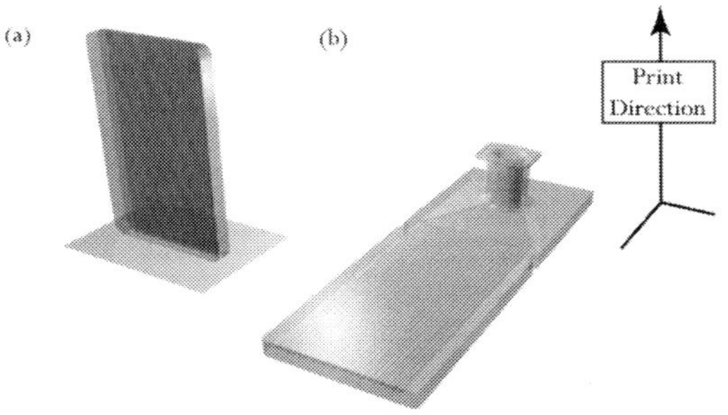

Figure 2. Perspective view of embedded channel designs, (a) vertical and (b) horizontal, which is representative of their orientation in the 3D-printer.

Vertically printing the channels ensured that only 3D-printed material formed the channels with no wax occupying the channels. Horizontally printing the channels, involved filling the channel with the support material, wax. To clean the horizontal channels, the arrays were then heated to 57°C in a oil bath to melt the wax within the embedded features. It was evident that wax melted when the channels were optically transparent. The channels were flushed out several times with syringes containing clean hot oil free from contaminants, hot water mixed with detergent and clean water. This process was repeated several times, as the presence of any remaining wax, oil or detergent would be diminished every time the process was repeated.

3.2. Channel measurement

The samples were cross-sectioned with a dicing saw and potted in epoxy. Samples were aligned perpendicular to the pot base to ensure a true cross-section of the channel (<1°). The epoxy encasing each sample was polished to a mirror finish and images of every channel were taken using a Hitachi TM1000 Tabletop Scanning Electron Microscope (SEM). The polished cross-sections were sputtered with a submicron layer of gold to eliminate charging while in the SEM. A custom MATLAB routine was used to measure area, perimeter, width, and height

Eurotherm Seminar 102: Thermal Management of Electronic Systems IOP Publishing
Journal of Physics: Conference Series **525** (2014) 012009 doi:10.1088/1742-6596/525/1/012009

of each cross-section. To minimize uncertainty in channel measurement, the scale bar was calibrated to pixel length to a resolution of 0.52 and 0.69 μm per pixel for 200 and 300 μm scale bars respectively. A representative polygon of the cross-section was taken, by selecting a series of points along the cross-section. The number of points selected was varied with shape complexity, with a minimum of 200 points taken for each image. A sensitivity analysis showed that a minimum of 200 points was needed to measure the channel area to within 0.9 %. A custom function was used to calculate area, and the sum of the length between points was used to calculate perimeter. Characteristic heights and widths were measured using the scaled images. This measurement methodology was similar to that of Eason *et al.* [2].

Channel length profile measurements were taken with a Hommel T500 surface profilometer to investigate the variance in channel wall roughness and, subsequently, CSA variance in the lengthwise direction. The samples were precision machined to expose the area of interest. The profile probe measured a length of 4.8 mm with the horizontal and vertical resolution of the scan was 2.4 μm and 0.02 μm respectively. The probe tip diameter was 2 μm. Each measurement was repeated three times to ensure that no damage was done to the channel and a consistent reading was taken.

4. Dimensional characteristics

This section discusses the CSA, channel dimensions and profilometry measurements for the vertical and horizontal 3D-printed microchannel arrays. CSA measurements for both Silicon microchannel arrays were also reported.

4.1. Vertical channels

Figure 3 depicts typical cross-sections of the rectangular and trapezoidal vertical 3D-printed microchannel arrays. The rectangular cross-section shows good shape conformity, with curvature present at the corner and base wall. This cross-section is similar to conventional DRIE microchannel cross-section, as shown by other authors [2][10]. The trapezoidal cross-section shows good shape conformity, with curvatures present at all corners and a base width larger than the nominal. The reason for the curvatures and larger base width is due to material shrinkage during the printing process [11].

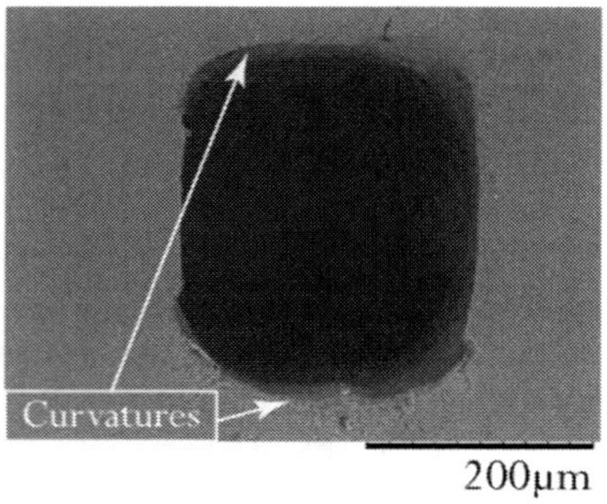

(a) Rectangular

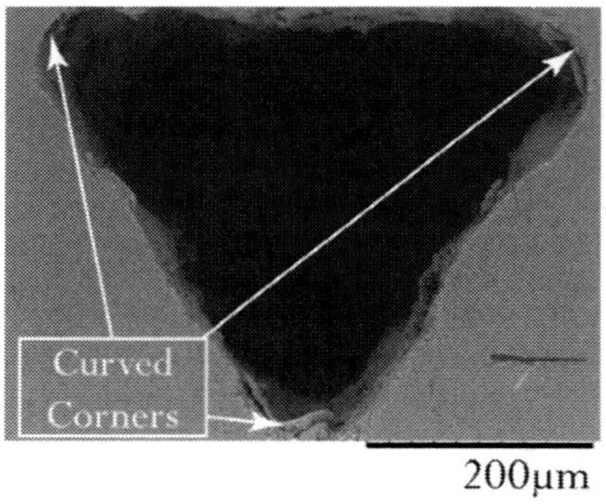

(b) Trapezoidal

Figure 3. Cross-sections of the vertical 3D-printed microchannel array shapes:(a) Rectangular, and (b)Trapezoidal.

Table 1 contains the measured average CSA and perimeter, the channel–to–channel variance and deviation from the nominal dimensions for each microchannel array.

[1] The labelling system describing each 3D-printed channel examined here, are labelled in terms of shape, orientation and number in sample size. The letters used are explained as follows: R = rectangular, T = trapezoidal,

Table 1. Comparison of the nominal and measured CSAs (mm^2),and perimeter (mm)for the vertical 3D-printed channels. Channel-to-channel variance per array and percentage deviation from the nominal dimension are presented.

	Nominal		Measured		Variance (%)		Change (%)	
Channel Type [1]	CSA	Perimeter	CSA	Perimeter	CSA	Perimeter	CSA	Perimeter
DRIE	0.101	1.27	0.099	1.20	2.7	2.1	-2.3	-0.1
KOH	0.123	1.50	0.117	1.50	4.9	4.3	-4.96	-5.8
RV1	0.101	1.20	0.087	1.10	7.7	4.9	-13.9	-8.3
RV2	0.101	1.20	0.086	1.10	10.2	6.8	-14.5	-8.4
RV3	0.101	1.20	0.097	1.20	11.7	4.5	-3.7	-0.5
TV1	0.123	1.50	0.115	1.40	7.5	6.6	-6.5	-6.6
TV2	0.123	1.50	0.115	1.41	6.0	4.6	-6.6	-6.0
TV3	0.123	1.50	0.104	1.31	9.0	5.1	-16.2	-12.6

All CSAs for both the rectangular and trapezoidal microchannel arrays were less than the nominal dimensions of the Silicon channels (-3.7– -16.2 %) with deviating channel-to-channel variances of 6.0–11.7 %. The perimeter measurements were less than the nominal dimensions; this is due to the curvature features present at the corners of each shape which reduce the perimeter. These deviations have been documented in other studies [5]. A possible reason for these deviations would be the effect of printer resolution, combined with material shrinkage associated with 3D-printing, which is a non-linear and a highly sensitive parameter [11]. In comparison to the Silicon channels, the variances associated with area and perimeter were 2.7 and 2.1 % for DRIE and 4.9 and 4.3 % for KOH, which are comparable to the 3D-printed channels.

Table 2 presents a comparison of the measured channel dimensions against the nominal, with associated deviations from nominal and variances of both dimensions.

Table 2. Comparison of the nominal and measured channel dimensions for the vertical 3D-printed channels. Channel-to-channel variance per array and percentage deviation from the nominal dimensions are presented.

	Nominal (mm)		Measured (mm)		Change (%)		Variance (%)	
Channel Type	Height	Width	Height	Width	Height	Width	Height	Width
DRIE	0.304	0.332	0.304	0.332	n/a	n/a	n/a	n/a
KOH	0.413	0.577	0.413	0.577	n/a	n/a	n/a	n/a
RV1	0.304	0.332	0.319	0.284	5.0	-14.4	5.3	6.6
RV2	0.304	0.332	0.314	0.300	3.2	-9.6	8.4	5.6
RV3	0.304	0.332	0.357	0.310	17.4	-6.5	7.8	10.8
TV1	0.413	0.577	0.362	0.456	-12.3	-20.9	7.3	7.7
TV2	0.413	0.577	0.382	0.479	-7.4	-16.9	2.9	5.6
TV3	0.413	0.577	0.363	0.442	-12.1	-23.4	7.1	7.6

The measured channel height fluctuated about the nominal (-12.3–17.4 %), while the channel widths were consistently smaller than the nominal. Reasons for such variance in the channel

V = vertical, H = horizontal. This labelling system will be used in all tables in this paper.

*** Formatting Issue - Best Available Paper/Graphic ***

Journal of Physics: Conference Series **525** (2014) 012009 doi:10.1088/1742-6596/525/1/012009

dimensions are due to both limited resolution and material shrinkage reducing both sharp corner features.

Vertical printing does vary in dimension and deviates from the nominal in both cases studied, however the variances range between two to four times larger than that of conventional Silicon fabrication. Other effects, such as shrinkage and resolution, should be accounted for in the design stage, as decreasing scale will result in larger variance. Although vertical printing is limited by its design flexibility, it does provide a swift and relatively cheap fabrication solution.

4.2. Horizontal channels

Figure 4 shows sample cross-sections for each horizontal 3D-printed microchannel array. A notable and common feature shared by both channels is the inherent roughness present at the side-walls. The rectangular channel cross-section, figure 4(a), has a rectangular profile, but the side-wall roughness distorts the channel shape. The trapezoidal cross-section, figure 4(b), shows a highly distorted trapezoidal shape. The side-wall roughness is more pronounced than the rectangular cross-section and, as a result, the base width dimension is not visible. There is an arching feature present at the top of the trapezoidal cross-section which reduces the sharp corner of the nominal shape. It is difficult to pinpoint this side-wall roughness feature origin; however one explanation would be an interaction between both the wax and 3D-printer material while being printed layer by layer during fabrication.

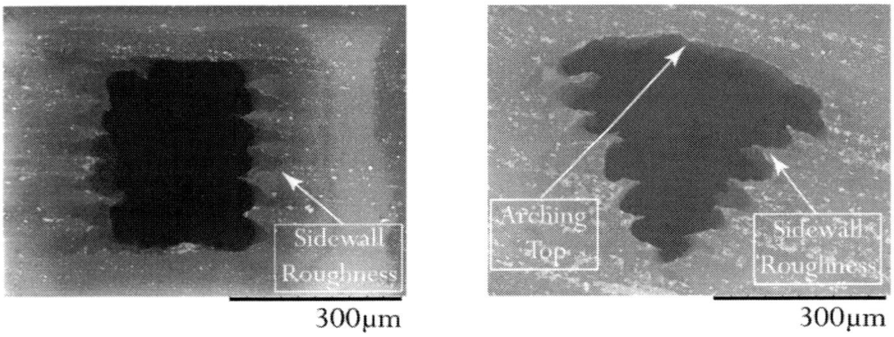

(a) Rectangular

(b) Trapezoidal

Figure 4. Cross-sections of the horizontal 3D-printed microchannel array shapes (a) rectangular and (b) trapezoidal.

Table 3 details the measured CSA and perimeter of each microchannel array, with the variance and deviation from the nominal.

A large variance in CSA and deviation from the nominal exists for both horizontal printed shapes (-31.7–11.9 %), primarily due to the side-wall roughness. The roughness resulted in a larger perimeter measurement (33.3–75%). However, there is a noticeable discrepancy in RH3. This channel array is an example where residual wax can remain within the side-wall reducing both CSA and perimeter. This is a shortcoming of the cleaning technique as it cannot consistently remove all wax from the channels due to the pronounced side-wall roughness. This also highlights an issue with scale when designing horizontally printed channels: i.e., at larger channel scales, the side-wall roughness will have a reduced effect on channel characteristics.

Table 4 is a comparison of the measured channel heights and widths with the nominal dimensions. This table also contains the variation of heights and widths for each microchannel array.

The measured rectangular channel heights have a lower variance than the measured widths. The reason for such high variance in width is the side-wall roughness which makes it difficult to discern the characteristic width of the channels. The width measurements presented here are measured from the base of each side-wall. The deviations from the nominal dimensions vary in both height and width. The rectangular channels, RH1 & RH2, are greater than the nominal height, while the widths are less than the nominal. A reason for this bias is the side-wall

Eurotherm Seminar 102: Thermal Management of Electronic Systems IOP Publishing
Journal of Physics: Conference Series **525** (2014) 012009 doi:10.1088/1742-6596/525/1/012009

Table 3. Comparison of the nominal and measured channel CSAs (mm^2) and perimeter (mm) for the horizontal 3D-printed channels. Channel-to-channel variance per array and percentage deviation from the nominal dimension are presented.

	Nominal		Measured		Variance (%)		Change (%)	
Channel Type	CSA	Perimeter	CSA	Perimeter	CSA	Perimeter	CSA	Perimeter
RH1	0.101	1.20	0.110	2.10	9.3	21.7	8.9	75.0
RH2	0.101	1.20	0.113	2.10	9.6	14.6	11.9	75.0
RH3	0.101	1.20	0.069	1.18	15.7	27.1	-31.7	-1.6
TH1	0.123	1.50	0.125	2.00	15.2	20.4	1.6	33.3
TH2	0.123	1.50	0.121	2.30	5.5	20.1	-1.6	53.3
TH3	0.123	1.50	0.121	2.30	9.7	19.3	-1.6	52.6

Table 4. Comparison of the nominal and measured channel dimensions for the horizontal channels. Channel-to-channel variance per array and percentage deviation from the nominal dimensions are presented.

	Nominal (mm)		Measured (mm)		Change %		Variance %	
Channel Type	Height	Width	Height	Width	Height	Width	Height	Width
RH1	0.304	0.332	0.358	0.282	17.6	-15.2	6.6	19.3
RH2	0.304	0.332	0.364	0.259	19.7	-22.1	6.2	11.7
RH3	0.304	0.332	0.303	0.237	-0.4	-28.6	4.2	10.9
TH1	0.413	0.577	0.439	0.468	6.3	-18.8	10.3	16.1
TH2	0.413	0.577	0.420	0.472	1.8	-18.1	13.6	18.7
TH3	0.413	0.577	0.412	0.491	-0.4	-14.9	16.5	7.1

roughness present in the channels. The measured trapezoidal channel heights were close to that to the nominal dimension (0.4–6.3 %). This result is due to the bowing feature present in the trapezoidal channel, which compensates for loss of the base width (21 μm). Although horizontal printing embedded channels yields good design flexibility, the degree of shape conformity and dimensional fidelity associated with the printing process are poor in comparison to both Silicon and vertical channels; as a result, this technique does not lend itself to creating dimensionally accurate channels. Similar to the vertical printing method, larger scale channels may yield higher dimensional accuracy as the side-wall roughness would have a diminishing effect on channel characteristics.

4.3. Surface profiles

This section compares the surface characteristics of the vertically and horizontally printed microchannel arrays fabricated and in order to determine effect of surface roughness along the channel length. To this end, SEM imaging of the exposed channel types, complemented with surface profiles, are presented. Finally, a spectral analysis of the surface profile plot was performed to detect the presence of spatially periodic roughness features along the channel length. Figure 5 illustrates the surface characteristics for both rectangular and trapezoidal vertical microchannel arrays with SEM images, sample profilometry measurements and a histogram of the distribution of profile height along the channel length.

Visual inspection of the rectangular channel profile, figure 5(a), shows a rough surface. The

*** Formatting Issue - Best Available Paper/Graphic ***

profile measurement of the channel length shows no clear periodic features, however a spectral analysis revealed a strong feature to be present every 200 μm along the channel length. The surface profile shows that the maximum feature size ranges between ± 20 μm, which is close to the printer resolution of ± 16 μm. The side-wall appears to be perpendicular to the base wall which implies a rectangular cross-section is maintained along the channel length. Finally, the side-wall roughness along the channel length appears to be similar to the base wall throughout the channel length, which implies a uniform roughness and surface profile along the channel length. Figure 5(b) shows the SEM image of the exposed trapezoidal channel accompanied with a surface profile and histogram of the same surface profile. Roughness features were found to be present at the base and side-wall of the trapezoidal channel. The resolution of the base width feature of 21 μm is lost with the average measurement being visually larger that the nominal value (approx. 100 μm). There are significant steps changes along the trapezoidal walls which suggests that the angled walls are not smooth. The metrology of the trapezoidal channel highlights the limit of the printer in creating embedded trapezoidal or triangular features. The surface profile shows no clear periodic features and is only presented up to 3 mm. This is due to the probe moving up the side-wall. It is important to note that adjusting the probe alignment correctly within the channel is difficult as the tolerances are small (approx. ± 50 μm). As a result, the histogram plot only uses data taken up to this point to prevent skewing of the distribution. A spectral analysis revealed two strong peaks occur at every 250 μm and 500 μm along the channel length. The histogram plot shows that the change in surface profile height is approximately ± 16 μm. Overall, the surface profile measurements for each vertical channel shape are approximately the resolution of the printer.

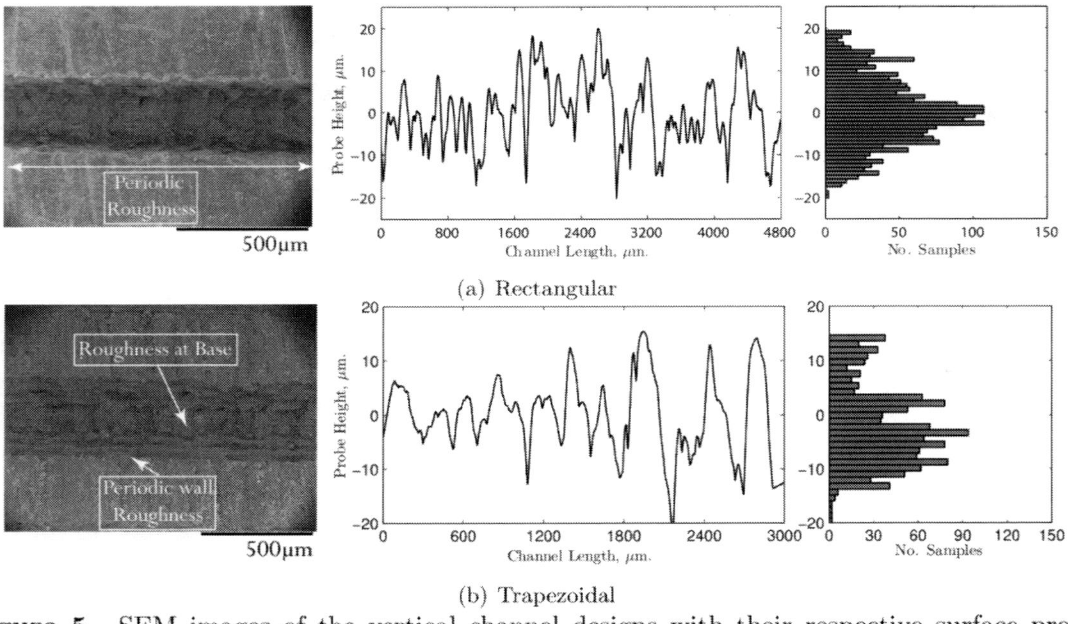

(a) Rectangular

(b) Trapezoidal

Figure 5. SEM images of the vertical channel designs with their respective surface profile measurements and related statistics.

Figure 6 depicts SEM images of the exposed horizontal 3D-printed microchannel arrays complete with both a sample surface profile measurement and histogram plot. A spectral anaylsis of the profile was carried out to determine if there were any periodic features present along the channel.

Figure 6(a) shows the exposed channel and surface profile information along the channel

Eurotherm Seminar 102: Thermal Management of Electronic Systems IOP Publishing
Journal of Physics: Conference Series **525** (2014) 012009 doi:10.1088/1742-6596/525/1/012009

length. The SEM imaging shows a strong patterned feature along the channel length. The pattern produced in the horizontal rectangular microchannel array is more consistent than that of the vertical rectangular microchannel arrays. The surface profile measurement shows no evident pattern, however the spectral plot revealed periods every 150 and 300 μm. The histogram plot revealed the change in profile height to be within ± 10 μm which is within the printer's highest resolution limit (± 16 μm)and performs better than the vertically printed channel. The side-wall roughness which was highlighted in the channel cross-section is evident along the channel sides due to protrusions from the side-wall.

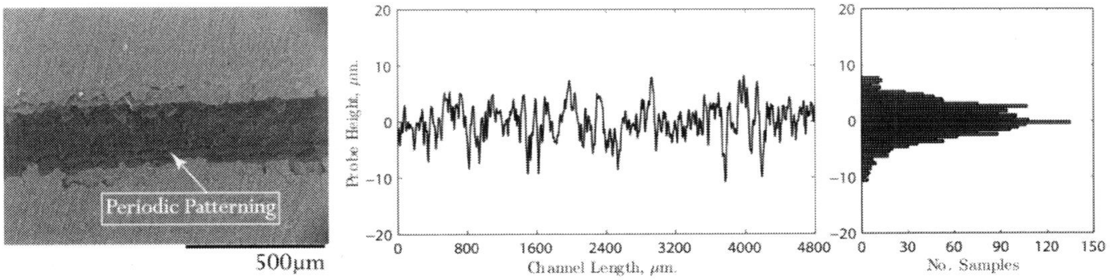

(a) Rectangular

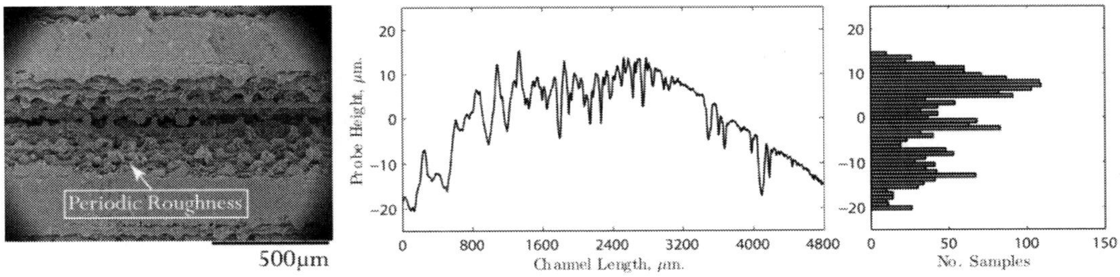

(b) Trapezoidal

Figure 6. SEM images of the horizontal channel designs with their respective surface profile measurements and related statistics.

Figure 6(b) shows the exposed horizontal trapezoidal microchannel, and depicts extensive side-wall roughness. Periodic patterning is evident along the channel length, however it is not as clear as the horizontal rectangular channel (Figure 6(a)). The surface profile depicted is not entirely representative of the total surface roughness of the channel. This is due to the coupled effect of the side-wall roughness and trapezoidal shape of the channel, as the probe was possibly travelling between the side-wall roughness layers along the channel length. This yields a visible bias in the histogram profile. Overall, the roughness present in the trapezoidal channel is extensively greater than the other channel measured and is not within the printer resolution. A spectral analysis of the profile revealed no periodic features present along the channel length.

The surface profile analysis shows that the profile along the channel length was approximately the resolution of the printer for the vertical channel. This is important when fabricating microchannels via vertical 3D-printing as this is the minimum roughness that will be present in the channels. At larger channel scales, the effect of surface roughness will be diminished: for example, a channel of 1000 μm hydraulic diameter would have a roughness of 3.2 % of the diameter for the given printer resolution. The horizontal channel featured a significant differing surface roughness features, with the side-wall present larger roughness than the base-wall. The trapezoidal channel surface profile presents a complex profile which is due to the side-wall roughness feature. Again, printing at a larger scale will diminish the side-wall roughness on embedded horizontally printed channels. Although the vertical printing approach yielded superior quality channels than the horizontal approach, printing vertically has more design constraints than horizontally printing, in particular that no overhanging feature can be incorporated into the design. This should be taken into consideration when fabricating embedded microchannel arrays. The paper highlights the capabilities of 3D-printing embedded microchannels; however, a comprehensive error analysis is needed to fully quantify the impact surface roughness has on channel characteristics.

5. Conclusions

A direct dimensional comparison between Silicon and equivalent 3D-printed microchannel arrays was conducted which considered the dimensional fidelity and shape conformity of the arrays. The following was concluded:

- The Silicon microchannel arrays produced superior quality channels compared to the 3D-printed equivalent. The vertical 3D-printed channel produced good shape conformity and lends itself to microchannel manufacture when compared to the horizontal printing approach.
- The CSA of the vertical channels were consistently less than the nominal dimensions due to curvature present at corners. This feature was a function of material shrinkage during the printing process and should be accounted for in its design.
- The horizontal channels demonstrated inferior shape conformity which was due to the additional side-wall roughness. This feature reduces the quality of the channel produced but would have a reduced effect for channels printed of larger scale.
- The surface profile varied about the resolution of the printer for the vertical 3D-printed channels (approx. 16 μm), however the surface roughness produced by the horizontal channels was larger than the printer resolution. Printing at larger scales would reduce this effect. A spectral analysis revealed a spatial surface periodicity along the vertical printed channels and horizontal printed rectangular channels. No notable patterns were evident in the horizontally printed trapezoidal channel.

In summary, for microfluidics research, vertically 3D-printed microchannels can yield suitable shape conformity and dimensional fidelity with variances two to four times larger than conventional fabrication techniques such as DRIE and wet-etching.Dimensional characterization is an important parameter as the variances measured with all fabrication methods can have direct implications on the hydrodynamic performance due to its sensitivity to channel dimensions.

Acknowledgments

The authors acknowledge the financial support of Science Foundation Ireland (SFI) under grant number 10/CE/I1853. Bell Labs Ireland would like to thank the Industrial Development Agency (IDA) Ireland for their continued support.

References

[1] Agostini B, Fabbri M, Park J E, Wojtan L, Thome J, R, and Michel B 2007 *Heat Transfer Eng.* **28** 4 p 258
[2] Eason C, Dalton T, Davies M, O'Mathna C, and Slattery O 2005 *Heat Transfer Eng.* **26** 3 p 79
[3] Martynova L, Locascio L E, Gaitan M, Kramer G W, Christensen R G, MacCrehan W A 1997 *Anal. Chem.* **69** 23 p 4783
[4] Paydar O H, Paredes C N, Hwang Y, Paz J, Shah N B, Candler R N 2014 *Sensor. Actuat. A-Phys.* **205** 0 p 199
[5] McDonald J C, Chabinyc M L, Metallo S J, Anderson J R, Stroock A D and Whitesides G M 2002 *Anal. Chem.* **74** 7 p 1537
[6] Moore J L, McCuiston A, Mittendorf I, Ottway R, and Johnson R D 2010 *Microfluid. Nanofluid.* **10** 4 p 877
[7] Bonyár A, Sántha H, Ring B, Varga M, Gábor Kovács J, Harsányi G 2010 *Procedia Engineering* **5** p 291
[8] Krejcova L, Nejdl L, Rodrigo M, Angel M, Zurek M, Matousek M, Hynek D, Zitka O, Kopel P, Adam V, and Kizek R 2013 *Biosens. Bioelectron.* **54** p 421
[9] McCullough E J and Yadavalli V K 2013 *J. Mater. Process Technol.* **213** 6 p 947
[10] Steinke M E and Kandlikar S G 2006 *Int. J. Therm. Sci.* **45** p 1073
[11] Wang R J, Wang L, Zhao L, and Liu Z, 2007 *Int. J. Adv. Manuf. Technol.* p498-504

Hydrodynamic characterization of a passive shape memory alloy valve

A M Waddell[1], **J Punch**[1], **J Stafford**[2] **and N Jeffers**[2]

[1] CTVR, Stokes Institute, University of Limerick, Limerick, Ireland
[2] Thermal Management Research Group, Bell Labs, Alcatel-Lucent, Dublin, Ireland

E-mail: alistair.waddell@ul.ie

Abstract. Next generation high-performance electronics will have large heat fluxes (>10^2 W/cm^2) and an alternative approach to traditional air cooling is required. An attractive solution is micro-channel cooling and micro-valves will be required for refined flow control in the supporting micro-fluidic systems. A NiTi Shape Memory Alloy (SMA) micro-valve design was hydrodynamically characterized in this work to obtain the valve loss coefficient (K) from pressure measurements. The hydrodynamic characterization was important as in the flow regime of the micro-fluidic system K is sensitive to Reynolds number (Re) and geometry. Static replicas of the SMA valve geometry were studied for low Reynolds numbers (110 – 220) in a 1x1 mm CSA miniature channel. The loss coefficients were found to be sensitive to flow rate and decreased rapidly with an increase in Re. The SMA valve was compared to a similar gate micro-valve and loss across both valves was of the same order of magnitude. The valve loss coefficients obtained in this work are important parameters in the modeling and design of micro-fluidic cooling systems.

1. Introduction

Electronic devices are constantly evolving towards compact form factors and higher performance. Accompanying this trend is an increase in the heat load (fluxes over 10^2 W/cm^2) that must be dissipated by internal circuitry. Traditional forced convection air cooling methods are reaching their limits and new approaches are required for future devices. A promising approach is the development of an integrated micro-channel cooling system on a chip. A vital component in this micro-fluidic system is a valve structure to control and regulate the flow of coolant.

A normally-closed passive Shape Memory Alloy (SMA) micro-valve design was described in previous work [1] and the developed prototype is shown in figure 1. The valve sits on an Acetal base (1) that slots into the (8x11 mm CSA) miniature-channel test-rig. The main body of the valve and source of obstruction when closed is a 127 μm thick curved stainless steel shim (2). Inserted through the steel shim is a ⌀0.4 mm NiTi actuator wire with an A_f temperature of 40°C. The valve is actuated by the SMA wire's shape memory effect when heated by the flow in the channel. A discussion on the function and response of the working SMA prototype valve is outside the scope of this paper. The focus is instead on the hydrodynamic characteristics of the valve geometry.

Content from this work may be used under the terms of the Creative Commons Attribution 3.0 licence. Any further distribution of this work must maintain attribution to the author(s) and the title of the work, journal citation and DOI.
Published under licence by IOP Publishing Ltd

Journal of Physics: Conference Series **525** (2014) 012010 doi:10.1088/1742-6596/525/1/012010

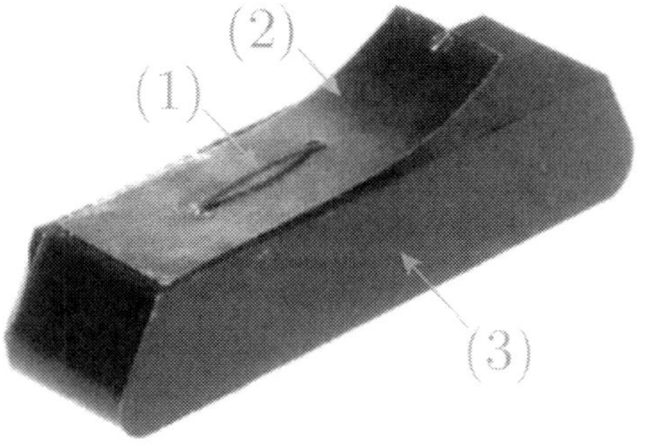

Figure 1. Prototype SMA valve [1]. (1) NiTi actuator wire; (2) stainless steel spring; (3) acetal base.

Figure 2(a) shows an illustration of the closed SMA prototype valve in elevation and end views. Also shown are fluid pathlines around the structure based on flow visualization performed by Waddell *et al.* [1]. On approach to the valve the flow area is gradually reduced. The fluid flows along the perimeter of the valve and subsequently through the two side-orifices in a complex and unsteady motion.

The hydrodynamic characteristics of the SMA valve will be compared with those of a similar gate valve [3]. The gate valve geometry is shown in figure 2(b). It consists of a vertical obstructing wall that occupies a certain fraction of the channel. The flow adjusts around the obstruction accordingly.

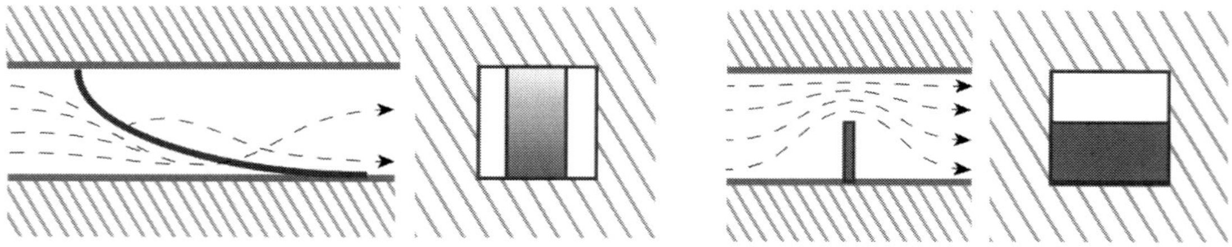

(a) SMA valve (b) Gate valve

Figure 2. Valve geometry in elevation and end view

The head loss coefficient (K) of a valve is an important parameter of the overall system design as the choice of pump must be able to overcome the associated pressure drop. In the turbulent flow regime K is practically constant and independent of Re. For laminar flows (Re $<10^3$), however, K is sensitive to Re and can vary greatly with geometry [2]. For this reason existing correlations could not be used to model the SMA prototype valve and a hydrodynamic characterization study was therefore performed. In the following sections the hydrodynamic characterization experiment on static valve-replicas will be detailed and results presented and discussed.

2. Hydrodynamic characterization method

Static test-pieces replicating the prototype SMA valve geometry were tested and details of these experiments will now be presented.

2.1. Test facility & samples

Reynolds scaling was used to characterize the macro-scale prototype valve geometry in micro-channel like conditions. Characterization was performed on replica geometry of the closed valve in a 1x1 mm CSA channel. The experimental setup is shown in figure 3. The working fluid (water) and was drawn from a reservoir (1) by a Cole-Parmer gear pump with a micro-fluidic head attachment. The reservoir remained at an ambient temperature of 22±1°C during the tests and the temperature of the flow entering the system was recorded with a K-type thermocouple. After the pump the flow entered the test rig (3) and an Omega PX2300 1 PSI transducer (4) recorded pressure drop across the test section. The transducer output was recorded with a 6009 Ni-DAQmx 14-bit data acquisition card.

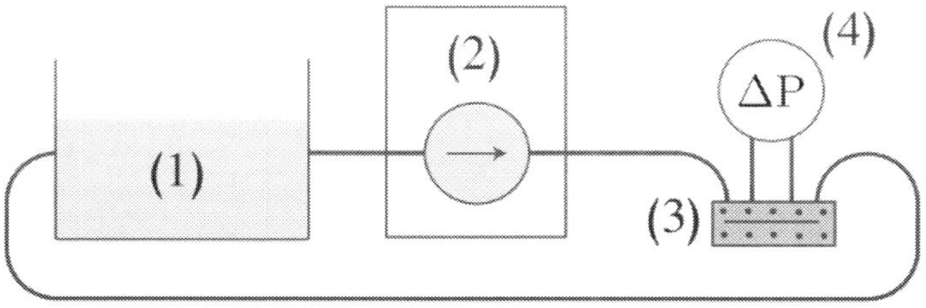

Figure 3. Pressure measurement facility. (1) reservoir; (2) gear pump; (3) test rig; (4) pressure transducer.

The 1 mm miniature channel rig is shown in figure 4. Fluid enters the channel at the inlet (1) and flows across two ⌀0.5 mm pressure taps (2 & 3) before exiting the rig (4). The pressure taps are located one flow development length ($x_{fd} \approx 0.05Re$, $Re = 240$) downstream of the inlet and test-section. The test section (5) consists of a slot where different replica valves can be inserted and tested. The test rig consists of an acetal body and polycarbonate window bolted together and sealed with an o-ring (6).

The valve geometry investigated in the current study was determined from the functional experiments involving the SMA valve shown in figure 1. Figure 5 shows a replica of this valve used in the characterization experiments. The sample was constructed from aluminium shim 100μm thick (1), bonded to a 3D-printed base (2) that slotted into the test rig. Six replica valve samples were constructed and had blockage ratios (w/W) ranging from 0.5 to 0.83.

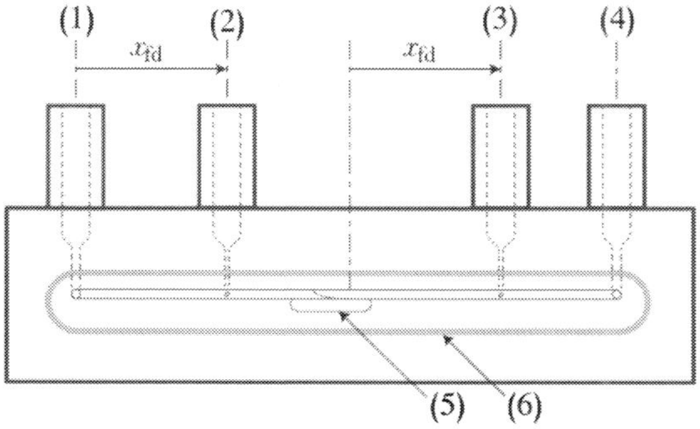

Figure 4. Pressure measurement rig. (1) flow inlet; (2) pressure port 1; (3) pressure port 2; (4) flow outlet; (5) removable test piece; (6) o-ring seal.

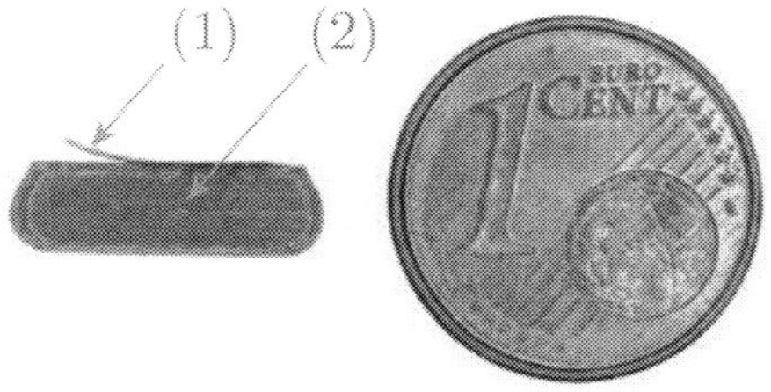

Figure 5. Miniature static valve replica. (1) 100 μm thick aluminium shim; (2) 13 mm long base.

2.2. Method

For each sample the test rig was assembled underwater to remove air, and the bolts were torqued to 5 Nm for consistent compression. The samples were tested over a range of Reynolds numbers in the channel between 110 and 220. After the system reached steady state for each flow-rate setting 10,000 pressure measurements were recorded and averaged.

3. Data analysis

The primary measurement recorded in the experiments was pressure drop across the test section. The valve loss coefficient (K) was then derived from this data. Analysis was performed on the results by the method of Kline and McClintock to determine the propagation of uncertainty within the results.

3.1. Data reduction

Equation 1 describes pressure-drop in the miniature-channel between the pressure ports. Head loss between two measured points is ΔP, the working fluid density is ρ, mean velocity in the miniature-channel is u, the valve loss coefficient is K, the friction factor f, the distance between the two pressure ports is L and the hydraulic diameter of the miniature-channel is D_h [4].

Head loss in the rig due to friction between the pressure ports was accounted for in the experimental results. Pressure-drop across an empty test section was recorded for the flow rates of interest yielding loss due to friction. These values were then subtracted from pressure-drop measurements across static valves to isolate the head loss contribution from the valve. Rearranging equation 1 for K (the valve loss coefficient) yields equation 2.

$$\Delta P = \frac{1}{2}\rho u^2\left(K + \frac{fL}{D_h}\right) \tag{1}$$

$$K = \frac{2\Delta P}{\rho u^2} \tag{2}$$

3.2. Uncertainty

Potential sources of uncertainty in the pressure measurements included the height-difference between two columns of liquid used in the transducer calibration and the transducer signal noise. A large sample size of 10,000 measurements was recorded and averaged to minimize uncertainty in the signal noise. A Newport optics stage was used to control the differential height between the two columns of water during the calibration. A total uncertainty in the pressure measurements was calculated as ±5.1 Pa by the method of Kline and McClintock [5]. Equation 2 was analyzed to evaluate the uncertainty in the derived variable K. The uncertainties of the primary variables are shown in table 1 and these values were combined by the method of Kline and McClintock, yielding an uncertainty in K that ranged between ± 4.7 and 10.34 %.

Table 1. Experimental uncertainties

Variable	Uncertainty
ΔP	± 5.1 Pa
$\dot{M}$	± 0.043 μL/s
A	± 0.02 mm^2
u	± 2.26 %
K	± 4.7–10.34 %

4. Results & discussion

The hydrodynamic characterization (see section 2) results will now be presented and discussed. Figure 6 shows pressure drop across the 6 samples for a range of flow-rates. Plotted through each data set is a power law curve fit in the form ax^b. All curve fitting constants can be found in table 2. Uncertainty in the data points range from ±1.18 to 9.3 %.

Journal of Physics: Conference Series **525** (2014) 012010 doi:10.1088/1742-6596/525/1/012010

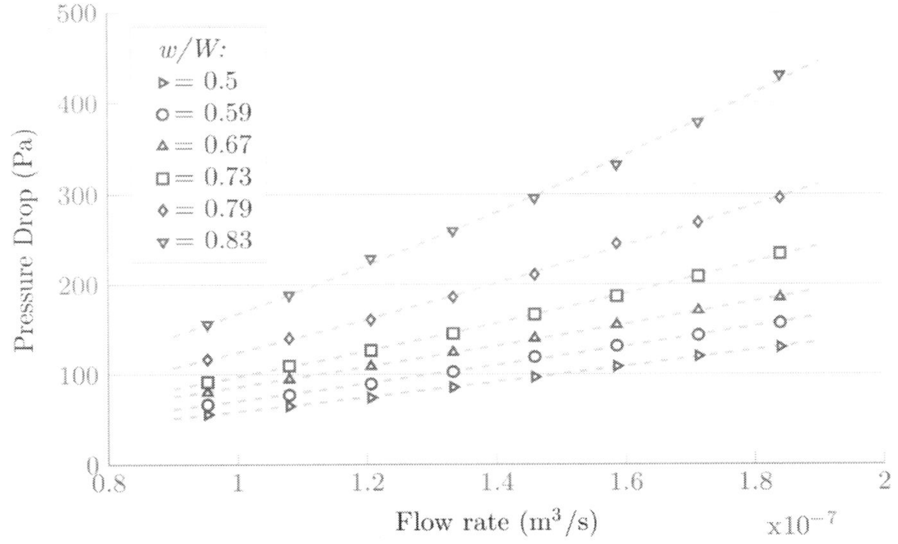

Figure 6. Pressure drop as a function of flow rate for various valve positions.

The data in figure 6 shows the pressure drop increases when w/W is increased and this is to be expected. Further, it can be seen that the pressure drop increases in a non-linear fashion as w/W is increased. More detailed information on the power law curve fits can be found in table 2. The power law exponent increases as w/W is increased, rising from 1.312 to 1.527 in the tests. The curve fits show a strong correlation in the data with an average R^2 of 0.9988.

Table 2. Power law (ax^b) curve fit constants

		ΔP			K	
w/W	a	b	R^2	a	b	R^2
0.5	$8.938\text{x}10^{10}$	1.312	0.9993	290.7	-0.6976	0.9968
0.59	$1.356\text{x}10^{11}$	1.326	0.998	298.3	-0.6662	0.9932
0.67	$6.928\text{x}10^{10}$	1.272	0.9996	450.7	-0.7119	0.9992
0.73	$1.009\text{x}10^{12}$	1.431	0.9995	282.9	-0.5861	0.9979
0.79	$1.29\text{x}10^{12}$	1.431	0.9986	347.6	-0.5781	0.9935
0.83	$8.116\text{x}10^{12}$	1.527	0.998	426.3	-0.5462	0.9915

K as a function of Re for the tested samples is shown in figure 7. Again, a power law curve fit in the form ax^b is plotted through each data set with the curve fitting constants located in table 2. Uncertainty in Re is ±3.5 %. The uncertainty in K ranges from ±4.7 to 10.34 %.

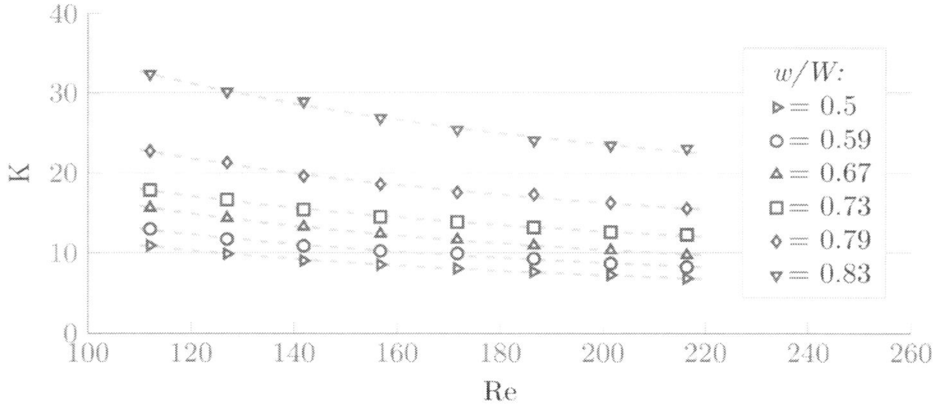

Figure 7. Valve loss coefficient (K) as a function of Re.

As Re is increased K is observed to decrease. The negative exponent of the power law curve fit increases with flow-rate and this implies that K is more sensitive to Re for the greater blockage ratios. For direct comparison with the gate valve (see section 1) characterized by Carretero-Benignos [3] a surface was plotted through the data and used to interpolate values for w/W=0.45. This comparison is plotted on a log-linear graph in figure 8.

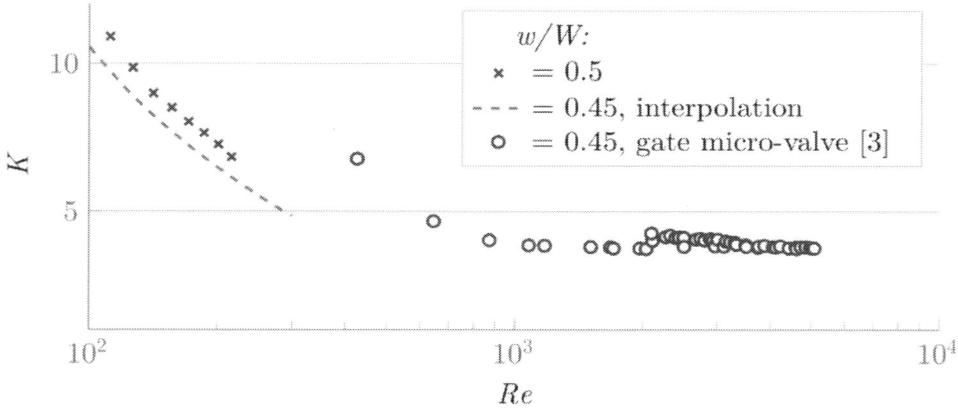

Figure 8. K factors for SMA valve replica compared with gate micro-valve

The K values for both the SMA and gate micro-valves are of the same order magnitude and reduce rapidly as Re is increased. The interpolated SMA valve K-factor curve is offset from the gate valve curve. Consider the geometry of the SMA valve (shown in section 1, figure 2(a)). The valve body is curved and this reduces the flow area gradually. This gradual reduction in area funnels the flow through the two side-orifices. The gate valve (figure 2(b)) consists of a wall perpendicular to the flow that blocks the channel. Unlike the SMA valve there is an abrupt reduction in flow area with no gradual reduction. These disparities in the two valve geometries may explain the difference in K; however the SMA valve has not been tested above Re >220.

In figure 8 K is relatively unaffected by the increase in Re above 10^3. There is a strong dependence as Re is decreased below 10^3 and the loss coefficients for both valves varies by a factor of two. This highlights the importance of understanding the losses associated with micro-scale valves as they operate in this flow regime. In systems incorporating micro-fluidic cooling the losses across micro-valves will need to be overcome by pumping. As K is sensitive to Re the overall work requirement of the system can vary significantly and needs to be considered in the design. The data presented in this work can be used as a reference in the design and modelling of valve components in future micro-fluidic cooling systems.

5. Conclusions

Hydro-dynamic characterization of a novel SMA micro-valve for application in passive thermal management of high heat flux devices was presented in this work. This analysis was performed on scaled down static replicas of the prototype SMA valve to derive K, the loss coefficient for the valve geometry. Key findings of the work were:

- The valve loss coefficients decrease rapidly with an increase in Re ($Re < 10^3$).
- The valve loss coefficients are lower than those of an equivalent gate valve.

This body of work presents information useful in the design of larger micro-fluidic cooling systems. The head loss curve fit's obtained for the range of experiment parameters can be used to predict the pressure drop across micro-valves in the system and provide a design specification for a micro-pump. Future work will include investigation of the flow structures around the valve and analysis of the heat transfer benefits that can be achieved.

Acknowledgements

The authors acknowledge the financial support of Science Foundation Ireland under grant number 10/CE/I1853. Bell Labs Ireland would like to thank the Industrial Development Agency (IDA) Ireland for their continued support.

Nomenclature

A	Area, miniature-channel	m^2	$\dot{M}$	Mass flow rate	kg/s
A_f	Austenite finish temperature	°C	Re	Reynolds number	-
D_h	Hydraulic diameter	m	u	Velocity	m/s
E	Young's Modulus	GPa	w	Width of valve	m
f	Darcy friction factor	-	W	Width of channel	m
K	Loss coefficient	-	ΔP	Differential pressure	Pa
L	Distance between pressure ports	m	ρ	Density	kg/m^3

References

[1] Waddell A M, Punch J, Stafford J and Jeffers N 2014 *Proc. 7th Int. Heat Transfer Conf. (Kyoto)* (New York: Begell House)

[2] Idelchik I E 1960 *Handbook of Hydraulic Resistance: Coefficients of Local Resistance and of Friction* (Washington: NTIS) p 351

[3] Carretero-Benignos J 2001 *Measurement and Modeling of the Flow Characteristics of Micro Disc Valves* (Massachusetts Institute of Technology: Masters Thesis) p 66

[4] Massey B 2006 *Mechanics of Fluids* (Oxon: Taylor & Francis) pp 245-253

[5] Kline S J and McClintock F A 1953 *Mech Eng.* **75** pp 385-387

Thermal Management Using Pulsating Jet Cooling Technology

S Alimohammadi[1,2], P Dinneen[1], T Persoons[1], D B Murray[1]

[1] Department of Mechanical and Manufacturing Engineering, Parsons Building, Trinity College, Dublin, Ireland

E-mail: alimohas@tcd.ie

Abstract. The existing methods of heat removal from compact electronic devises are known to be deficient as the evolving technology demands more power density and accordingly better cooling techniques. Impinging jets can be used as a satisfactory method for thermal management of electronic devices with limited space and volume. Pulsating flows can produce an additional enhancement in heat transfer rate compared to steady flows. This article is part of a comprehensive experimental and numerical study performed on pulsating jet cooling technology. The experimental approach explores heat transfer performance of a pulsating air jet impinging onto a flat surface for nozzle-to-surface distances $1 \leq H/D \leq 6$, Reynolds numbers $1{,}300 \leq Re \leq 2{,}800$ pulsation frequency $2Hz \leq f \leq 65Hz$, and Strouhal number $0.0012 \leq Sr = fD/U_m \leq 0.084$. The time-resolved velocity at the nozzle exit is measured to quantify the turbulence intensity profile. The numerical methodology is firstly validated using the experimental local Nusselt number distribution for the steady jet with the same geometry and boundary conditions. For a time-averaged Reynolds number of 6,000, the heat transfer enhancement using the pulsating jet for $9Hz \leq f \leq 55Hz$ and $0.017 \leq Sr \leq 0.102$ and $1 \leq H/D \leq 6$ are calculated. For the same range of *Sr* number, the numerical and experimental methods show consistent results.

1. Introduction

The existing methods of heat removal from compact electronic devises are known to be deficient as the growing technology demands more power and accordingly better cooling techniques with time. Impinging jets can be used as a satisfactory method for thermal management of electronic devices with limited space and volume. Pulsating flows can produce an additional enhancement in heat transfer rate compared to steady flows. This article represents parts of a comprehensive study performed on pulsating jet cooling technology both experimentally and numerically.

Potential heat transfer enhancement using pulsating jet impingement has not been broadly studied to date. A few of studies have identified a considerable enhancement in heat transfer coefficient [3]-[4]. Zumbrunnen and Aziz have experimentally examined the intermittent pulsation with square-wave form [3]; using square-wave pulsation at high frequency, a large heat transfer enhancement of up to 100% is measured by them. Herwig and Middleberg [4] have studied different waveform signals including triangular, sinusoidal and square-wave forms. They have shown that square-wave form velocity signals are able to generate higher heat transfer coefficients compared to other types. A criterion based on the Strouhal number is partly recognized by different findings. Herwig and Middelberg [4] observed an increasingly positive enhancement for $Sr > 0.029$ for a circular jet from a long pipe nozzle, up to $\delta Nu_0 = 30\%$ at $Sr = 0.076$ for square wave excitation. Conversely, some researchers have concluded a decrease of heat transfer using pulsating jet impingements [5]-[7].

[2] Corresponding Author: Phone= +353-876145739

Content from this work may be used under the terms of the Creative Commons Attribution 3.0 licence. Any further distribution of this work must maintain attribution to the author(s) and the title of the work, journal citation and DOI.
Published under licence by IOP Publishing Ltd

Azevedo [5] reported that the introduction of pulsation only served to reduce stagnation point heat transfer in all instances and for all *H/D*. The reductions reported equal to less than -25% degradation in heat transfer. Liu and Sullivan [6] have measured the relative heat transfer enhancement changing between -10% to +10% with pulsation frequency and *H/D*. Using an acoustically exited air jet from a contoured nozzle, O'Donovan and Murray [7] have also reported a heat transfer reduction up to -21% for $10{,}000 \leq Re \leq 30{,}000$ and $0.5 \leq H/D \leq 2$. The very different finding by various sources as described is main reason for which the unanimity about the effect of flow pulsation on magnitude of the heat transfer enhancement has not reached yet. This lack of consistency can partly be partly related to the wide range of operating conditions used in different studies.

Both numerical and experimental capabilities in investigation of the cooling effect of pulsating jets are presented in this paper. The impinging jet test facility has been previously fabricated for fluid dynamics and local heat transfer coefficient measurements by Persoons et al. [1] and O'Donovan and Murray [8]-[9]. A couple of modifications are performed in order record and process the fluctuating velocity signal at the exit of pipe nozzle. Also, the fluctuating velocity signals at the nozzle exit are processed in order to derive the values of turbulence intensities, as a major intention of current study. The experimental approach explores heat transfer performance of a pulsating air jet impinging onto a flat surface for nozzle-to-surface distances $1 \leq H/D \leq 6$, Reynolds numbers $1{,}300 \leq Re \leq 28{,}00$ pulsation frequency $2Hz \leq f \leq 65Hz$, and Strouhal number $0.0012 \leq Sr \leq 0.0839$.

The CFD methodology has been earlier validated using the experimental local Nusselt number distribution for the steady jet in the first step [10]-[12]. An essential requirement for CFD investigation of pulsating flow is the correct numerical modelling of laminar-turbulent boundary layer transition is. This paper employs a transition turbulence model called the Gamma-Theta model. This approach captures the laminar-turbulent transition in the wall jet for small nozzle-to-surface distances ($H \leq 2D$), and thus the intensity and extent of the secondary peak in the radial Nusselt number distribution. As another important motivation of this paper, it is aiming to generate a reliable CFD approach for heat transfer analysis of pulsating jet cooling technology for thermal management of compact electronic devices. The numerical methodology is firstly validated using the experimental local Nusselt number distribution for the steady jet with the same geometry and boundary conditions. As a main object, for the time-averaged Reynolds number of 6,000, the heat transfer enhancement using pulsating jet for $9Hz \leq f \leq 55Hz$ and $0.017 \leq Sr \leq 0.102$ and $1 \leq H/D \leq 6$ are calculated. This study can be considered as the second step toward a numerical methodology for unsteady impinging flows such as synthetic or pulsating jets [10]-[12].

2. Experimental approach

The schematic representation of the experimental setup is displayed in figure 1.

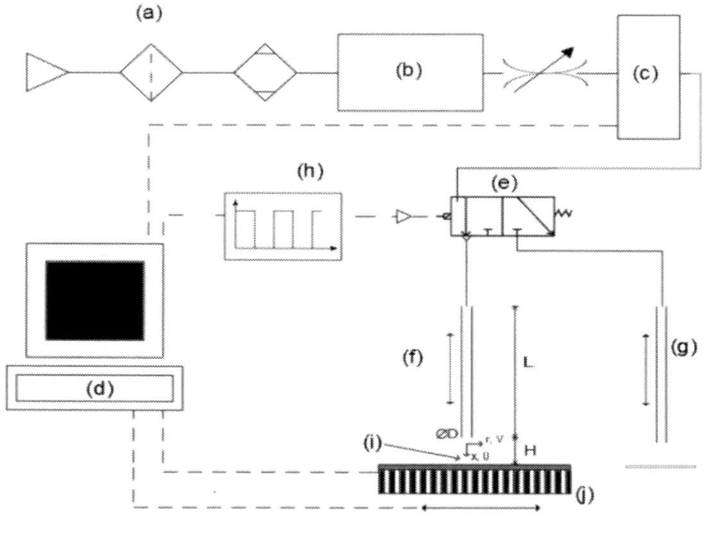

Figure 1. Experimental Setup (a) Air Compressor, Filter and Dryer, (b) Plenum Chamber, (c) Mass Flow Meter, (d) Data Acquisition System, (e) Solenoid Valve, (f) Testing Pipe Nozzle, (g) Secondary Nozzle, (h) Function Generator, (i) Microfoil Sensor Flush with Hot Plate, (j) Adjustable Hot Plate.

The air was supplied from the buildings air supply having passed through a number of filters (a) and a plenum chamber (b) in order to rid the flow of any fluctuations and impurities. The air was then passed through a digital mass flow controller (c) with an accuracy of ± 1%, controlled by a computer algorithm at (d), through to the solenoid valve (e). This valve divides the flow between the testing pipe (f) and the secondary pipe (g), at a frequency specified by the amplified signal from the function generator (h). Both the testing and secondary nozzles are identical, straight edged pipes with a diameter D = 5mm and length L = 80D. The secondary pipe is used so as to allow for air to be expelled to the ambient when the valve is closed, preventing pressure build up in the system. The height of the testing pipe, H, is adjusted as required with the height of the dummy pipe in order to maintain the same boundary conditions at both pipes, providing equal the back pressures at each pipe. The testing pipe was centred over an RDF Microfoil™ sensor at (i) which was installed flush with the hot plate at (j). This sensor was connected by the DAQ system and computer at (d) along with three T-type thermocouples allowing for the ambient, plate, and flow to be measured. This hot plate could be positioned radially using a stepper motor controlled through the computational algorithm at (d).

The plate was heated using a DC power supply to provide the electrical power to the electrical heaters stored in a 1.1mm silicon matt beneath the 5mm thick copper plate. This mat is glued to the underside of the copper plate, with the topside of the plate being the impingement surface. The plate was maintained at a constant temperature of 60°C with the ambient remaining around 22°C. The plate has been found to maintain a uniform wall temperature boundary condition according to previous studies [8]. The angle of impingement for all tests was 90°, maintaining the same experimental setup from previous studies [1]. The relationship used in this study to define heat transfer achieved is that of Shadlesky [13] who utilised laminar flow theory and assumptions in order to develop the following expression; $Nu_o = 0.585Re^{0.5}Pr^{0.4}$ for H/D < 2. Using this expression and a 13mm jet the relationship between heat flux and voltage from the Microfoil™ was correlated. This is performed to provide an accurate calibration for the 5mm straight pipe nozzles.

The heat transfer was approximated using $h = q''(T_{surf} - T_{ref})$ where the heat flux was provided by the RDF Microfoil™ before converting to the Nusselt number $Nu = hD/k$.

Hot wire measurements were carried out using the one-dimensional Dantec MiniCTA™ 54T30 hot wire (response time of 1-2 ms). This was clamped to the plate in order to give accurate radial temperature measurements of the flow utilising to the spatial resolution of the stepper motor. Measurements were carried at a sampling frequency proportional to the pulsation frequency being implemented. The uncertainty in the velocity measurements obtained is estimated at $\pm$2.7%.

The Nusselt number was measured radially for Reynolds number of 1300 and 2800. These low Reynolds numbers were selected as a result of previous studies that had suggested that the valve was struggling to maintain a square wave pulse at high frequencies. The duty cycle, assumed to be 0.5, was not reached and the study to achieve similar values to that of Persoons et al. [1]. It was therefore decided to reduce the volumetric flow rate by introducing 5mm piping. This decrease in piping lead to a reduction in the Reynolds numbers tested so as to maintain a similar order of Strouhal number to Persoons et al. [1].

3. Numerical approach

The CFD model reproduces the exact configuration used in the accompanying experiments for steady jet which used a smooth pipe nozzle with inner diameter D = 13mm and length L = 32D. The initial computational mesh is refined and adapted iteratively in regions with large velocity, pressure, temperature and turbulence gradients, such as the impingement wall region and the mixing region of jet and surrounding entrained air, in order to attain a stable solution. Figure 2 represents the computational domain together with the generated mesh used for simulations.

As a meticulous numerical uncertainty analysis, the Grid Convergence Index (*GCI*) study is performed in order to show that solutions are in asymptotic range of convergence [10]-[11]. The results which represent the discretization error in the solution can be plotted in the form of error band on the numerical results. As shown in figure 3, for steady jet with Re = 6,000 and H/D = 1, Alimohammadi et. al [10] have shown that local distribution of numerical uncertainty has some peaks with the maximum of smaller than 5% in the region of $1 < r/D < 2$.

The validation of heat transfer results for the steady jet for larger values of nozzle-to-surface distances ($2D \leq H \leq 6D$) is performed using a comparison of numerical results for local Nusselt number distribution over the impingement wall and experimental data. The statistical deviations of numerical results from experimental data for Re = 14,000 are quantified as 5%, 3% and 2% in terms of local, area-averaged and stagnation point Nusselt numbers, respectively. Conclusively, the results confirms that the developed numerical steady jet model, as the baseline model, can be extended for simulation of unsteady jet impingement problems such as pulsating flows. Further details about the computational procedure and the comprehensive sensitivity analysis performed on the effect of grid density, spatial discretization scheme, turbulent turbulence model and turbulent Prandtl number can be found in Alimohammadi et al. [10]-[12].

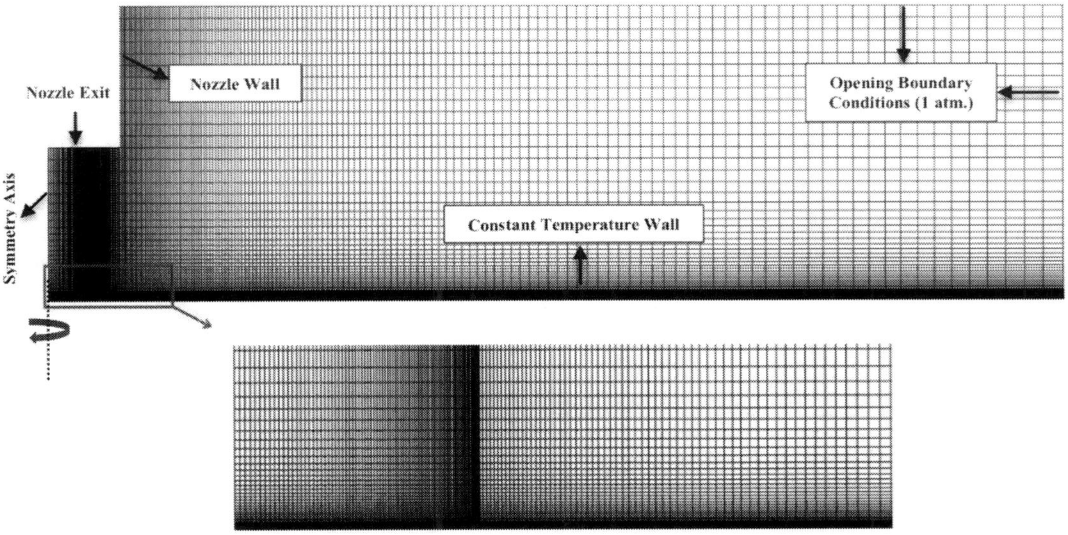

Figure 2. Computational domain and mesh for numerical simulation of axisymmetric pulsating jet impingement.

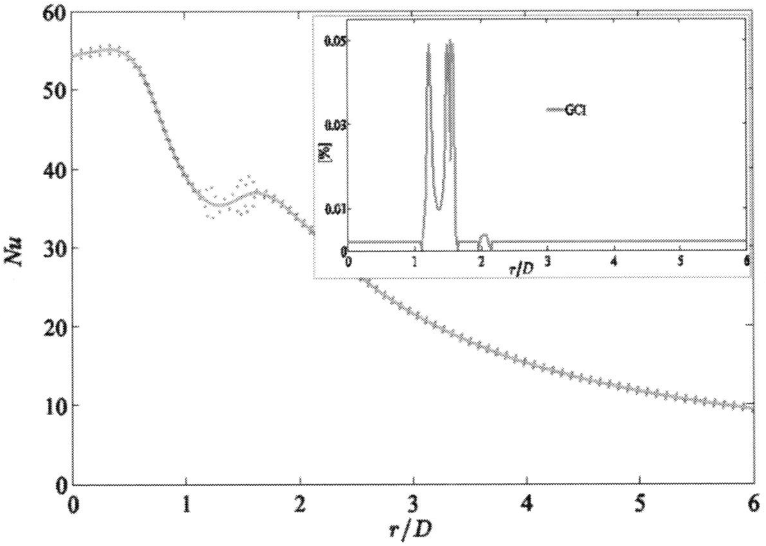

Figure 3. Radial distribution of Nusselt number with local numerical uncertainties (GCI (%)).

At the next step, the steady model is modified with an aim to investigate the pulsating jet flows. Detailed adjustments are implemented to account for the intermittent pulsation with a square-wave form with exponential decay/growth as the inlet velocity signal [11]. The exponential decay/growth ($\pm e^{-t/\tau}$; where τ is a decay constant) between extrema of the pulsating signal is used to take into consideration the response time of the pneumatic valve which provides the pulsating flow component and has a response time of approximately 2 milliseconds (see section 2).

4. Results and Discussion

4.1. Experimental results

4.1.1. Radial heat transfer results. The following experimental results for the radial Nusselt number distributions are presented in Figures 4. These display the effects of pulsation frequency (($2Hz \le f \le 65Hz$; $0.0012 \le Sr \le 0.0839$) and nozzle to plate spacing ($1D \le H \le 6D$) on the heat transfer for time averaged Reynolds numbers of 1,300 and 2,800.

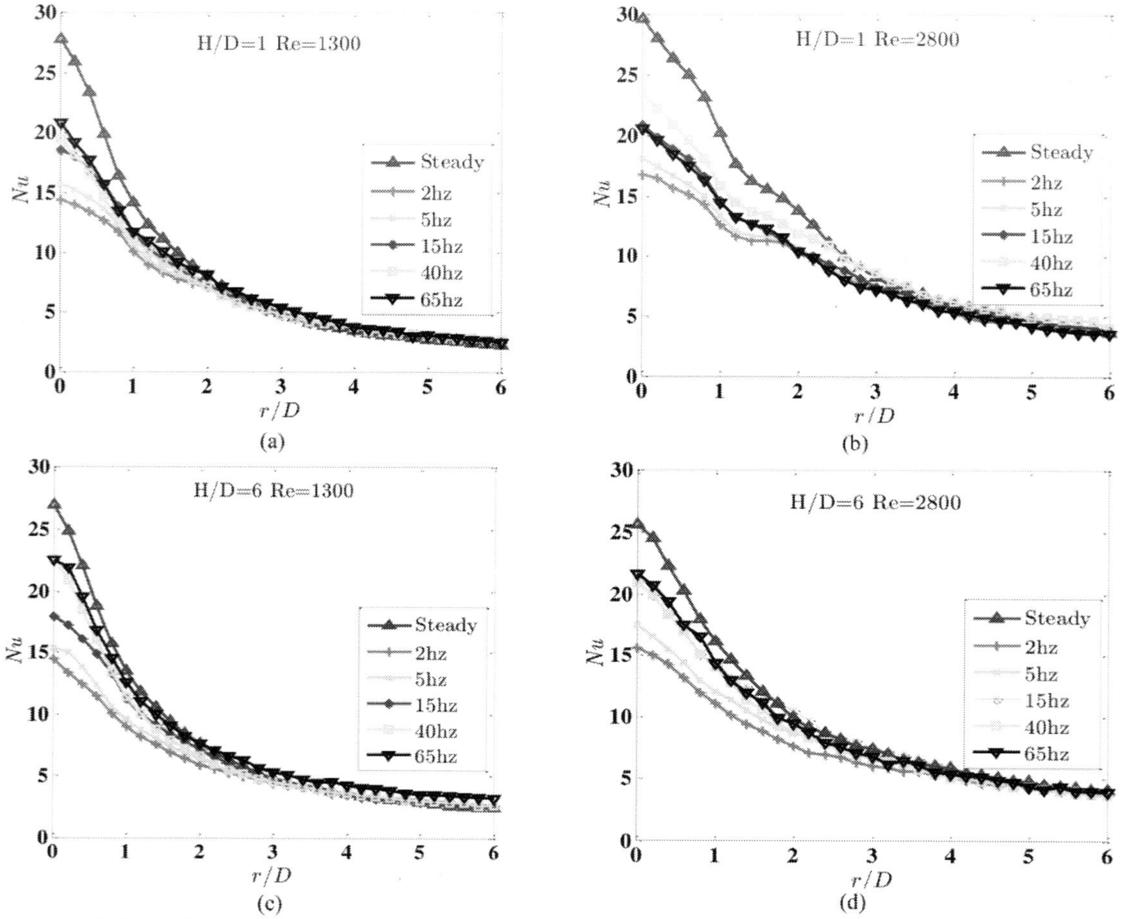

Figure 4. Radial heat transfer for $2Hz \le f \le 65Hz$ and $0.0012 \le Sr \le 0.0839$ at (a). $H/D = 1$, $Re = 1300$; (b). $H/D = 1$, $Re = 2800$; (c). $H/D = 2$, $Re = 1300$; (d). $H/D = 2$, $Re = 2800$.

The impinging jet Nusselt number correlation employed in the following analysis is that of Shadlesky [13], $Nu_o = 0.585Re^{0.5}Pr^{0.4}$,, which provides a lower bound estimate for the stagnation Nusselt number due to the use of laminar flow theory and assumptions in its conception. This provides a lower bound estimate for the stagnation Nusselt number due to its use of laminar flow theory and assumptions during its formulation.

As shown in figures 4, the highest heat transfer is achieved at the stagnation point. The highest velocity and lowest jet temperature is observed at this point, causing a corresponding increase in the Nusselt number due to the well documented dependencies on Re and ΔT [14].

The stagnation Nusselt number Nu_0 decreases for all Reynolds numbers and *H/D* values when pulsation is introduced. The results above were reflected by Azevedo et al. [5] who reported that the introduction of pulsation only served to reduce stagnation point heat transfer in all instances and for all *H/D*. The reductions reported equate < -25% degradation in heat transfer. This however was found at higher Strouhal numbers than those in this study, hence why such a significant degradation is observed above. It displays a marked decrease for pulsating jets at low *H/D*, caused by the increased entrainment and mixing provided by the introduction of pulsation. This leads the jet to become shorter and broadening its velocity profile, reducing the length for which the potential core persists and degrading the mean velocity of the jet at the core, confirmed in the studies of Hofman et al. [15]. This increased mixing and rate of decay of the potential core of the jet is accelerated by increasing the frequency of pulsation and therefore Strouhal number [14], [16]. However, the distance is not large enough to allow the increased mixing provided by the pulsating jet to impact the potential core of the jet. The impact of the ordered, periodic vortices of a pulsating jet do not rival the stagnation point heat transfer of a steady jet until they have penetrated the core of the jet at a higher *H/D*, shown figure 4 [14]-[15]. This results in increased turbulence intensity and mixing compensating for the loss in mean velocity when compared to a steady jet. It should also be noted that Nu_0 decreases for a steady jet with increasing *H/D* as compared to pulsating jets, which remain relatively constant, increasing in some isolated incidences.

Interestingly there are no sizeable "saddle back" profiles observed in these distributions. This is a phenomenon specific to impinging jets, where by a secondary peak is observed around $r/D = 1.5$ where the boundary transitions from a laminar to a turbulent regime at low H/D, increases the momentum transport [17]. This was observed in the previous set of results obtained using a 13.3mm contoured pipe by Persoons et al. [1]. This has been attributed to the low Reynolds number of the flow, in contrast to the considerably higher Reynolds numbers which were analysed by the previous studies using this experimental set up and the majority of the authors in the review of the literature above. The lower Reynolds number and therefore slightly higher Strouhal number also means that the potential core of the jet is diminished at a quicker rate as outlined above. This is shown by Lytle and Webb [18] who display that the secondary maximum only occurs at higher Reynolds number when H/D is increased. Conversely this secondary maximum becomes evident at lower Re with a corresponding decrease in *H/D*. Therefore such a saddle back profile may be evident testing at $H/D < 1$. This secondary peak was mildly visible in this investigation however, at $H/D = 1$ when a Reynolds number of 2800 was utilised. This secondary maximum can also be observed at higher frequencies for low *H/D*, a result akin to that of Persoons et al. [1]. This is caused by the Reynolds number being doubled during pulsating flows so as to allow for the same time averaged velocities compared to a steady jet.

4.1.2. Average radial heat transfer results. To discern as to whether any improvements in heat transfer are obtained beyond the stagnation point, the area averaged Nusselt number was derived. This was defined utilising the length scale of the jet nozzle to define a circular impingement area of radius $R = 6D$ around the stagnation point. The integral of the Nusselt number with respect to r/D was then calculated and divided by the surface are to give $\overline{Nu} = \frac{1}{\pi R^2}\int_0^{2\pi}\int_0^R Nu(r)r\,dr\,d\theta$. The enhancement was then derived from the following expression: $\delta\overline{Nu} = \left(\frac{\overline{Nu}_{pulse}}{\overline{Nu}_{steady}} - 1\right) \times 100\%$. The stagnation and area-averaged heat enhancements, when compared to steady jet, are quantified as follows $-49\% \leq \delta Nu_0 \leq -8\%$ and $-21\% \leq \delta\overline{Nu} \leq 29.5\%$. Variation of $\delta\overline{Nu}$ with *H/D* is shown in figure 5.

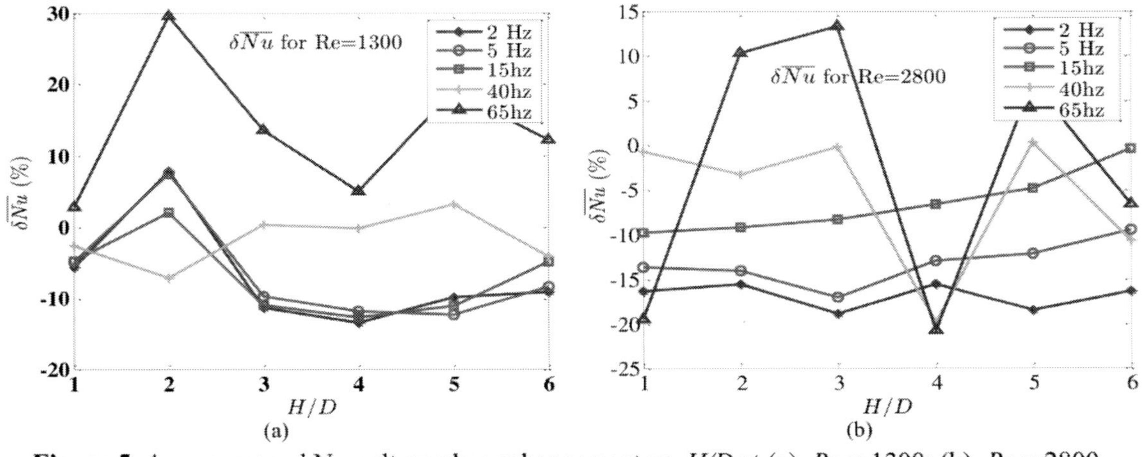

Figure 5. Area-averaged Nusselt number enhancement vs. *H/D* at (a). *Re* = 1300; (b). *Re* = 2800.

The enhancements in the area-averaged Reynolds number increase substantially at *H/D* > 2 for both Reynolds numbers, as displayed in figure 5. This is reflective again of the normally clear secondary maximum found at low *H/D*. The enhanced mixing and turbulence intensity found to exist as the velocity decreases radially and the turbulence intensity increases; this has been found by Mladin and Zumbrunnen [20], and therefore leads to increased heat transfer for *r/D* > 2. These higher turbulence intensities observed in pulsating flows have been noted to cause increases in the radial velocity of a pulsating jet improving the heat transfer in the wall jet region [21]. The peak in area-averaged heat transfer is found at *H/D* = 2 for the lower Reynolds number of 1300 and increases to *H/D* = 3 for *Re* = 2800. These peaks in area-averaged heat transfer coincided with a disappearance in of the small secondary maximum visible in the radial heat transfer for *H/D* = 1 *Re* = 2800 in figure 4, indicative of the existence of the potential core of the jet. This would suggest that at H/D=3, the potential core of the jet has been completely enveloped by the shear layer allowing for increased turbulence intensity across the jet [15]. This shortening and spreading of the jet also allows for an increased radial velocity increasing heat transfer away from the stagnation point. This results in a flatter radial heat transfer distribution [22]. Some area-averaged enhancement has already been observed at *Re* = 1300 and *H/D* = 1 at 65*Hz*, with no secondary maximum visible at this *H/D* spacing in the radial profiles in figure 4. The $\overline{Nu}$ profile in figure 5 suggests that this envelopment of the potential core has already occurred due to the higher Strouhal numbers in effect. The increases in heat transfer, as shown above, display a clear dependency on the reduction in the mean velocity flow and increasing the frequency of pulsation. This confirms that there is a definite dependency on the Strouhal number as suggested by a number of authors [3] and [15]-[16].

4.1.3. Turbulence Intensities. The compressibility of air results in pulsating jets of higher frequencies struggling to maintain a square wave pulsation. The theoretical analysis of the effective pulsation amplitude displays that at higher frequencies there is effectively a steady flow oscillating to a maximum. This observation of the flow never reaching a standstill is reflected in the predictions of Persoons et al. [1].

The following velocity signals, as shown in figure 6, were obtained using hot wire anemometry. It is the intention of this paper to provide an extended range of turbulence intensities occurring in a pulsating jet for varying pulsation frequencies and Reynolds numbers. The centreline turbulence intensity was evaluated for a selected number of frequencies using Reynolds decomposition. This was carried out by decomposing the flow into periodical signals, as shown in figure 6, before deriving the phase averaged velocity ($\bar{u}_\emptyset$) for each period. This phase averaged velocity was then subtracted from the true velocity ($u(t)$) of the flow to give the turbulent fluctuations ($u'(t) = u(t) - \bar{u}_\emptyset$). Turbulence intensity is defined as the root-mean-square of the turbulent velocity fluctuations $u'(t)$ with respect to mean averaged jet exit velocity $\bar{u}_m$; $Tu = \sqrt{\overline{u'(t)^2}/\bar{u}_m}$. Noise was eliminated on a case by case basis so as to ensure the accuracy of the values obtained.

Figure 6 diplays the dependence of turbulence intensity on the Reynolds number. This dependence is often assumed for a fully developed pipe flow, as is the case in this study, through the correlation $Tu = 0.16(Re)^{-\frac{1}{8}}$ [23]. The profiles above demonstrate this relationship with the turbulence decreasing for higher Reynolds numbers. Using this expression it is expected that the turbulence intensities for these flows would be ≈ 6% for *Re* = 2800. It is expected however that there be some deviation from this due to the pulsating nature of the flow, while the above expression is proposed for steady flows.

Hofman et al. [15] measured the turbulence intensity of steady jet, which was also found to be ≈ 4% for *Re* = 78,000, analogous to the turbulence intensities retrieved in the current investigation, reaffirming the quasi-steady assumption utilised in the analysis of these low frequency jets. The ability, to calculate these turbulence intensities will aid the accuracy of numerical studies being run concurrently with the current experimental setup allowing for more accurate predictions of heat transfer enhancement.

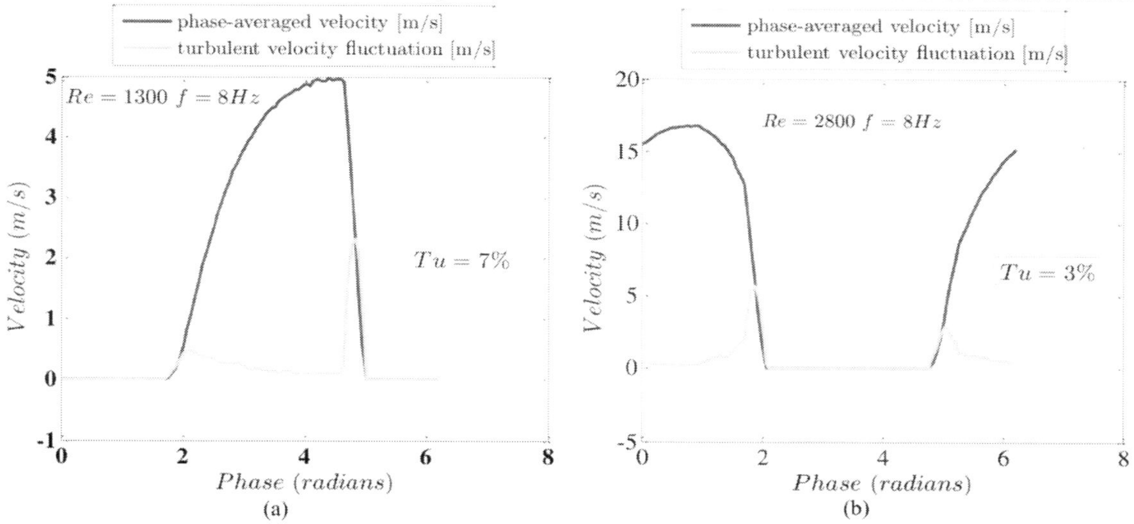

Figure 6. Velocity signal decomposition and $Tu(\%)$ at (a). $Re = 1300, f = 8Hz$; (b). $Re = 2800, f = 8Hz$.

4.2. Numerical results

4.2.1. Radial heat transfer results. The effect of pulsating inlet flow on the local Nusselt number distribution and thermal enhancement on stagnation (δNu_0) and area-averaged (δNu_{ave}) Nusselt numbers are discussed. As shown in Fig. 1, a degradation of Nusselt number is obvious in the stagnation zone. At higher frequencies ($27Hz \leq f \leq 55$Hz) and larger values of *H/D*, the pulsating jet flow raises the local Nusselt number distribution over the whole impingement surface. For the radial distances of $r/D > 2$, after the secondary peak, the results show a gradual reduction in heat transfer distribution, and interestingly, the difference between the results of various frequencies after this region becomes insignificant. In addition, as the frequency of the pulsation increases, the second peak in the Nusselt number distribution becomes less intense while remaining at the same location.

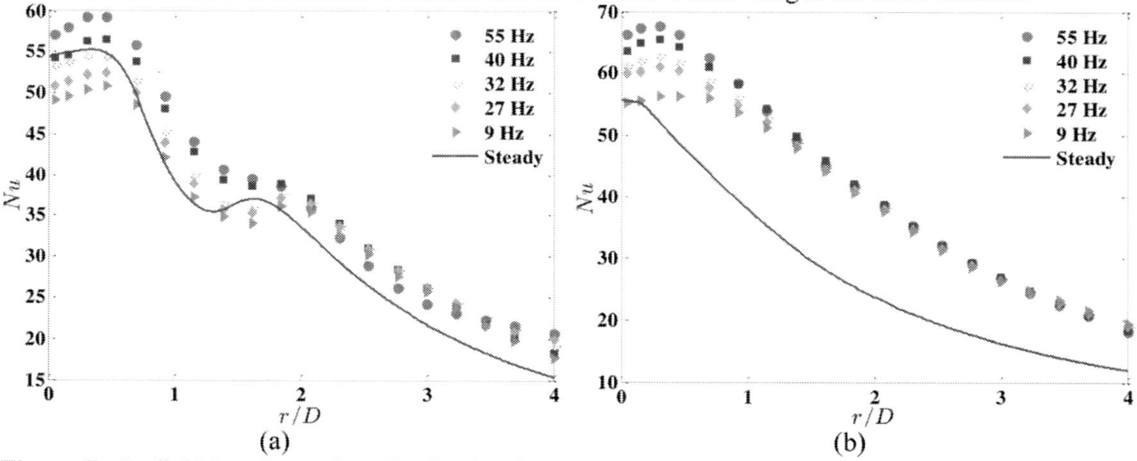

Figure 7. Radial Nusselt number distribution for steady and pulsating flows at different frequencies ($f = 9 – 55$ *Hz*; $Sr = 0.017$-0.102; $Re = 6{,}000$) for: (a). $H/D = 1$, (b). $H/D = 6$.

4.2.2. Area-averaged and Stagnation Nusselt Numbers. The enhancement effects of flow pulsation on area-averaged Nusselt (δN_{uave}) and stagnation Nusselt (δNu_0) numbers, when compared to corresponding steady jet, are shown in figure 8.

For $f = 9$ *Hz*, a reduction in the stagnation Nusselt number occurs when pulsation is used, followed by increases with increasing frequency from 9 to 55 *Hz*. For *H/D* of 1, the maximum reduction of -10% of the steady Nusselt number is observed (f = 9 *Hz*, $Sr = 0.017$), and only at the uppermost frequency of 55 *Hz*, is an increase of the stagnation point Nusselt number detectable.

Journal of Physics: Conference Series **525** (2014) 012011 doi:10.1088/1742-6596/525/1/012011

The level of area-averaged enhancement (δN_{uave}) increases with frequency (or Strouhal number). *H/D* strongly affects the area-averaged heat transfer. The results for two higher frequencies, 40 and 55 *Hz*, are very close; this becomes more evident at *H/D* = 6, where the maximum δNuave of 63% occurs (f = 55, Sr = 0.102). The lowest enhancement is equal to 10% for *H/D* =1 (f = 9 *Hz*, Sr = 0.017). The increased δN_{uave} outweighs the reductions in δNu_0. The stagnation and area-averaged Nusselt numbers exhibit similar trends with frequency and nozzle-to-surface distance. The area-averaged Nusselt numbers represent more consistent and larger enhancements showing a monotonic increase with frequency at different *H/D* values. The variation of enhancement in stagnation and area-averaged Nusselt numbers with f and *H/D* are consistent with experimental data of Persoons et al. [1].

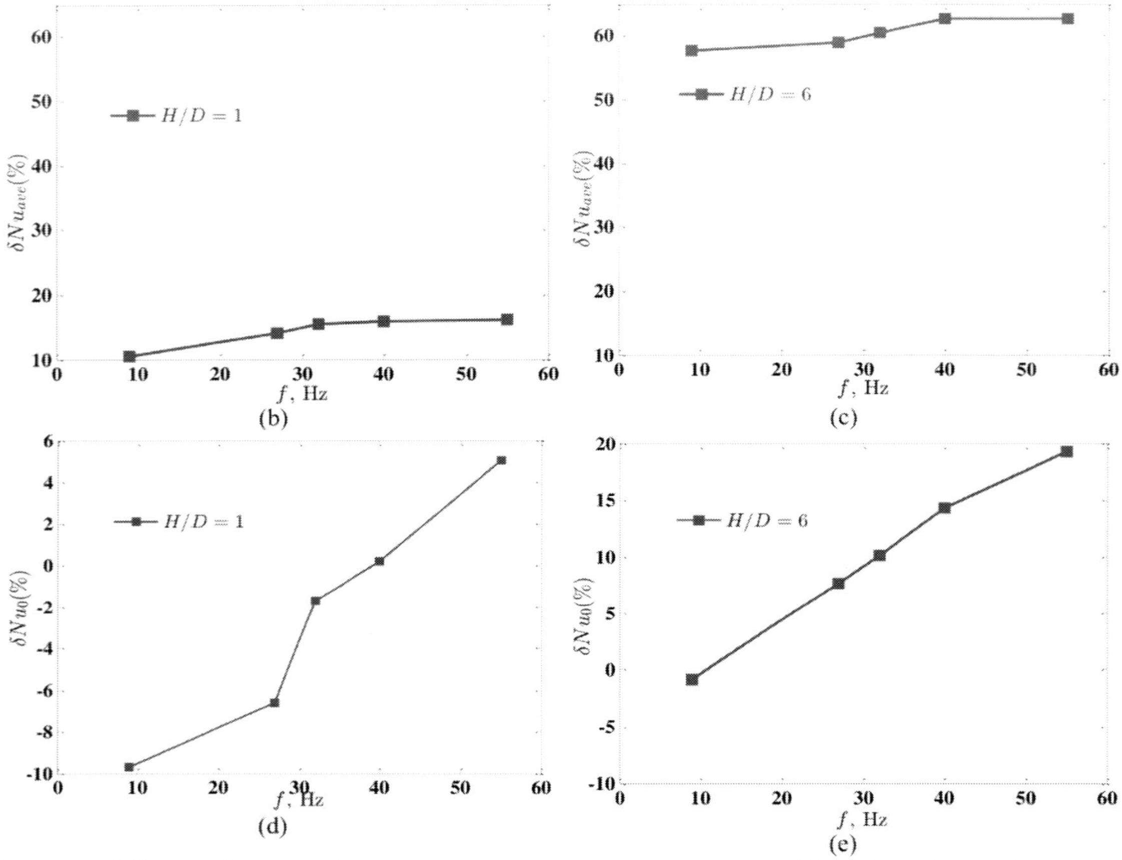

Figure 8. Effect of flow pulsation on area-averaged Nusselt number (a. *H/D* = 1, b. *H/D* = 6) and Stagnation Nusselt number (b. *H/D* = 1, b. *H/D* = 6) for f = 9 – 55 *Hz*; Sr = 0.017-0.102; Re = 6,000.

5. Conclusions

Using the numerical approach, the effect of flow pulsation on stagnation and area-averaged Nusselt numbers, when compared to steady jet, are respectively quantified as -10% ≤ δN_0 ≤ 19% and 10% ≤ δN_{uave} ≤ 63%. For experimental approach, the stagnation and area-averaged heat enhancements are quantified as follows -49% ≤ δNu_0 ≤ -8% and -21% ≤ δNu_{ave} ≤ 29.5%.

The results of numerical and experimental approaches are consistent for identical values of Sr numbers. A clear unanimity about the effect of flow pulsation on magnitude of the heat transfer enhancement can be achieved by adapting the range of the operating conditions in both approaches, as a further step of the current study. By adjusting the effective parameters appropriately, the employment of the pulsating jet flow can cause a significant enhancement compared to the equivalent steady jet flow. All in all, pulsating jet cooling technology shows great potentials for thermal management in compact electronic systems.

References

[1] Persoons T, Balgazin, K, Brown K and Murray D B 2013 Scaling of convective heat transfer enhancement due to flow pulsation in an axisymmetric impinging jet *J. Heat Transf.* **135** 111012.

[2] Garimella S V, Yeh L T and Persoons T 2012 Thermal Management Challenges in Telecommunication Systems and Data Centers *IEEE Trans. Compon. Packag. Manuf. Technol.* **2** 1307–16.

[3] Zumbrunnen D A and Aziz M 1993 Convective heat-transfer enhancement due to Iintermittency in an impinging jet *Trans. J. Heat Transf.* **115** 91–98.

[4] Herwig H and Middelberg G 2008 The physics of unsteady jet impingement and its heat transfer performance *Acta Mech.* **201** 171–184.

[5] Azevedo L F A 1994 Pulsed Air Jet Impingement Heat Transfer *Exp. Therm. Fluid Sci.* **8** 206–213.

[6] Liu T S and Sullivan J P 1996 Heat transfer and flow structures in an excited circular impinging jet *Int. J. Heat Mass Transfer* **39** 3695–3706.

[7] O'Donovan T S and Murray D B 2007 Effect of acoustic excitation on the heat transfer to an impinging air jet *Proc. ASME-JSME Thermal Eng. Summer Heat Transf. Conf.* HT2007–32800 (Vancouver: CANADA).

[8] O'Donovan T S and Murray D B 2007 Jet impingement heat transfer – Part I: Mean and root-mean-square heat transfer and velocity distributions *Int. J. Heat Mass Transf.* **50** 3291–3301.

[9] Persoons T, McGuinn A and Murray D B 2011 A general correlation for the stagnation point Nusselt number of an axisymmetric impinging synthetic jet *Int. J. Heat Mass Transf.* **54** 3900–08.

[10] Alimohammadi S, Murray D and Persoons T 2014 Experimental Validation of a CFD Methodology for Transitional Flow Heat Transfer Characteristics of a Steady Impinging Jet *J. Heat Transf.* HT-14-1049 (accepted; in press).

[11] Alimohammadi S, Persoons T and Murray D B 2014 A numerical-experimental study of heat transfer enhancement using unconfined steady and pulsating turbulent air jet impingement *Proc. 15th Int. Heat Transfer Conf.* IHTC15-8765 (Kyoto:Japan) (accepted; in press).

[12] Alimohammadi S, Persoons T, Murray D B, Tehrani M S, Farhanieh B and Koehler J 2013 A Validated Numerical-Experimental Design Methodology for a Movable Supersonic Ejector Compressor for Waste-Heat Recovery *J. Therm. Sci. Eng. Appl.* **6** 021001:1-10.

[13] Shadlesky P S 1983 Stagnation point heat transfer for jet impingement to a plane surface *AIAA J.* **21** 1214–15.

[14] Farrington R B and Claunch S D 1994 Infrared Imaging of Large-Amplitude, Low-Frequency Disturbances on a Planar Jet *AIAA J.* **32** 317–323.

[15] Hofmann H M, Movileanu D L, Kind M and Martin H 2007 Influence of a Pulsation on Heat Transfer and Flow Structure in Submerged Impinging Jets *Int. J. Heat Mass Transf.* **50** 3638–3648.

[16] Mladin E and Zumbrunnen D 1997 Local Convective Heat Transfer To Submerged Pulsating Jets *Int. J. Heat Mass Transf.* **40** 3305–21.

[17] Lee J and Lee S 1999 Stagnation Region Heat Transfer of a Turbulent Axisymmetric Jet Impingement *Exp. Heat Transf.* **12** 137-156.

[18] Lytle D and Webb B W 1994 Air jet impingement heat transfer at low nozzle-plate spacings **37** *Int. J. Heat Mass Transf.* 1687–97.

[19] Sheriff H S and Zumbrunnen D A 1994 Effect of Flow Pulsations on the Cooling Effectiveness of an Impinging Jet *J. Heat Transfer* **116** 886–895.

[20] Mladin E C and Zumbrunnen D A 2000 Alterations to Coherent Flow Structures and Heat Transfer Due to Pulsations in an Impinging Air-Jet *Int. J. Therm. Sci.* **39** 236–248.

[21] Zulkifli R and Sopian K 2007 Studies on Pulse Jet Impingement Heat Transfer: Flow Profile *Int. J. Eng. Technol.* **4** 86–9.

[22] Camci C and Herr F 2002 Forced Convection Heat Transfer Enhancement Using a Self-Oscillating Impinging Planar Jet *J. Heat Transfer* 124 770-782.

[23] Ansys Inc. 2011 Ansys Fluent User Guide Release 14.0 (Pennsylvania:USA) pp 263-4.

The effects of stroke length and Reynolds number on heat transfer to a ducted confined and semi-confined synthetic air jet

D.I.Rylatt*, T.S.O'Donovan

Heriot Watt University Edinburgh

Email: dir1@hw.ac.uk

Abstract. Heat transfer to three configurations of ducted jet and un-ducted semi-confined jets is investigated experimentally. The influence of the jet operating parameters, stroke length (L_0/D) and Reynolds (Re) number on the heat transferred to the jet is of particular interest. Heat transfer distributions to the jet are reported at $H/D = 1$ for a range of experimental parameters Re (*1000* to *4000*) and L_0/D (*5* to *20*). Secondary and tertiary peaks are discernable in the heat transfer distributions across the range of parameters tested. It is shown that for a fixed Re varying the L_0/D has little effect on the magnitude of the stagnation region heat transfer but does effect the position and magnitude of the secondary and tertiary peaks in the heat transfer distribution. It is also shown that for a fixed L_0/D increasing the Re has a significant effect on the magnitude of the stagnation region heat transfer but has little impact on the position of the secondary and tertiary peaks in the heat transfer distributions. Ducting is added to the configuration to improve heat transfer by drawing cold air from a remote location into the jet flow. Ducting is shown to increase stagnation region and area averaged heat transfer across the range of jet parameters tested when compared with an un-ducted jets of equal confinement. Increasing the stroke length from $L_0/D = 5$ to *20* for a Reynolds number of *2000* reduces the enhancement in stagnation region heat transfer provided by the ducting from *35%* to *10%;* the area averaged heat transfer provided by the ducting also changes from a *42%* to a *21%* enhancement. This is shown to be partly due to relative magnitude of the peaks in heat transfer outwith the stagnation region; at low stroke lengths, the difference in the magnitude of these peaks is large and reduces with increasing L_0/D. It is also shown that as L_0/D is increased the stagnation region heat transfer to the un-ducted jets increases while for the ducted jets stagnation region heat transfer decreases with increasing L_0/D. Increasing Reynolds number from *1000* to *4000* for a stroke length from $L_0/D = 10$ increases the increase in stagnation region heat transfer provided by the ducting from *10 %* to over *50 %* and increases the increase in area averaged heat transfer provided by the ducting from *15 %* to *45 %*. This is shown to be primarily due to the magnitude of the stagnation region heat transfer. While the heat transfer increases with Re for all configurations of jet the increase is much more significant for the ducted jets.

Content from this work may be used under the terms of the Creative Commons Attribution 3.0 licence. Any further distribution of this work must maintain attribution to the author(s) and the title of the work, journal citation and DOI.
Published under licence by IOP Publishing Ltd

1. Introduction

Due to the high stagnation region and area averaged heat transfer coefficients jet impingement can achieve it has, for many years, been used in industrial applications. Synthetic air jets are a promising new technology which has been shown by both Smith and Swift [1] and Pavlova and Amitay [2] to be capable of providing more than twice the cooling of continuous air jet impingement for a similar range of Reynolds numbers.

The processing power of computers has increased greatly in recent years leading to higher demands on thermal management systems. This increase in processing power has been accompanied by a steady reduction in size and the available space in computers due to the desire for increased functionality in the same or smaller housing. Kercher et al. [4] have shown that a synthetic air jet using the same power as a standard commercial fan can provide more than twice the cooling. The current CPU Thermal Design Power (TDP) for an Intel® Core™ i7-860 is 95W and the TDP of the Pentium 3 circa 1999 was around 35W. While this shows that Gunther's [5] prediction in 2001 of a twofold growth in the TDP every two years was not realised, it is clear that the fan-fin heat exchangers will not meet the growing cooling requirement of electronic component. New technologies, such as synthetic air jets have the potential to meet this requirement. Chaudhari et al [6] show that a synthetic jet has a maximum heat transfer performance at $H/D = 10$. However when the jet is confined to $H/D = 3$ this is reduced by *80%*. Persoons et al [7] present heat transfer data for an axisymmetric jet similar to that used in this study; and report a reduction in cooling efficacy when the jet operates below its optimal spacing of 3.4 jet diameters. Smith and Glezer [8] and Smith and Swift [3] attribute the high rates of heat transfer achieved by synthetic jets to high rates of growth in both the jet column width and volume flux due to entrainment. This is limited at lower nozzle to impingement surface spacings. In addition to this, at low H/D, the fluid between the heated impingement surface and the jet orifice plate is highly confined and is re-circulated into the jet flow. The elevated temperature of this fluid above that of the ambient air reduces the temperature difference between the synthetic air jet and the impingement surface and therefore results in substantially reducing cooling performance. Several studies have investigating the effects of semi-confinement on the heat transfer to steady impinging jets [9-14]. Youn et al [13] investigated the effects of semi-confinement of a microscale steady jet for a wide range of confining plate lengths, L_{op} (*3 to 48* jet diameters) and nozzle to impingement surface spacings, H (*1 to 20* jet diameters). Overall, confinement is shown to reduce the heat transfer. At low nozzle to impingement surface spacings, even relatively minor semi-confinement ($L_{op}/D=6$) can reduce the stagnation point heat transfer by more than 50%. In general, at low H/D however, further increasing the confinement does not lead to substantial further reductions in the surface heat transfer, but it does affect the height at which the effect of the increased plate length will be overcome. Herein lays one significant difference between steady and synthetic air jet impingement heat transfer. Chaudhari et al [6] investigated the effect of increasing the semi-confinement of a synthetic jet from $L_{op}/D = 7.86$ to *13.75* and report a *50%* reduction in heat transfer for the more confined jet across the full range of nozzle to impingement surface spacings tested ($H/D=1$ *to 14*). Therefore, it is clear that heat transfer to semi-confined impinging synthetic jets is significantly different from steady jets, which only display the effects of increased confinement at low spacings. The motivation for this work is to ameliorate the effects of confinement on heat transfer to a synthetic jet at low H/D by adding ducting to the orifice plate of the jet. This allows air to be drawn into the jet from some remote location. Rylatt and O'Donovan [14] have shown that ducting can significantly increase the heat transfer in both the stagnation region and on an area averaged basis. This paper focuses on the effect of the stroke length and Reynolds number of the jet on the efficacy of the ducting.

2. Experimental setup;

A schematic of the experimental setup is presented in figure 1; the rig consists of two main components: 1) the synthetic air jet assembly, and 2) a heated impingement surface. The jet is mounted directly above the heated surface. As can be seen in figure 1, the synthetic air jet consists of a cavity with an acoustic speaker operating as the jet diaphragm on one end and the jet orifice opposite it. A detailed description of the experimental rig used in this investigation is shown by Rylatt and O'Donovan [14].

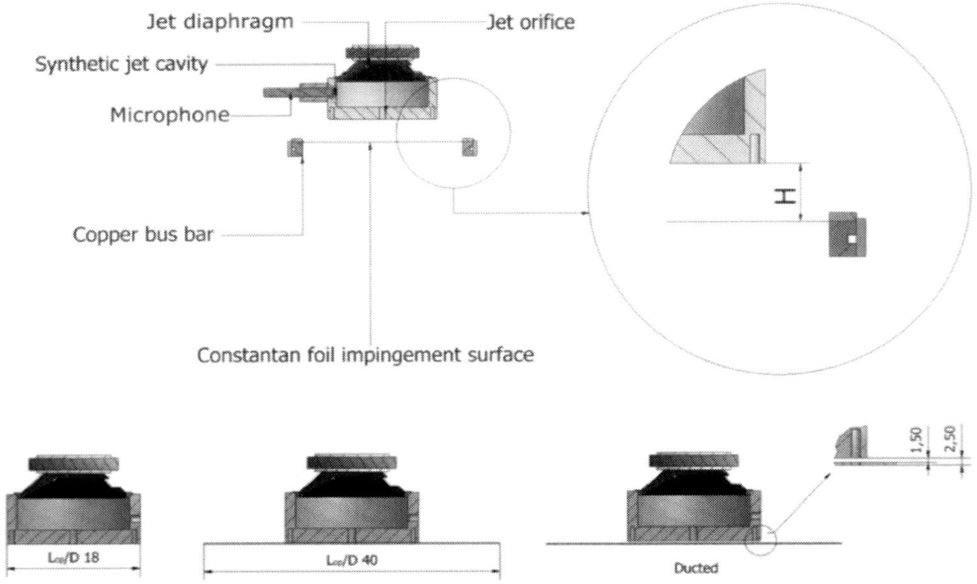

1: Experimental facility and jet configurations

Heat transfer data are presented in the form of the time-average and time-varying Nusselt number which uses the diameter of the jet to normalise the surface convective heat transfer coefficient; this is calculated using equations 1 to 3. Equation (2) is based on the assumption that heat generated within the foil is conducted in 1 dimensional through the thickness of the foil before convecting to the impinging jet. However, Patil and Narayanan [15], who also investigated the application of the thin foil infrared thermography technique for heat transfer measurements in convective flows, found that "lateral conduction heat flux rate contributed a significant portion to the net heat entering the control volume". Therefore, in order to accurately assess the local heat transfer taking place from the foil to the impinging jet, lateral conduction must be considered.

$$Nu = \frac{hD}{k} \tag{1}$$

$$h = \frac{q''_{gen}}{(T_{surf} - T_{jet})} \tag{2}$$

$$q''_{gen} = \frac{VI}{A_{surf}} \tag{3}$$

A correction for lateral conduction as used by Golobic et al [16] was used in this study to calculate the mean and time-varying surface heat transfer coefficient. As with the rig description a more detailed explanation of this process has been presented by Rylatt and O'Donovan [14]. The Reynolds number, *Re* is defined using the velocity scale propose by Smith and Glezer [8] shown in equation 4:

$$U_0 = \frac{L_0}{\tau} \tag{4}$$

Where L_0 is the stroke length, given by equation 5:

$$L_o = \int_0^{\tau/2} u_0(t)dt \tag{5}$$

The synthetic jet Reynolds number, *Re*, may be calculated as in equation 6:

$$Re = \frac{U_0 D}{\nu} \tag{6}$$

In this investigation the jet outlet velocity is not measured directly rather the time varying relative pressure of the cavity is measured and the jet exit velocity calculated via the method proposed by Persoons and O'Donovan [17]. By using the conservation of momentum in the orifice, the time varying jet exit velocity can be calculated as follows:

$$m\frac{dU}{dt} + F_D(U_0) = pA \tag{7}$$

where $F_D(U_0)$ represents a damping force, and $m = \rho AL'$ is the mass of gas in the orifice. The effective length L is the sum of the geometric length L and end corrections $L'-L=2\beta D$, where $\beta=0.425$ [18] for a sharp-edged circular orifice. The uncertainty in the Nusselt number is calculated using the method outlined by the ASME [19] and found to be less than *3%*.

3. Results and Discussion

Heat transfer to a ducted and un-ducted synthetic jet is presented in this section. Also assessed is the influence of confinement on heat transferred to the jet. Various configurations of the jet were compared over a range of Reynolds numbers (*1000 to 4000*) and stroke lengths (*5 to 20* L_0/D) at a jet impingement surface spacing of $H/D=1$. The ducting added to the jet is a *1.5 mm* thick steel plate measuring *20 mm* by *20 mm* and was attached at a distance *1 mm* from the orifice plate of the jet. Three different duct configurations are studied, each with a hole in the centre of the plate which is centred on the jet orifice; the holes were *6, 8* and *10 mm* in diameter or *1.2, 1.6* and *2* times the jet diameter.

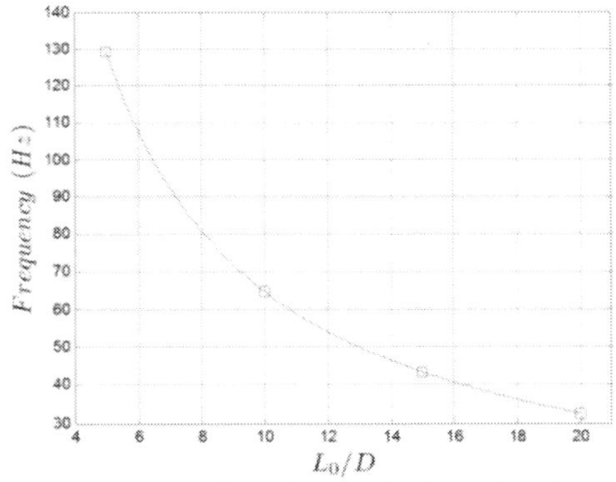

2: Frequency variation with stroke length $H/D = 1$, $Re=2000$

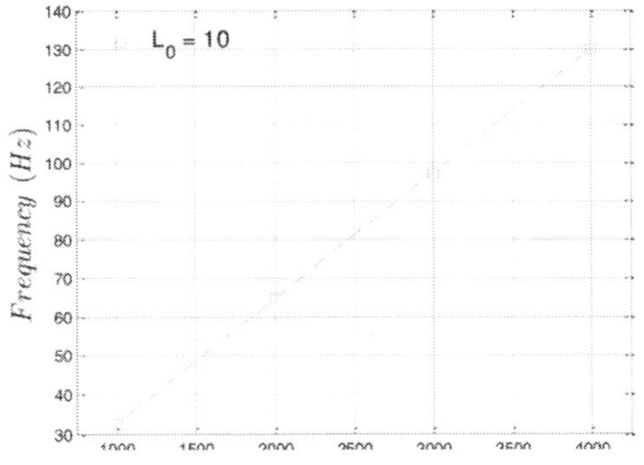

3: Frequency variation with Reynolds number $L_0/D=10$, $H/D=1$

To hold the Reynolds number constant and vary the stroke length the amplitude and frequency of the excitation signal to the jet diaphragm is varied. Figure 2 shows the excitation frequency variation with Reynolds number at $H/D=1$. The constant Reynolds number jets used in this investigation has a Reynolds number $Re = 2000$. With increasing stroke length the excitation frequency of the jet is reduced which increases the period of the jets expulsion and suction strokes, generating a jet of constant velocity that expels a larger volume of air in a single cycle. The stroke lengths tested in this investigation were *5, 10, 15* and *20* which had corresponding excitation frequencies of *129, 65, 43* and *33* at $H/D = 1$.

In order to hold the stroke length constant and vary the Reynolds number, the amplitude and frequency of the excitation signal to the jet diaphragm is varied. Figure 3 shows that the excitation frequency varies linearly with Reynolds number at $H/D = 1$. The constant stroke length jets used in this investigation have a stroke length $L_0/D = 10$. As the stroke length is held constant the Reynolds number is varied by changing the frequency which reduces the time period in which the same volume of fluid passes through the jet orifice on both the expulsion and suction stroke of the jet cycle. This has

Journal of Physics: Conference Series **525** (2014) 012012 doi:10.1088/1742-6596/525/1/012012

the effect of increasing the fluid velocity and therefore the Reynolds number. The Reynolds numbers tested in this investigation were *1000, 2000, 3000* and *4000* which had corresponding excitation frequencies of *33, 43, 65* and *129* at $H/D=1$.

3.1 Constant Reynolds number

Figure 4 shows the time averaged heat transfer to the three configurations of ducted jet and the jet confined by $L_{op}/D = 18$ and *40*. At $L_0/D = 5$, $Re = 2000$ (figure $4_{(a)}$) it can be seen that heat transfer to the more confined jet ($L_{op}/D = 40$) is lower than that of the less confined jet ($L_{op}/D = 18$). This is most noticeable in the stagnation region and the region where peaks in the heat transfer distributions occur, $r/D \approx 3$. It can also be seen that heat transfer to all there configurations of ducted jet is higher in magnitude than that of the two jets confined by $L_{op}/D = 18$ and *40*. It can also be seen that the peaks in heat transfer that occur outwith the stagnation region for all configurations of jet occur are closer to the stagnation region when ducting has been added to the jet; the peaks occur at $r/D \approx 2$ for the ducted jets and at $r/D \approx 3$ for the un-ducted jets. It is apparent that as L_0/D is increased the difference in magnitude of the peaks outwith the stagnation region between the ducted and un-ducted jets reduces. For the jet setting of $L_0/D = 10$, $Re = 2000$ (figure $4_{(b)}$) it is again noticeable that heat transfer to the more confined jet ($L_{op}/D = 40$) is lower than that of the less confined jet ($L_{op}/D = 18$). This is most noticeable in the stagnation region and the region where peaks in the heat transfer distributions occur, $r/D \approx 3$.Again it is observed that heat transfer to all three configurations of ducted jet is higher in magnitude than that observed for the confined jets. When comparing figures $4_{(a)}$ and $4_{(b)}$ it can be seen that the increase in heat transfer provided by ducting the jet when compared with the jets semi-confined to $L_{op}/D = 18$ and *40* is lower for the higher stroke length jet.

Comparing the performance of the jet confined by a plate, $L_{op}/D = 18$ to that confined by $L_{op}/D = 40$ across the range of stroke lengths tested, it is seen that heat transfer to the jet begins to converge both in the stagnation region and in the areas where peaks occur outwith the stagnation region, with increasing stroke length. Comparing the performance of the three different configurations of ducted jet across the range of stroke lengths tested it can be seen that the increase in stagnation heat transfer offered by the ducting reduces with increasing stroke length. It can also be seen that the level of heat transfer provided by the three different configurations of ducted jet varies little across the range of stroke lengths tested. It is observed that at $L_0/D = 5$ (figure $4_{(a)}$) the peaks in heat transfer outwith the stagnation region occur in close proximity but vary in magnitude for the different ducts significantly; as L_0/D is increased, the peaks become similar in magnitude. It can be seen that the peaks in heat transfer outwith the stagnation region for the un-ducted jets very little in position across the range of stroke lengths tested ($r/D \approx 3$ to *3.5*) whereas the ducted jets vary more ($r/D \approx 2$ to *3*).

Heat transfer to the jet in two regions is quantitatively accessed; these regions are 1) the stagnation region which extended over an area *1D* in diameter centred around the stagnation point, to account for local minima at the stagnation point and 2) the area averaged heat transfer which is accessed over a circular area of diameter *14 D*. Stagnation region heat transfer is an indication of the maximum heat transfer rate the jet can provide and is used in many investigations [7, 13, 20-22] to assess a jets performance in both steady and synthetic jet studies. Area averaged heat transfer gives an indication of the practical heat transfer rate available to the jet in real world applications; this area is selected because it is comparable with the size of commonly used CPU heat exchangers [23].

Figures $5_{(a)}$ and $5_{(b)}$ show how heat transfer to the various jet configurations varies with L_0/D at $Re = 2000$ in the stagnation region and on an area averaged basis respectively. By comparing the performance of the un-ducted jet with the ducted jet in the stagnation region (figure $5_{(a)}$), it is noticeable that Nu_{stag} for the ducted jets decrease with increasing stroke length while the opposite is true of the un-ducted jets.

For the two differing levels of confinement it can be seen that the stagnation region heat transfer to the less confined jet is higher in magnitude across the range of stroke lengths tested, but that as L_0/D is increased the difference in their performance reduces. It is clear that the two larger diameter ducts perform best across the range of stroke lengths tested, it is also clear that the *1.6D* ducting performs best at $L_0/D = 5$ and *10* while the *2D* ducting performs best at $L_0/D = 15$ and *20*.

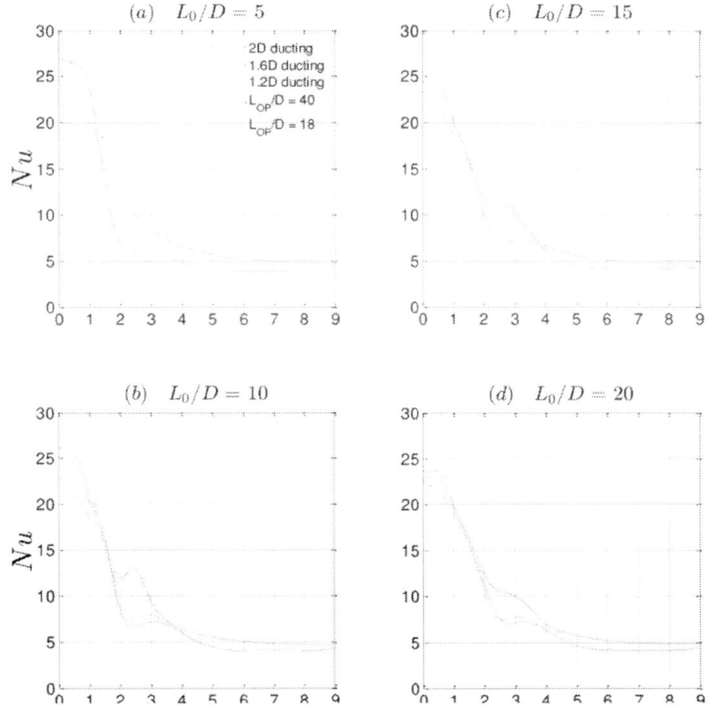

4: Heat transfer distributions for ducted and un-ducted jets for fixed Reynolds number of *Re = 2000, H/D = 1,* *($L_0/D = 5, f_e = 129$ Hz).), ($L_0/D = 10, f_e = 65$ Hz), ($L_0/D = 15, f_e = 43$ Hz), ($L_0/D = 20, f_e = 33$ Hz),*

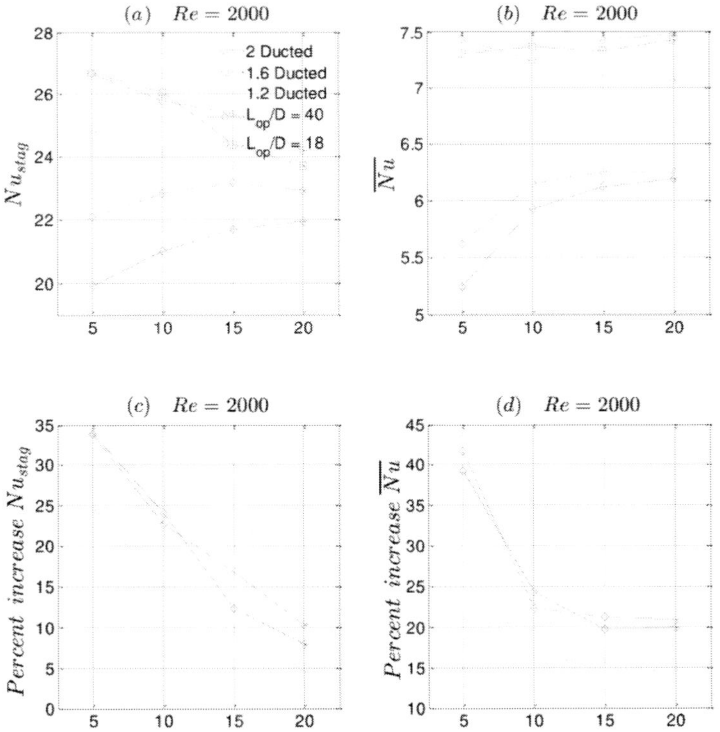

5: stagnation region, area averaged heat transfer and percentage increase in heat transfer for ducted and un-ducted jets *H/D = 1, Re = 2000* and *($L_0/D = 5, f_e = 129$ Hz).), ($L_0/D = 10, f_e = 65$ Hz), ($L_0/D = 15, f_e = 43$ Hz), ($L_0/D = 20, f_e = 33$ Hz),*

Journal of Physics: Conference Series **525** (2014) 012012 doi:10.1088/1742-6596/525/1/012012

Figure $5_{(c)}$ shows the percentage increase in heat transfer when compared to the more confined jet of $L_{op}/D = 40$; this level of confinement has been selected as it is the same level of confinement provided by the duct. As can be seen in Figures $5_{(c)}$ the *1.6 D* ducting is most effecting at $L_0/D = 5$, providing a *34 %* increase in heat transfer; the efficacy of the ducting reduces with increasing L_0/D and at $L_0/D = 20$ the *2 D* ducting is only increasing the stagnation region heat transfer by *10 %*
Figure $5_{(b)}$ shows the area average heat transfer provide by all jet configurations across the range of stroke lengths tested. Again it is observed that the jet semi-confined to $L_{op}/D = 18$ delivers higher rates of area averaged heat transfer across the range of stroke lengths tested than the more confined jet of $L_{op}/D = 40$. It is apparent that the difference in magnitude of the area averaged heat transfer delivered for the two different levels of confinement is less than is observed for stagnation region heat transfer and by $L_0/D = 20$ both levels of confinement deliver similar heat transfer rates. This is due the secondary peaks been similar in magnitude for both levels of confinement across the range of stroke lengths tested with the difference in the magnitude of the peaks reducing with increasing stroke length. It is noticeable that unlike the stagnation region, the area averaged heat transfer to the ducted jets increases with stroke length. This is due the position of the peaks in heat transfer outwith the stagnation region. As L_0/D increases these peaks reduce in magnitude but increase in radial position from the stagnation point and therefore act over a larger surface area and increase the area averaged heat transfer. It is clear that the *1.6* and *2 D* ducting provide the greatest increase in area averaged heat transfer across the range of stroke lengths tested with the *2 D* ducting performing best at all L_0/D except $L_0/D = 10$.
Figure $5_{(d)}$ shows the percentage increase in area averaged heat transfer achieved by adding the ducting to the jet when compared with the un-ducted jet confined to $L_{op}/D = 40$. It is noticeable that the efficacy of the ducting varies differently on an area averaged basis when compared with the stagnation region. The effectiveness of the ducting reduces almost linearly with increasing L_0/D in the stagnation region whereas on an area averaged basis the efficacy of the ducting reduces significantly from $L_0/D = 5$ to *10* but then remains relatively constant for the rest of the stroke lengths tested. This is due to the peaks outwith the stagnation region acting over a larger area with increasing L_0/D; this would seem to not only compensate for their reduction in magnitude but also the reduction in the efficacy of the ducting in the stagnation region with increasing stroke length. It is clear that the *2 D* ducting provides the greatest increase in heat transfer cross the range of stroke lengths tested except at $L_0/D = 10$ where the *1.6 D* ducting performs better. The *2 D* ducting increase area averaged heat transfer by 41% at $L_0/D = 5$ and delivers 215 at $L_0/D = 20$.

Constant stroke length
Figure 6 shows the time averaged heat transfer to the three configurations of ducted jet and the jet semi-confined to L_{op}/D *18* and *40* for a range of Reynolds when the stroke length is held constant at $L_0/D = 10$. Comparing the jets with differing levels of semi-confinement across the range of Reynolds numbers tested it can be seen that the less confined jet ($L_{op}/D = 18$) delivers higher rates of stagnation region heat transfer but that the peaks in heat transfer outwith the stagnation region are of similar magnitude for both levels of confinement and occur at the same radial location at each Reynolds number. As with the constant Reynolds number jets radial location of the peaks in the heat transfer outwith the stagnation region for the undusted jets vary little across the range of Reynolds numbers tested ($r/D \approx 3$ to *3.8*). It is noticeable that the peaks in heat transfer outwith the stagnation region for the ducted jets vary little across the range of stroke lengths tested in figure 6 occurring at $r/D \approx 2.5$, comparing this with the constant Reynolds number jets shown in figure 4, it is clear from this that it is the variation in stroke length which influences the location of these peaks. It is observed that as the Reynolds number is increased the difference in the heat transfer delivered by the three ducted configurations of jets becomes more distinct. In figure 4 it was observed that the difference in heat transfer delivered by the three different configurations of ducting reduces with increasing stroke length. It is also noticeable that magnitude of the heat transfer varies more when the Reynolds number if varied for a fixed stroke length than when the stroke length is varied for a fixed Reynolds number. From this we can see that the stroke length of a synthetic jet influences the magnitude and location of peaks in the heat transfer outwith the stagnation region whereas the Reynolds number of a synthetic jet influences the magnitude of the heat transfer to the jet.

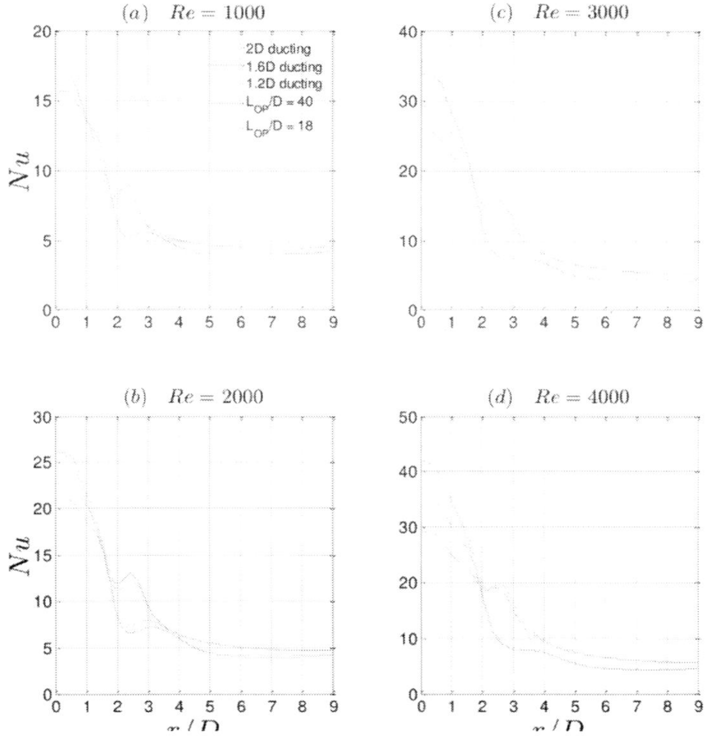

6: heat transfer distributions for ducted and un-ducted jets for fixed stroke length of $L_0/D = 10$, $H/D = 1$ ($Re = 1000, f_e = 33$ Hz), ($Re = 2000, f_e = 65$ Hz), ($Re = 3000, f_e = 96$ Hz), and ($Re = 4000, f_e = 129$ Hz),

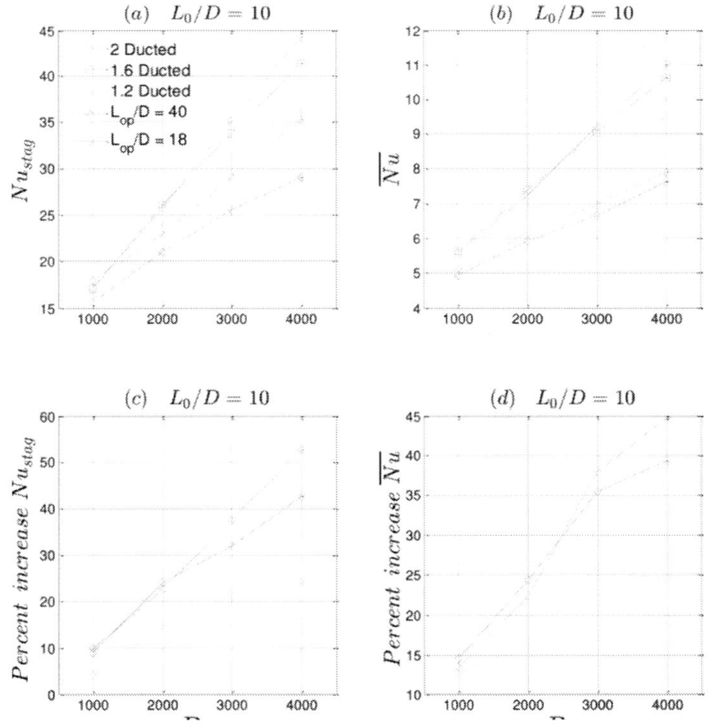

7: stagnation region, area averaged heat transfer and percentage increase in heat transfer for ducted and un-ducted jets $H/D = 1$, $L_0/D = 10$, ($Re = 1000, f_e = 33$ Hz), ($Re = 2000, f_e = 65$ Hz), ($Re = 3000, f_e = 96$ Hz), and ($Re = 4000, f_e = 129$ Hz),

Figures $7_{(a)}$ and $7_{(b)}$ show how heat transfer to the various jet configurations varies with L_0/D at Re = *2000* in the stagnation region and on an area averaged basis respectively. By comparing the performance of the un-ducted jet with the ducted jet in the stagnation region (figure $7_{(a)}$), it is noticeable that Nu_{stag} for the ducted jets increase with increasing Reynolds number.

For the two differing levels of confinement it can be seen that the stagnation region heat transfer to the less confined jet is higher in magnitude across the range of stroke lengths tested, but that as the Reynolds number is increased the difference in their performance increases showing that the effects of semi-confinement is more significant with increasing Reynolds number. It is clear that the two larger diameter ducts perform best across the range of stroke lengths tested, the *1.6D* and *2D* ducting performing identically at Re = *1000* and *2000* with the *2D* ducting performs better at Re = *3000* and *4000*. Figure $7_{(c)}$ shows the percentage increase in heat transfer when compared to the more confined jet of L_{op}/D =*40*; this level of confinement has been selected as it is the same level of confinement provided by the duct. As can be seen in Figures $7_{(c)}$ the *1.6 D* and *2D* ducting is most effecting at Re = *1000* providing a *10 %* increase in heat transfer; the efficacy of the ducting increases with increasing Re and at Re = *4000* the *2 D* ducting increases the stagnation region heat transfer by *53 %*

Figure $7_{(b)}$ shows the area average heat transfer provide by all jet configurations across the range of stroke lengths tested. Again it is observed that the jet semi-confined to L_{op}/D =*18* delivers higher rates of area averaged heat transfer across the range of stroke lengths tested than the more confined jet of L_{op}/D = *40*. It is apparent that the difference in magnitude of the area averaged heat transfer delivered for the two different levels of confinement is less than is observed for stagnation region heat transfer. This is due the secondary peaks been similar in magnitude for both levels of confinement across the range of stroke lengths tested with the difference in the magnitude of the peaks reducing with increasing stroke length. It is clear that the *1.6* and *2 D* ducting provide the greatest increase in area averaged heat transfer across the range of Reynolds numbers tested with the *1.6 D* ducting performing best at all Re = *1000* and *2000* and the *2 D* ducting performing best Re = *3000* and *4000*. It is

Figure $7_{(d)}$ shows the percentage increase in area averaged heat transfer achieved by adding the ducting to the jet when compared with the un-ducted jet confined to L_{op}/D = *40*. It is shown that *1.6 D* ducting performs best at Re = *1000* and *2000* offering *15* and *24 %* increase in area averaged heat transfer respectively. The *2 D* ducting performs best at Re = *1000* and *2000* increasing area averaged heat transfer by *35%* and *45%* respectively.

4. Conclusion

Increasing confinement has been shown to reduce heat transfer to a synthetic air jet. The reduction in heat transfer caused by increased semi confinement is most significant in the stagnation region, and this is most pronounces for low stroke lengths for a fixed Reynolds number and high Reynolds numbers for a fixed stroke length.

Ducting is shown to increase stagnation region and area averaged heat transfer across the range of jet parameters tested when compared with an un-ducted jet of equal confinement. Increasing the stroke length from L_0/D = *5* to *20* for a Reynolds number of *2000* reduces the enhancement in stagnation region heat transfer provided by the ducting from *35%* to *10 %* and reduces the enhancement in area averaged heat transfer provided by the ducting from *42%* to *21 %*. This is shown to be partly due to relative magnitude of the peaks in heat transfer outwith the stagnation region; at low stroke lengths the difference in the magnitude of these peaks is large and reduces with increasing L_0/D. It is also shown that as L_0/D is increased, the stagnation region heat transfer to the un-ducted jets increases while for the ducted jets the stagnation region heat transfer decreases with increasing L_0/D. Increasing Reynolds number from *1000* to *4000* for a stroke length from L_0/D = *10* increases the enhancement in stagnation region heat transfer provided by the ducting from *10%* to over *50%* and increases the enhancement in area averaged heat transfer provided by the ducting from *15%* to *45%*. This is shown to be primarily due to the magnitude of the stagnation region heat transfer, while it increases with Re for all configurations of jet the increase is much more significant for the ducted jets.

5. References

[1] Smith, B.L. and G.W. Swift. *Synthetic jets at large Reynolds number and comparison to continuous jets*. in *31st AIAA Fluid Dynamics Conference and Exhibit*. 2001. Anaheim.

[2] Pavlova, A. and M. Amitay, *Electronic cooling using synthetic jet impingement*. ASME Journal of Heat Transfer, 2006. **128**: p. 897 - 907.

[3] Smith, B.L. and G.W. Swift, *A comparison between synthetic jets and continuous jets*. Experiments in Fluids, 2003. **34**: p. 467 - 472.

[4] Kercher, D.S., et al., *Microjet cooling devices for thermal management of electronics*. IEEE Transactions on Components and Packaging Technologies, 2003. **26**: p. 359 - 366.

[5] S. Gunther, F.B., D. M. Carmean, J. C. Hall, *Managing the Impact of Increasing Microprocessor Power Consumption* Intel Technology Journal, 2001. **Vol. 5,** (No. 1).

[6] Chaudhari, M., B. Puranik, and A. Agrawal, *Heat transfer characteristics of synthetic jet impingement cooling*. International Journal of Heat and Mass Transfer, 2010. **53**(5–6): p. 1057-1069.

[7] Persoons, T., A. McGuinn, and D.B. Murray, *A general correlation for the stagnation point Nusselt number of an axisymmetric impinging synthetic jet*. International Journal of Heat and Mass Transfer, 2011. **54**(17–18): p. 3900-3908.

[8] Smith, B.L. and A. Glezer, *The formation and evolution of synthetic jets*. Physics of Fluids, 1998. **10**: p. 2281 - 2297.

[9] Ashforth-Frost, S. and K. Jambunathan, *Effect of nozzle geometry and semi-confinement on the potential core of a turbulent axisymmetric free jet*. International Communications in Heat and Mass Transfer, 1996. **23**: p. 155 - 162.

[10] Garimella, S.V. and R.A. Rice, *Confined and submerged liquid jet impingement heat transfer*. ASME Journal of Heat Transfer, 1995. **117**: p. 871 - 877.

[11] Colucci, D.W. and R. Viskanta, *Effect of nozzle geometry on local convective heat transfer to a confined impinging air jet*. Experimental Thermal and Fluid Science, 1996. **13**: p. 71 - 80.

[12] Gao, N. and D. Ewing, *Investigation of the effect of confinement on the heat transfer to round impinging jets exiting a long pipe*. International Journal of Heat and Fluid Flow, 2006. **27**(1): p. 33-41.

[13] Youn, Y.J., K. Choo, and S.J. Kim, *Effect of confinement on heat transfer characteristics of a microscale impinging jet*. International Journal of Heat and Mass Transfer, 2011. **54**(1-3): p. 366-373.

[14] Rylatt, D.I. and T.S. O'Donovan, *Heat transfer enhancement to a confined impinging synthetic air jet*. Applied Thermal Engineering, 2013. **51**(1–2): p. 468-475.

[15] Patil, V.A. and V. Narayanan, *Application of heated-thin-foil thermography technique to external convective microscale flows*. Measurement Science and Technology, 2005. **16**: p. 472-476.

[16] Golobic, I., et al., *Experimental determination of transient wall temperature distributions close to growing vapor bubbles*. Heat and Mass Transfer, 2007.

[17] Persoons, T. and T.S. O'Donovan, *A pressure-based estimate of synthetic jet velocity*. Physics of Fluids, 2007. **19**.

[18] Beranek, L.L., *Acoustics*. Physics Today, 1955. **8**(10): p. 27-28.

[19] *Policy on reporting uncertainties in experimental measurements and results*. ASME Journal of Heat Transfer Policy.

[20] Hoogendoorn, C.J., *The effect of turbulence on heat transfer at a stagnation point*. International Journal of Heat and Mass Transfer, 1977. **20**: p. 1333 - 1338.

[21] Shadlesky, P.S., *Stagnation point heat transfer for jet impingement to a plane surface*. AIAA Journal, 1983. **21**: p. 1214 - 1215.

[22] Choo, K.S., et al., *Heat transfer characteristics of a micro-scale impinging slot jet*. International Journal of Heat and Mass Transfer, 2009. **52**(13-14): p. 3169-3175.

[23]. Saini, M. and R.L. Webb, *Heat rejection limits of air cooled plane fin heat sinks for computer cooling*. IEEE Transactions on Components and Packaging Technologies, 2003. **26**(1): p. 71-79.

High fidelity phase locked PIV measurements analysing the flow fields surrounding an oscillating piezoelectric fan

Nicholas Jeffers, Kevin Nolan, Jason Stafford, and Brian Donnelly
Thermal Management Research Group, Bell Labs, Alcatel-Lucent Ireland, Blanchardstown, Dublin 15, Ireland
E-mail: nick.jeffers@alcatel-lucent.com

Abstract. Piezoelectric fans have been studied extensively and are seen as a promising technology for thermal management due to their ability to provide quiet, reliable cooling with low power consumption. The fluid mechanics of an unconfined piezoelectric fan are complex which is why the majority of the literature to date confines the fan in an attempt to simplify the flow field. This paper investigates the fluid mechanics of an unconfined fan operating in its first vibration frequency mode. The piezoelectric fan used in this study measures 12.7mm x 70mm and resonates at 92.5Hz in air. A custom built experimental facility was developed to capture the fan's flow field using phase locked Particle Image Velocimetry (PIV). The phase locked PIV results are presented in terms of vorticity and show the formation of a horse shoe vortex. A three dimensional λ_2 criterion constructed from interpolated PIV measurements was used to identify the vortex core in the vicinity of the fan. This analysis was used to clearly identify the formation of a horse shoe vortex that turns into a hairpin vortex before it breaks up due to a combination of vortex shedding and flow along the fan blade. The results presented in this paper contribute to both the fluid dynamics and heat transfer literature concerning first mode fan oscillation.

1. Introduction

Piezoelectric cooling devices are seen as a promising air moving technology that bridge the gap between natural convection heat sinks and axial fan forced convection heat sinks. The main advantages of piezoelectric fans are high reliability, low power consumption and minimal noise generation. A typical piezoelectric fan is comprised of a piezoelectric material bonded to a flexible cantilever beam or fan blade. The piezoelectric material expands and contracts when an AC voltage is applied to it causing the fan blade to deflect. The fan achieves its maximum deflection when it is rigidly secured at one end, for example to a metal pillar, and the electrical excitation frequency matches the resonant frequency of the fan blade and fluid. This oscillating motion induces an air flow that can subsequently be used to cool adjacent heated surfaces.

Different configurations of piezoelectric fans have been studied since 1979 when Toda and Osaka [1] first proposed their use for cooling applications. However, only in the last ten years has the technology gained popularity amongst the scientific and industrial community since device reliability, energy consumption, and miniaturisation have become critical design parameters. A number of articles found in the literature present experimental and modelled

Content from this work may be used under the terms of the Creative Commons Attribution 3.0 licence. Any further distribution of this work must maintain attribution to the author(s) and the title of the work, journal citation and DOI.
Published under licence by IOP Publishing Ltd

data for confined piezoelectric fans. These confinements are used to simplify the complex fluid mechanics generated by a fan operating in its first vibration mode.

Kim et al. [2] studied a confined piezoelectric fan using phase-resolved Particle Image Velocimetry (PIV) and smoke visualisation techniques. The fan used in this study had a height to length aspect ratio of 1:1.2 and was confined by plates positioned on the lateral edges of the fan in an attempt to make the flow quasi-two-dimensional. They showed that a pair of counter rotating vortices were generated at the fan tip with a maximum velocity almost four times that of the maximum fan tip velocity. They also showed that the velocity fields off-centre were considerably weaker than those in the centre, they attributed this to a three dimensional flow field resulting from the fluid interaction with the no-slip boundary condition at the confining planes. They postulated that this fluid interaction contributed to a breakdown in the vortex formation from the fan tip in these areas.

To gain further insight, Choi et al. [3] employed the experiments of Kim et al. [2] to validate a 2D numerical simulation of a piezoelectric fan. They numerically investigated the fluidic mechanism generating the vortices observed in experiments. They reported that there were four distinct vortex formation stages induced by the oscillation of the fan blade: initiation, development, separation, and propagation. A vortex is *initiated* when the fan blade begins to move causing high and low pressure zones on either side of the blade. The air surrounding the fan blade travels from the high pressure zone over the fan tip to the low pressure zone thus *initiating* a vortex. As the fan blade continues to advance, more air travels from the high pressure zone to the low pressure zone, thus *developing* the vortex. This vortex continues to develop until the fan blade reaches its maximum deflection and thus zero velocity. As the fan begins to accelerate in the opposite direction, another counter-rotating vortex is initiated causing the original vortex to *separate* away from the fan blade. As this secondary vortex develops it causes the primary vortex to *propagate* down stream. The two dimensional fluid flow analysis presented by Choi et al. [3] shows the basic fluid mechanics of an oscillating piezoelectric fan at its centre and does not account for strong out of plane vortex formation.

Kim et al. [4] performed Continuous Wavelet Transform and Proper Orthogonal Decomposition on a piezoelectric fan's flow field to reveal data hidden by phase-averaging. They analysed the size, strength and distribution of the vortices generated by a fan confined along its lateral edges. They found from their instantaneous results that the size of all the vortices at each phase angle had a similar velocity magnitude, however, the location of the vortex cores were scattered suggesting a somewhat irregular motion of the vortex structures once they separate from the blade tip. This further supports the opinion that the fluid mechanics surrounding a piezoelectric fan are very complex and difficult to predict.

To date the literature has predominantly reported on confined fans restricted by a number of confinement plates in an attempt to simplify the fan's flow structures. However, there is a gap in the literature reporting on the fluid mechanics resulting from an unconfined fan in free space. For this reason it is important to understand how an unconfined fan generates air flow. This paper aims to help fill this gap in the literature by assessing the fluid mechanics of an unconfined fan. A custom measurement facility was built to non-intrusively measure the fluid mechanics of the fan using PIV. The flow field measurements obtained with the PIV system are used to explain how the oscillating fan interacts with the surrounding fluid.

2. Experimentation

This paper primarily examines the aerodynamic effects of a piezoelectric fan operating in its first mode of oscillation. In order to achieve this objective, a rigorous regime of PIV measurements were carried out at different planes to obtain a phase averaged 3D flow field in a volume surrounding the fan. The complete apparatus used in this work is discussed in greater detail by Jeffers et al. [5], nevertheless this section discusses the experimental apparatus and details

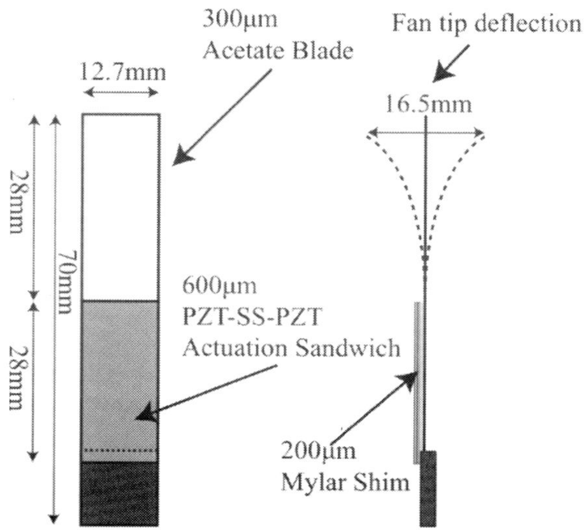

Figure 1. A schematic of the piezoelectric fan used in this investigation.

pertaining to this study.

Figure 1 illustrates the piezoelectric fan used in this investigation. The device is adapted from a commercially available fan in order to aid with PIV measurements. The opaque mylar blade was replaced with a transparent 300μm acetate blade to allow the PIV laser to pass through unobstructed. The new blade was thicker, stiffer and 6mm shorter resulting in a shift in the first harmonic from 60Hz to 92.5Hz in light of the observations of Kimber and Garimella [6] who found that elevated frequency had a greater influence on heat transfer rates compared to fan tip amplitude. Additionally the peak fan tip deflection amplitude was reduced from 25mm to 16.5mm. The power consumption of the modified fan was 100mW which is 70mW higher than the unmodified fan. The fan is driven with a 110V (rms) sine wave produced by a TTi TG1000 function generator and a linear amplifier model. The fan was fixed to a three-axis stage to allow precise positioning.

The PIV measurement apparatus used in this study is the same as that used by Jeffers et al. [5] apart from the TSI PowerView Plus 4MP PIV Camera used in this work. The camera and the laser were on two traversing stages mechanically held perpendicular to each other, allowing precise placement of the camera and laser in the x, y and z directions.

To obtain the 3D flow field surrounding the oscillating fan, measurements were acquired in two perpendicular planes as illustrated in figure 2. In configuration A, the laser is positioned perpendicular to the fan and the PIV camera records images in the $x - y$ plane. Measurements were taken at 5mm increments from 5mm beyond the blade tip towards the blade root. Configuration B positions the laser parallel to the fan's major axis and the PIV camera records images in the $x - z$ plane. The five planes taken were located 3mm below and above the long side of the fan, one at each of the fan's edges and one in the centre of the fan.

The periodic sine wave used to drive the fan facilitates the use of phase locked PIV measurements. Here, PIV images are acquired at precise times, synchronised with the blade's deflection. For the present measurements, a total of 11 positions along a half period of the fan's input signal were acquired as illustrated in figure 3. By assuming symmetry, half of a fan's oscillation is sufficient to describe the flow fields for a full oscillation. In the uppermost plot of figure 3, open circles show the measured fan deflection at equally spaced phases of the fans oscillation while closed circles represent those positions plotted herein. For the present data the

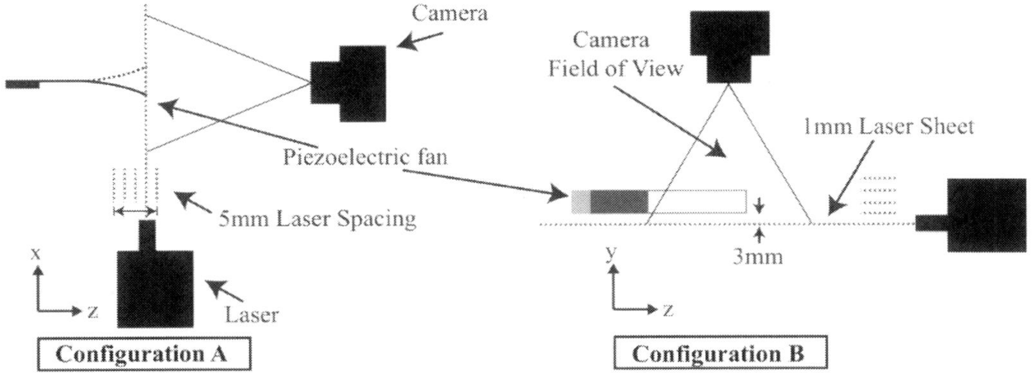

Figure 2. PIV measurement plane configurations. Configuration A comprises measurements spaced at 5mm increments in $x - y$ planes from 5mm in front of the fan toward the root. Configuration B comprises measurements in the $x - z$ plane corresponding to 3mm above and below the fan geometry and at the extrema and centreline of the fan.

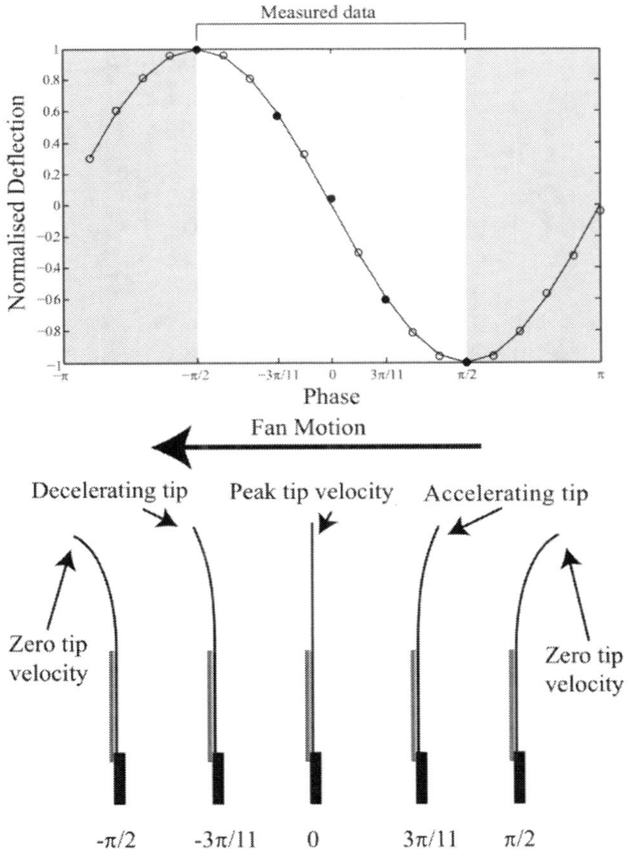

Figure 3. Illustration of the phase locked measurements. (top) Measured fan tip deflection and the driving sine wave, the closed circles represent the data presented herein; (bottom) deflection of the fan for a half period.

fan moves from right to left, i.e. from $\pi/2$ to $-\pi/2$. At these extrema the fan has zero tip velocity accelerating to its peak velocity at the midpoint and back to zero again.

The PIV test procedure used to produce the results in this paper was the same as that used by Jeffers et al. [5]. 300 image pairs were acquired at each phase and spatial position ensuring that the ensemble averages of velocity magnitude and vorticity had sufficiently converged towards a representative time-average. A full-field analysis, as described by Stafford et al. [7], was conducted on the flow fields generated by this piezoelectric fan. As the sampling rate constitutes random sampling of unique flow structures, standard error estimates were used to determine the uncertainty in the ensemble average velocity magnitude compared to a time-average flow field. The current sample size results in an uncertainty in the full-field first order data of less than 6.5%. Therefore, the velocity data presented in this paper sufficiently represents the time-average flow fields.

Choi et al. [3] utilised vorticity, $\omega = \nabla \times \mathbf{v}$, to identify vortices in their data. This is a common and simple technique to identify vortices in flow fields in the literature. However it is also susceptible to error in the presence of strong shear and can vary significantly in magnitude along the length of a vortex in 3D flow fields hindering structure identification.

Jeong and Hussain [8] proposed a technique for vortex structure identification, commonly referred to as the λ_2 criterion, as a good estimate to capture circulation in a vortex core. Here the velocity gradient tensor $\nabla\mathbf{u}$ is split into symmetric and anti symmetric parts corresponding to the strain rate tensor $\mathbf{S} = \frac{1}{2}(\nabla\mathbf{u}+\nabla\mathbf{u}^T)$ and the rotation tensor $\mathbf{\Omega} = \frac{1}{2}(\nabla\mathbf{u}-\nabla\mathbf{u}^T)$ so that λ_2 is the second eigenvalue of $\mathbf{S}^2+\mathbf{\Omega}^2$. Selecting a negative value of λ_2 and computing an isosurface can reveal the vortical structure of a flow field. Both identification methods will be used herein.

In order to compute λ_2 for the 2D PIV measurements the data were interpolated into a 3D field using linear interpolation owing to the high spatial density of velocity vectors. This made available the phase averaged velocity gradient tensor and thus λ_2 for each phase measured.

3. Results and Discussion

This section presents and discusses the fluid mechanics of a piezoelectric fan in free space. Figure 4 shows the phase averaged flow field contoured with vorticity normal to the measurement plane surrounding the fan. Two views of the fan blade's deflection are shown: the front view which is taken 5mm from the fan tip in the $x-y$ plane; and the top view that is taken in the centre of the fan in the $x-z$ plane. The dashed line in each subplot indicates the intersection of these two measurement planes. At $\pi/2$ the fan velocity is zero. As the fan accelerates to $3\pi/11$ the front view shows two vortices are formed on the top and bottom of the trailing side of the fan blade. The simulations of Choi et al. [3] have shown that the formation of these vortices is caused by the fan blade generating positive pressure on its leading surface and a negative pressure on its trailing surface. This pressure differential and the shear with the ambient is what drives the vortex formation at both top and bottom of the fan blade.

Simultaneously the top view shows that a vortex rolls up over the fan tip while the fan is accelerating, resulting in a vortex detached via a shear layer. This comma shaped structure appears to result from the vortex separating from the rapidly accelerating fan tip. This is analogous to vortex shedding over a bluff body. As the piezoelectric fan is effectively a flapping cantilever, it has zero displacement at one end (the fixed end) and maximum displacement at the other (the free end). Therefore the rate of acceleration, and therefore velocity, increases from the fixed to free end. As a result of this, it is expected that the flow behaviour will change with distance along the fan blade from root to tip (consider how the flow over a bluff body changes as the Reynolds number increases). The measurements in the $x-y$ plane are located 5mm before the blade tip, therefore the local blade velocity at this position is lower than the blade tip velocity. The results show that vortices in close proximity to the fan are entrained behind the blade as it moves through a half oscillation and they resist detachment. However, in the $x-z$

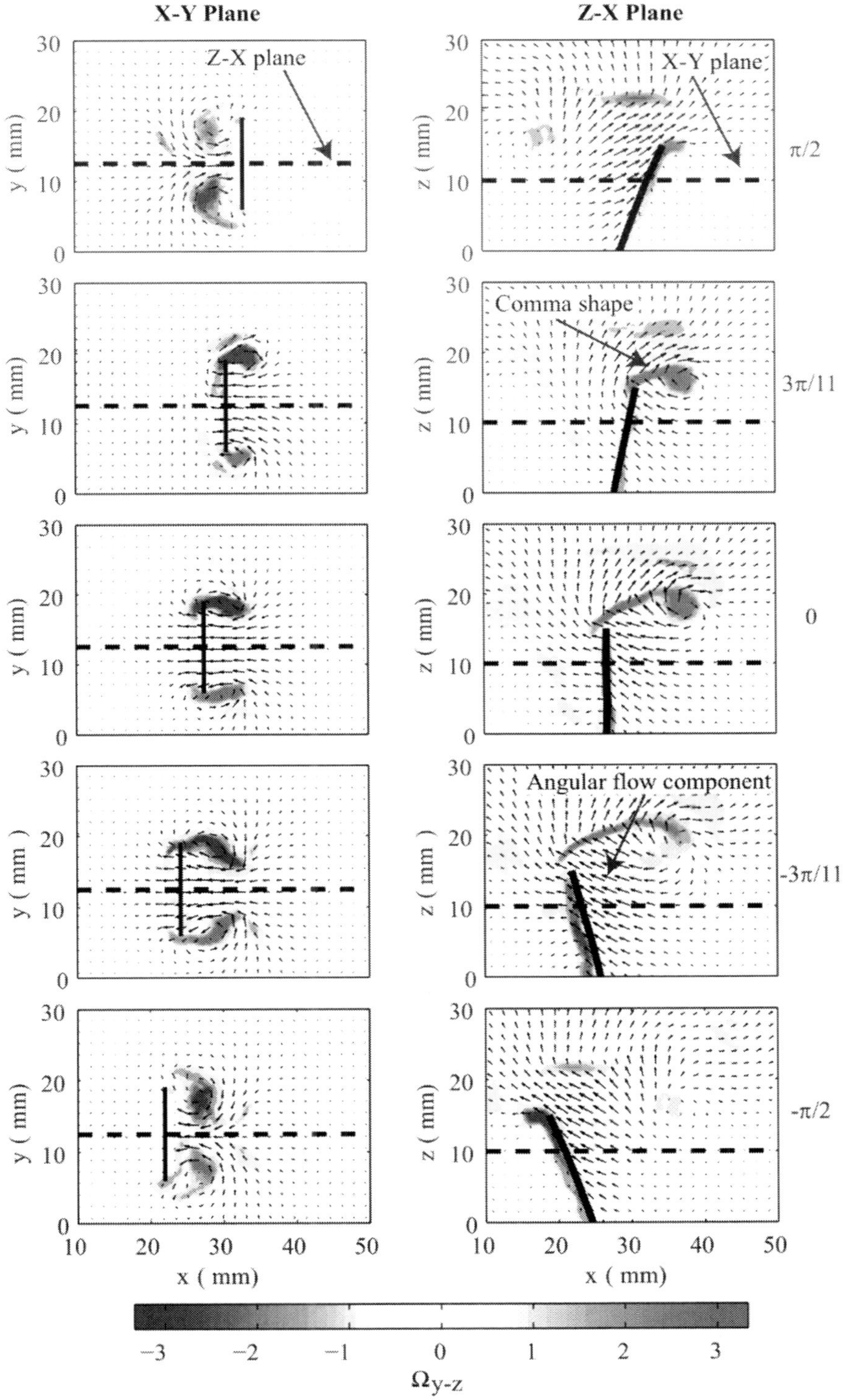

Figure 4. Phase locked PIV measurements for the vorticity of an unconfined fan in free space corresponding to the filled circles in figure 3. The left column shows measurements in the $x - y$ plane while the right column shows measurements in the $x - z$ plane. The dashed lines show the intersection of the $x - y$ and $x - z$ planes.

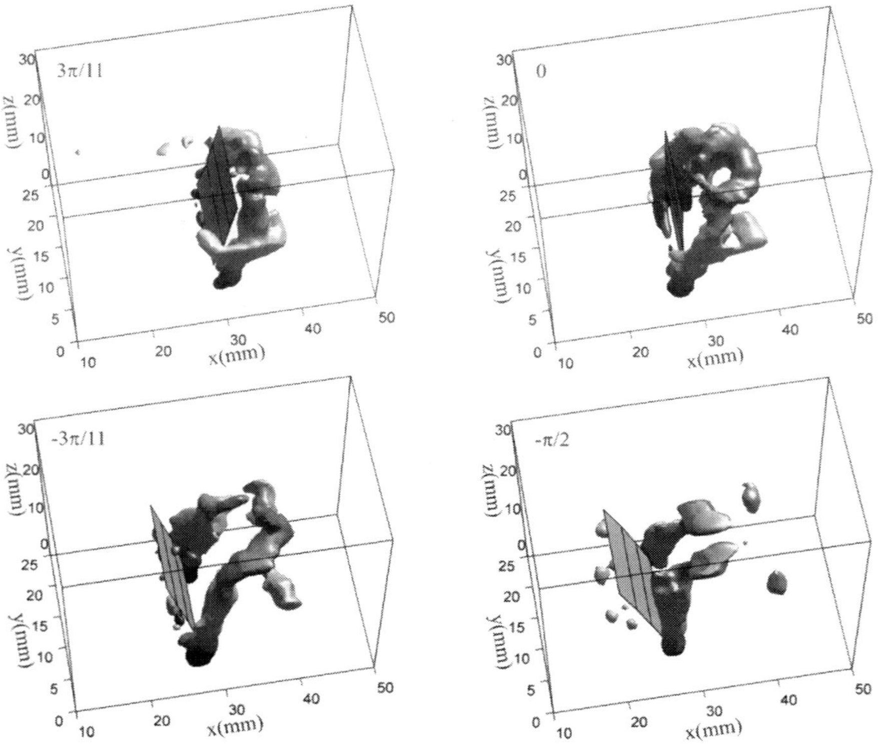

Figure 5. Phase locked interpolated PIV measurements showing isosurfaces of $\lambda 2$. Time is increasing from top left to top right and bottom left to bottom right.

plane, the vortex formed behind the blade is shed due to the higher velocity and thus remains in approximately the same $x - z$ location ($x \approx 37mm$, $z \approx 20mm$). This is to be expected from observation of the flow over bluff bodies; as the Reynolds number increases, vortices form behind the body that eventually begin to detach and shed downstream. This behaviour is similar to the observations of Kim et al. [4] and Choi et al. [3] for confined fans, however their results only consider a two dimensional $x - z$ plane.

The influence of the top and bottom vortices on this unconfined configuration can be better understood by looking at the 3D flow field, in particular isosurfaces of λ_2, as shown in figure 5. As discussed in section 2, the λ_2 criterion is used to visualise the vortical structures in the phase locked data. The colours plotted in figure 5 represent distance from the blade root in the positive z direction. Four phases are shown in figure 5 between $3\pi/11$ and $-\pi/2$ as the fan moves from right to left ($\pi/2$ is omitted for clarity). As the fan accelerates from zero velocity to $-3\pi/11$, a horse shoe type vortex is formed from both sets of vortices previously described and illustrated in figure 4. Figure 5 illustrates that for all phases, sections of the vortex remain close to the fixed end of the blade, these are coloured blue in the figure. Moving along the length of the fan in the z direction, towards the blade tip, the vortex core increasingly lags behind the blade throughout its motion.

Beyond the tip of the fan, as it moves through zero deflection (0 to $-3\pi/11$), the vortex structure is observed to pull away from the blade more aggressively than along the blade edges. This region corresponds to the detached vortex and shear layer observed on the right side of figure 4 and appears to result from an inertial resistance to entrainment behind the retreating blade. As the fan decelerates to its point of maximum deflection (from $-3\pi/11$ to $-\pi/2$), the source of this inertia becomes apparent. The fluid behind the blade has momentum from the acceleration phase and as a result it is deflected towards the fan tip by the slower moving, decelerating fan. This results in an ejection of fluid outwards form the blade tip. The direction of this fluid is at about twice the deflection angle of the blade tip measured relative to the z-axis. The competing entrainment of the retreating blade and the inertia of this ejection drive the development of the strong tip vortex.

At $-\pi/2$ the fan is at maximum deflection, the vortex structure indicated by the λ_2 isosurface in figure 5, appears to be split in half at the tip. Unlike the measurements of Kim et al. [2] and the simulations of Choi et al. [3] where a train of counter-rotating vortices are ejected from the tip of the fan, the upper and lower edge vortices appear to twist the tip vortex in line with their rotation. This can also be inferred in figure 4 as the vortex appears to suddenly vanish from the ω_y colormap at $-\pi/2$.

4. Conclusions

This paper presents the fluid mechanics of a piezoelectric fan operating in its first vibration mode. Phase locked PIV flow visualisation measurements for one half oscillation of the fan are presented. Using this data, three dimensional λ_2 isosurfaces were constructed from interpolated PIV data. These results showed a horse shoe type vortex initially forming around the unconfined fan blade as it accelerates from zero velocity at maximum deflection. As the fan blade advances beyond $-3\pi/11$ the horse shoe vortex begins to separate from the tip of the fan and a hairpin vortex is formed.

The segment of the vortex structure entrained along the upper and lower edges of the fan blade appears to remain attached to the fan blade unlike the region at the tip. This attachment is due to the strong coherence of the structure extending from the fan root along its length. The momentum of the flow resulting from the observed region of strong angular flow is enough to cause a strong rollup of the flow that resists entrainment to the retreating fan blade. This has the effect of peeling the edge vortices away from the blade with increasing z position. When the fan blade completes one half period of oscillation, the vortex structure ruptures and the two separated halves become aligned with the edge vortices expelling fluid outwards in the process.

The experimental results presented in this paper have provided new insights into the flow physics of piezoelectric fans. However, assessing the influence of fan geometry, confinement, frequency, and displacement is necessary to determine the best piezoelectric fan configuration.

5. Acknowledgments

Bell Labs would like to thank the Industrial Development Agency (IDA) Ireland for their continued support.

References

[1] M. Toda and S. Osaka, "Vibrational fan using the piezoelectric polymer pvf2," *Proceedings of the IEEE*, vol. 67, no. 8, pp. 1171–1173, Aug 1979.

[2] Y. H. Kim, S. Wereley, and C. H. Chun, "Phase-resolved flow field produced by a vibrating cantilever plate between two endplates," *Physics of Fluids*, vol. 16, no. 1, pp. 145–162, Jan 2004.

[3] M. Choi, C. Cierpka, and Y.-H. Kim, "Vortex formation by a vibrating cantilever," *Journal of Fluids and Structures*, vol. 31, no. 0, pp. 67 – 78, 2012.

[4] Y. H. Kim, C. H. Chun, and S. Wereley, "Flow field around a vibrating cantilever: coherent structure eduction by continuous wavelet transform and proper orthogonal decomposition," *Journal of Fluid Mechanics*, vol. 669, pp. 584–606, 2011.

[5] N. Jeffers, J. Stafford, and B. Donnelly, "Heat transfer and fluid mechanics from a piezoelectric fan operating in its second resonant frequency mode," 2014, to be published.

[6] M. Kimber and S. V. Garimella, "Measurement and prediction of the cooling characteristics of a generalized vibrating piezoelectric fan," *Journal of Heat and Mass Transfer*, vol. 52, pp. 4470–4478, 2009.

[7] J. Stafford, E. Walsh, and V. Egan, "A statistical analysis for time-averaged turbulent and fluctuating flow fields using particle image velocimetry," *Flow Measurement and Instrumentation*, vol. 26, pp. 1–9, 2012.

[8] J. Jeong and F. Hussain, "On the identification of a vortex," *Journal of Fluid Mechanics*, vol. 285, pp. 69–94, 2 1995.

Journal of Physics: Conference Series **525** (2014) 012014 doi:10.1088/1742-6596/525/1/012014

Heat Transfer and Flow Structure Evaluation of a Synthetic Jet Emanating from a Planar Heat Sink

Paul Manning[1], Tim Persoons[1], Darina Murray[1]

[1]Mechanical and Manufacturing Engineering, Parsons Building, Trinity College, Dublin 2, Ireland

E-mail: manninp@tcd.ie

Abstract. Direct impinging synthetic jets are a proven method for heat transfer enhancement, and have been subject to extensive research. However, despite the vast amount of research into direct synthetic jet impingement, there has been little research investigating the effects of a synthetic jet emanating from a heated surface, this forms the basis of the current research investigation. Both single and multiple orifices are integrated into a planar heat sink forming a synthetic jet, thus allowing the heat transfer enhancement and flow structures to be assessed. The heat transfer analysis highlighted that the multiple orifice synthetic jet resulted in the greatest heat transfer enhancements. The flow structures responsible for these enhancements were identified using a combination of flow visualisation, thermal imaging and thermal boundary layer analysis. The flow structure analysis identified that the synthetic jets decreased the thermal boundary layer thickness resulting in a more effective convective heat transfer process. Flow visualisation revealed entrainment of local air adjacent to the heated surface; this occurred from vortex roll-up at the surface of the heat sink and from the highly sheared jet flow. Furthermore, a secondary entrainment was identified which created a surface impingement effect. It is proposed that all three flow features enhance the heat transfer characteristics of the system.

1. Introduction

Over the past decade, heat dissipation in microprocessors has more than doubled [1]. The trend of electronic device miniaturization and increasing power density results in very high localised heat fluxes, this presents a thermal bottleneck in the electronics component cooling sector [2]. The basic requirement for a thermal management system is to efficiently dissipate heat from a heated surface to ensure safe operating conditions, as thermal overstressing is one of the major causes for failure of electronic devices [3]. However, as form factors of electronic devices decrease in size, there is an increase in the demand for confined space thermal management solutions.

Synthetic jets have been at the forefront of electronic component cooling research for the past number of years. The two stroke motion of a synthetic jet results in the periodic ejection and suction of a working fluid across an orifice, creating vortex roll-up at the surface of the orifice and subsequent vortex ejection into the surrounding flow field. Conventional synthetic jets create a train of vortices which impinge onto a heated surface resulting in enhanced heat transfer characteristics. The strength and size of the vortex ejected from the synthetic jet cavity is determined by the dimensionless stroke length, L_0/D, where the stroke length L_0 is determined by the following expression:

Content from this work may be used under the terms of the Creative Commons Attribution 3.0 licence. Any further distribution of this work must maintain attribution to the author(s) and the title of the work, journal citation and DOI.
Published under licence by IOP Publishing Ltd

$$L_0 \int_0^{\frac{T}{2}} u_m(t)dt = \frac{2U_0}{f} \qquad (1)$$

where U_0 is the average ejection velocity from the orifice and f is the forcing frequency. The dimensionless stroke length can be defined as L_0/D and is inversely proportional to the Strouhal number, therefore $\frac{L_o}{d} = \frac{1}{2}\left(\frac{fd}{U_o}\right)^{-1}$. This relationship determines the characteristics of the synthetic jet flow, since the stroke length parameter governs vortex detachment at the orifice [10]. The following criterion must be satisfied for vortex detachment from a sharp edged orifice of a synthetic jet, $\frac{L_o}{d} > 0.5$ [10] as below this threshold vortex rings do not escape the orifice and are re-entrained back into the cavity on the suction phase, resulting in reduced heat transfer [6]. For $2.56 < \frac{L_o}{d} < 8.9$ the vortices increase in size, and there is virtually no interaction between successive vortices [11]. Increasing the vortex strength will therefore increase the volumetric entrainment of the surrounding flow field; this has a positive effect on heat transfer due to the increased mass flux. Another method which increases the volumetric flow entrainment is by integrating multiple in-phase synthetic jets. Research conducted by Chaudari et al. [3] identified that the heat transfer coefficient for multiple synthetic jets is up to 30% higher as compared to that obtained with a conventional single orifice synthetic jet at the same excitation frequency. This heat transfer enhancement is attributed to the increase in fresh fluid entrainment resulting from the increased entrainment rates.

Current research into synthetic jets focusses on parameters such as stroke length [6,7,9], orifice to impingement surface distance [6,7,9], flow confinement and recirculation [4], orifice geometries [5] and multiple orifices [3]. However, this investigation evaluates the heat transfer characteristics of single and multiple synthetic jets *emanating* from a heated surface. It is hypothesised that the synthetic jet flow will periodically break down the thermal boundary layer and promote localised turbulence generation and mixing at the surface of the heat sink. Additionally, heated air from the thermal boundary layer is expected to be entrained into the synthetic jet and ejected from the surface. The heat transfer analysis provides an insight into the average Nusselt number resulting from the synthetic jet actuation. Through additional heat transfer analysis, the relative heat transfer contribution from the upper and lower surfaces of the heat sink is also identified. The overall heat transfer resulting from jet actuation is examined for a single orifice and multiple orifices and flow structures responsible for the heat transfer enhancements are identified using three flow structure analysis methods; flow visualisation, thermal imaging and thermal boundary layer analysis.

2. Experimental Set-up and Procedure

The unique feature about the heat sink design is that the synthetic jet is part of the heated surface, whereas conventionally synthetic jets direct cooler air at a heated surface. The ejection plate channels the flow by forcing the enclosed air through an orifice thus creating a synthetic jet. By attaching an oscillating diaphragm onto the base of the heat sink, a synthetic jet flow can be created. Two synthetic jet heat sink configurations are created for the heat transfer and flow structure analysis; a single orifice and a 2 X 2 array of orifices with a centre-to-centre spacing, Z, of 15 mm. The diameter for each orifice, d, is 5 mm as shown in *figure 1*. The multiple orifice heat sink therefore has a Z/d ratio of 3. Heat is provided by four cartridge heaters which fit into machined holes on the side of the heat sink. Pressure recovery slots are machined into the sides of the heat sink perpendicular to each heater slot to alleviate any difficulty associated with cartridge heater removal. Thermocouple grooves are machined into the upper and lower surface of the heat sink fitted with T-type thermocouples, to measure the heat sink temperature. Copper is the preferred material for the heat sink design for its excellent thermal conductivity, creating an isothermal boundary condition. Thermal imaging results indicated that the heat sink surface has a temperature deviation of 0.9°C for a surface heat sink temperature of 74°C, with the highest temperature variation occurring at the heat sink/insulation interface. Both heat sink designs are depicted in *figure 1*.

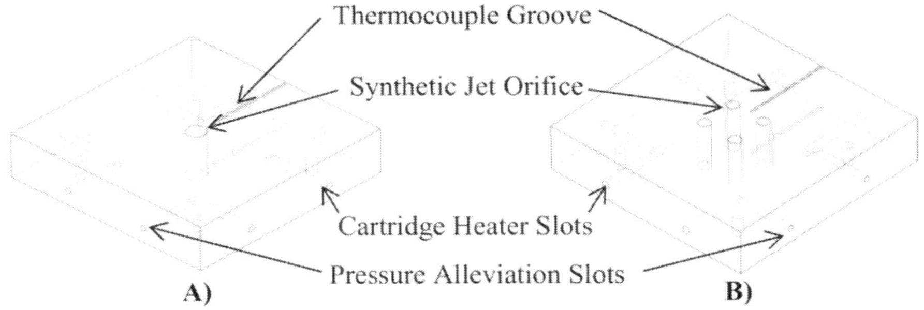

Figure 1. Copper heat sink designs A) Single orifice and B) Multiple orifices

The synthetic jet was generated using an electro mechanically driven actuator i.e. an acoustic speaker. An acoustic cavity, with embedded pressure microphone is used to measure the pressure generated by the speaker within the chamber. The inner diameter of the acoustic cavity conforms to the 71.5 mm diameter of the speaker ensuring no flow disruptions, which could affect the pressure measurements. The pressure measurements are used to calculate the Reynolds number and stroke length for the synthetic jet flow through the orifice. This was achieved using Persoons and O'Donovan [12] cavity pressure method for synthetic jet velocity estimation.

The experimental set-up was designed to minimize thermal conduction losses through the system. This was achieved using low thermal conductivity (~0.023 $\frac{W}{m.K}$) insulation. A cross section of the synthetic jet test section is shown in *figure 2*.

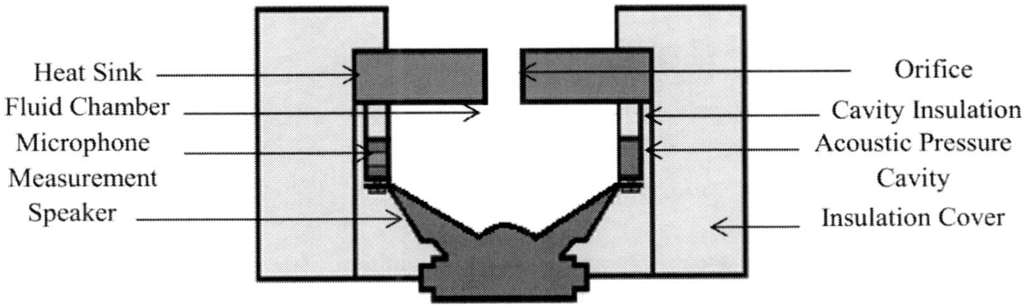

Figure 2. Schematic of synthetic jet set-up

4. Experimental Procedure

4.1 Heat Transfer

The cartridge heaters supply the copper heat sink with a constant power of 4 W via an HP 6033A power system supply (0 – 20 V / 0 – 20 A, 200 W). The Visaton FRS 8 speaker is actuated using LabView software which controls the amplitude and frequency of a sine wave output to a signal amplifier. The amplified signal is then passed into the speaker, which actuates the speaker diaphragm creating a synthetic jet through the heat sink orifice. Temperature measurements are taken on the upper and lower surfaces of the heat sink, inside the jet cavity, in the ambient air and at the base of the speaker. Once the temperature of the heat sink reaches steady-state, the temperatures, Reynolds number, Stroke Length and pressure are all recorded. The Reynolds number is calculated by the following expression:

$$Re = \frac{U\,d}{\nu} \qquad (2)$$

where U is the jet velocity, d is the orifice diameter and ν is the kinematic viscosity of air. For the multiple orifice case, the diameter of a single orifice (i.e. 5 mm) is used; therefore, the Reynolds number represents the jet flow through a single orifice. The testing procedure is repeated for a variety of frequencies and stroke lengths. The multiple orifice heat sink was tested at the same stroke length twice with two different speakers identifying repeatability of the test and assessing the influence of the speaker characteristics. The maximum variation between both data sets for Nusselt number and Reynolds number is 5.2% and 5.6% respectively.

The novel implementation of a synthetic jet integrated into a heat sink increases the complexity of the heat transfer analysis. Synthetic jet actuation will enhance the heat transfer from the heat sink, however, quantifying how much of the heat transfer enhancement is a direct result of jet actuation requires accounting for the system losses. The main losses which must be investigated are: conduction through the insulation housing, Q_{cond}, radiative heat losses from the surface of the heat sink, Q_{rad}, and the additional heat added to the copper heat sink induced from the heat generated due to ohmic heating in the voice coil driver in the loudspeaker, Q_{add}. The convective heat transfer is therefore a function of these parameters, as shown in following equation:

$$Q_{Total} = Q_{heater} + Q_{add} - Q_{cond} - Q_{rad} \tag{3}$$

Once the system losses have been accounted for, the heat transfer coefficient, h, is calculated from:

$$h = \left(\frac{Q_{Total}}{A\Delta T}\right) \tag{4}$$

where Q_{Total} is defined above [W], h is the heat transfer coefficient $\left[\frac{W}{m^2k}\right]$, A is the area exposed to the jet flow [m^2] and ΔT is the temperature difference between the surface heat sink temperature and ambient temperature [K]. The average Nusselt number, $\overline{Nu}$, for mixed convection is calculated as:

$$\overline{Nu}_{mix} = \left(\frac{h\,L_{char}}{k}\right) \tag{5}$$

where k is the thermal conductivity $\left[\frac{W}{mk}\right]$ and L_{char} is the length scale for the heat sink surface.

The Nusselt number defined by the experimental measurements incorporates natural convection from the heat sink. Therefore in order to obtain a representative Nusselt number for the synthetic jet actuation alone, natural convection heat transfer must be accounted for. The influence of natural convection on the total heat transfer is described by the following relationship:

$$Nu_{mix}{}^3 = Nu_{Forced}{}^3 + Nu_{Natural}{}^3 \rightarrow Nu_{Forced} = \left(Nu_{mix}{}^3 - Nu_{Natural}{}^3\right)^{\frac{1}{3}} \tag{6}$$

The direct heat transfer enhancements resulting from synthetic jet actuation is represented by the forced Nusselt number. The estimated relative uncertainty on the power input, thermocouples, Reynolds number and Nusselt number (based on the error propagation rules and 95% confidence level) does not exceed 13.4%, 1.7%, 6.4% and 14.9% respectively, and is typically below 7%, 1%, 6% and 9.8% for the majority of test conditions. The relative uncertainty for the actuation frequency was 0.006% and therefore was assumed negligible.

4.2 Flow Visualisation

Flow visualisation was implemented to capture high speed images of the synthetic jet flow. Flow visualisation operates by injecting seeding particles into the flow field, so that it describes the same flow pattern as the jet flow [13]. The particles can be visualised using a laser sheet and camera. The laser sheet was orientated vertically and positioned at the centre of the orifice. The camera was positioned perpendicular to the laser sheet, as depicted in *figure 3*. The camera and laser are controlled using LaVision software. The sample rate of the camera is slightly lower than the actuation frequency of the jet, therefore capturing vortex propagation over many periods. The seeding medium is introduced to the synthetic jet system and is highlighted by the laser allowing the identification of the flow features generated by the synthetic jet. The post processing of the results was undertaken using the LaVision software.

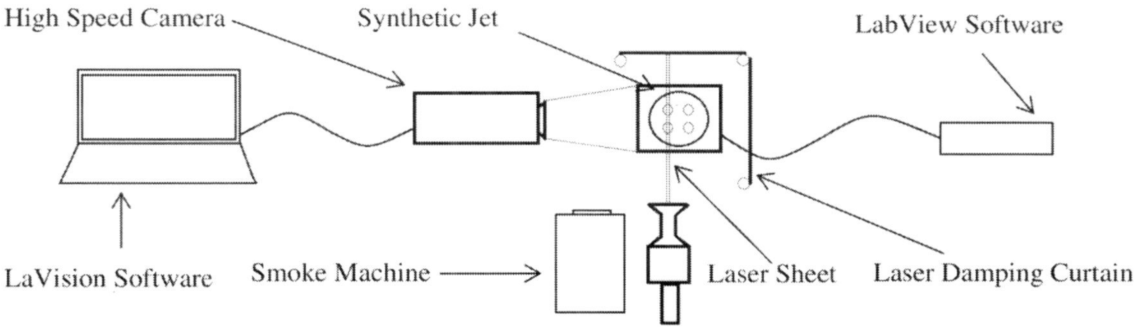

Figure 3. Schematic of the flow visualisation experimental set-up

4.3 Thermal Imaging

Thermal imaging gives another visual representation of the effects of the synthetic jet flow structures on surface heat transfer. The thermal imaging camera was mounted onto a tripod and orientated vertically downward, the heat sink was positioned underneath camera. The experimental set-up for the thermal imaging procedure is depicted in *figure 4*. The emissivity of the heat sink surface was experimentally obtained by in camera comparison to a reference thermocouple. The analysis and post processing was undertaken using FLIR software.

Figure 4. Schematic of the thermal imaging experimental set-up

4.4 Temperature/Boundary Layer Analysis

To establish a thermal boundary layer profile for the heat sink, temperature measurements were taken at and directly above the heat sink surface. Four measurement points which were identified along the radius of the multiple orifice heat sink as indicated in *figure 5 A)*. These are located at:

1) the edge of the insulation
2) 15 mm from the insulation edge
3) between the multiple orifice pair
4) at the centre of the heat sink

The temperature measurements recorded above the measurement positions were taken at 1 mm increments from 1 mm 10 mm above the heat sink, in 2 mm increments from 10 mm to 20 mm and 5 mm increments from 20 mm to 70 mm above the heat sink. This was done for natural convection and synthetic jet conditions. The experimental set-up is depicted in *figure 5 B)*.

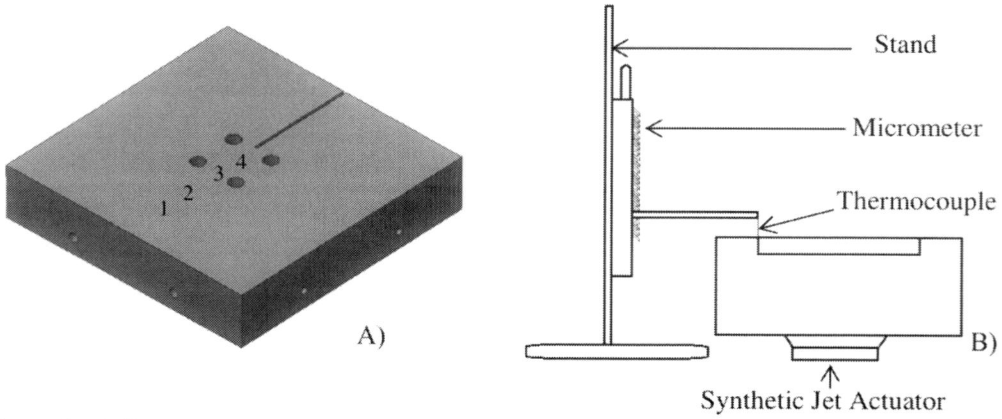

Figure 5. A) Temperature measurements used for boundary layer analysis, B) Schematic of the thermal boundary layer experimental set-up

5. Results and Discussion

5.1.Heat Transfer Analysis

Firstly, the single orifice heat sink was assessed with synthetic jet flow as a function of frequency. The total heat transfer was calculated, this included the heat transfer from the upper and lower surfaces of the heat sink. The results indicate that increasing the frequency of actuation, increases the rate of heat transfer in general. A higher frequency range was tested for the lower stroke length of 2.5, this indicated in a peak in heat transfer at 350 Hz, after which the heat transfer decreases as shown in *figure 6*. This may be due to a cavity resonant frequency or a harmonic of the resonant frequency at 350 Hz. The rate of heat transfer also increases substantially as the stroke length of the synthetic jet is increased, this is a consequence of an increase in the volumetric flow rate. As the volumetric flow rate through the orifice increases, the entrainment of the surrounding flow field is also increased. The jet flow becomes highly sheared increasing the rate of entrainment. This process is hypothesised to decrease the boundary layer thickness, reducing the temperature gradient between the surface and ambient.

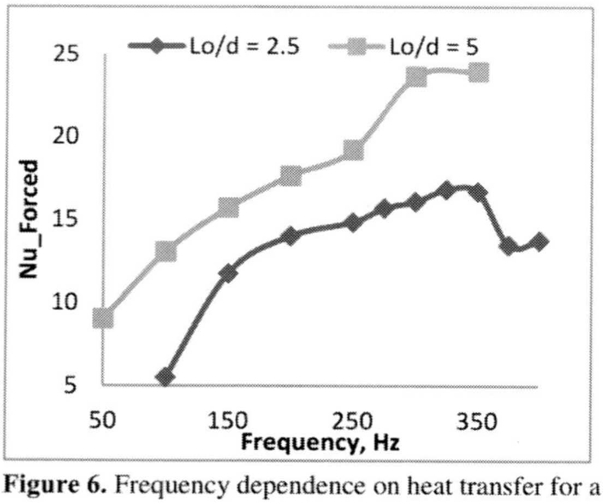

Figure 6. Frequency dependence on heat transfer for a single orifice at stroke lengths of 2.5 and 5

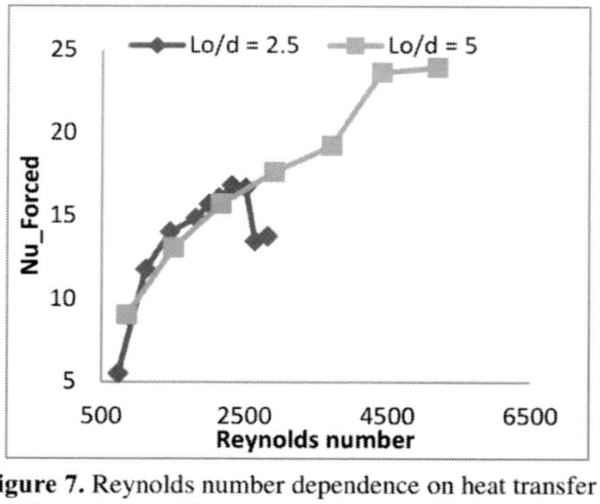

Figure 7. Reynolds number dependence on heat transfer for a single orifice at stroke lengths of 2.5 and 5

The dimensionless stroke lengths used during testing ensures that the vortex generated at the orifice detaches from the surface and is not re-ingested back into the cavity. To avoid the re-ingested process the minimum stroke length tested was 2.5, which is above the sharp edge orifice criterion for vortex detachment of $\frac{L_o}{d} > 0.5$ [10]. The higher stroke length substantially increases the rate of heat transfer, which has also been identified in the literature for direct synthetic jet impingement [6, 7, 14].

The heat transfer analysis on the single orifice also highlighted that the greatest heat transfer enhancement occurs at the upper surface of the heat sink. Both upper and lower surfaces were insulated, in turn, and the heat transfer enhancements for both surfaces were analysed. The results are presented in *figure 8*.

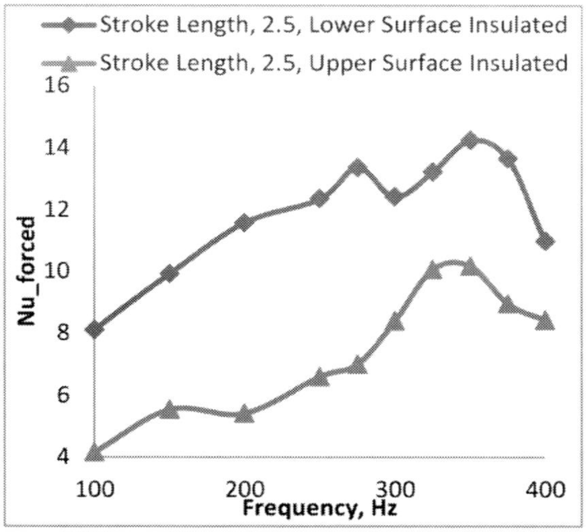

Figure 8. Comparison of heat transfer between the upper surface to the lower surface heat sinks at a constant stroke length of 2.5

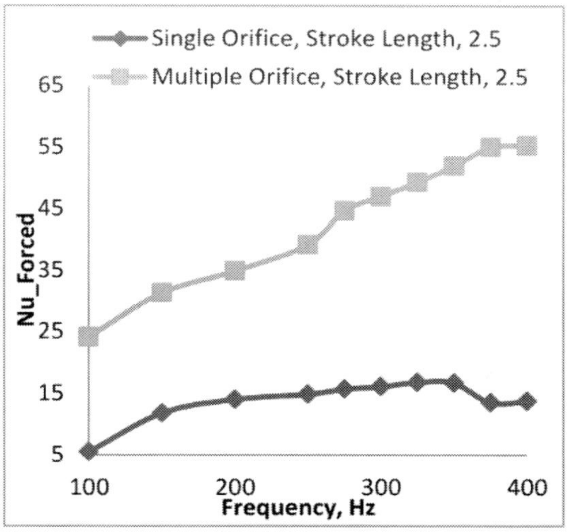

Figure 9. A comparison between the heat transfer of a single orifice and a multiple orifice heat sink at a constant stroke length of 2.5

Figure 8 indicates that the trends between both datasets are similar achieving peak heat transfer at 350 Hz. However, as the excitation frequency is increased the heat transfer on the lower surface increases at a higher rate compared to the upper surface. The physical cause of this is yet unknown and will require further investigation.

The multiple orifice heat sink resulted in a significantly larger heat transfer enhancement compared to the single orifice heat sink. The significant increase in heat transfer coincides with results found by Chaudari et al. [3] for multiple orifice impinging synthetic jets. However, a more significant finding in this research was that the increase in the number of orifices increased the volumetric entrainment of the surrounding flow field, this would indicate why the heat transfer is significantly higher. The difference in heat transfer between the single and multiple orifice heat sink is evident in *figure 9*. The peak heat transfer for the multiple orifice heat sink occurs at a higher frequency than for the single orifice, this may be due to the change in resonant frequency of the synthetic jet cavity or the increase in surface area due to extra orifices. The results obtained indicate that the frequency and Reynolds number increase linearly, therefore a Nusselt number correlation at a set stroke length is obtained. The Nusselt number corrections are presented in *Table 1*.

Table 1. Nusselt number correlations for single and multiple orifice synthetic jet actuation

Nusselt Number Correlations				
Jet Type	*Stroke Length*	*Reynolds Number*	*Correlation*	R^2
Multiple Orifice	2.5	750 < Re < 3,000	$0.6987\ Re^{0.5443}$	0.9831
Single Orifice	2.5	1,100 < Re < 2,500	$0.5877\ Re^{0.4309}$	0.9703
Single Orifice	5	750 < Re < 5,500	$0.2522\ Re^{0.5344}$	0.986

5.2. Flow Structure

The multiple orifice heat sink resulted in the highest heat transfer enhancement of all the orifice configurations tested, therefore the flow structures developed by the multiple orifice heat sink were investigated. The analysis of these flow structures is presented in *figure 10.*

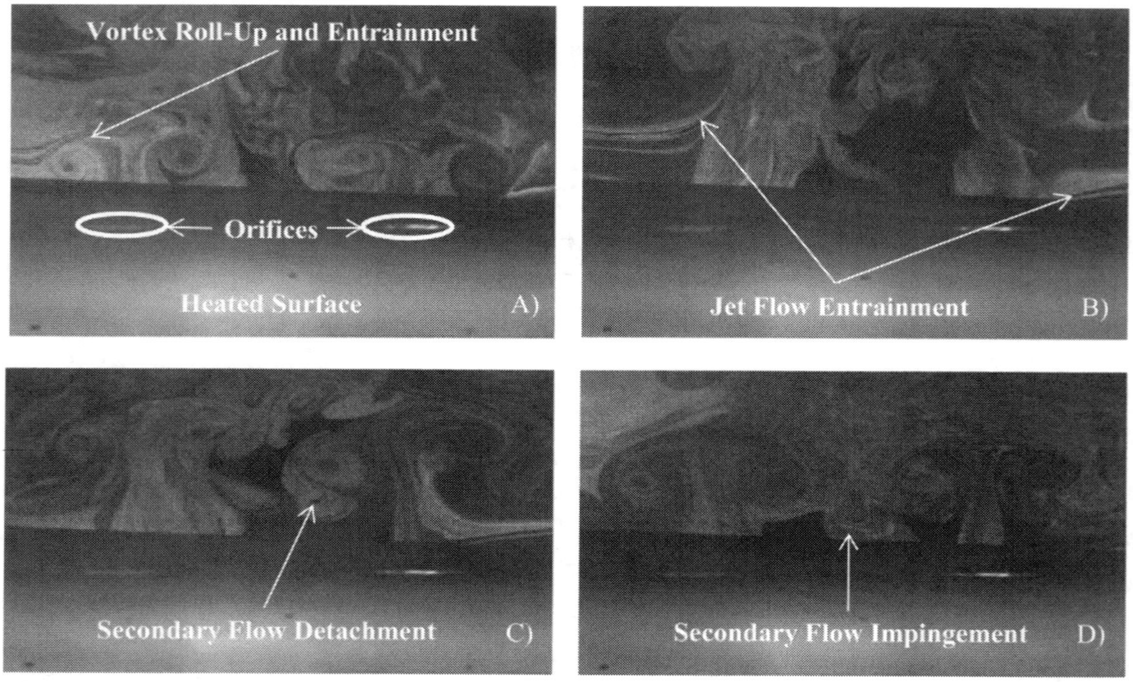

Figure 10. Flow visualisation of the multiple orifice heat sink at a stroke length of 5 A) Vortex Roll-Up, B) Jet Flow Entrainment, C) Secondary Flow Detachment and D) Secondary Flow Impingement

Vortex formation initiates at the edge of the orifice as the top of the fluid slug is ejected from the cavity. The vortex continues to roll-up during the ejection phase until it is finally advected downstream. This process is depicted in *figure 10 A)* which identifies vortex roll-up at the surface of the orifice, this is typical behaviour for a synthetic jet. The image provides sufficient evidence to indicate that local turbulence generation and flow field entrainment is occurring, again satisfying one of the initial hypotheses as a method responsible for increasing heat transfer. The figure also depicts both vortices rolling up simultaneously, which indicates that both of the synthetic jets are in phase, as expected.

The high momentum ejection stroke of the synthetic jet creates a highly sheared jet flow, these shear layers entrain the surrounding ambient flow, breaking down the thermal boundary layer. However, it is not only during vortex roll-up that the highly sheared flow entrains the ambient flow; entrainment induced by the jet flow as the vortex is advected downstream is also evident in *figure 10*

B). While the vortex is advected downstream, the synthetic jet is drawing the surrounding air into the cavity during the suction phase. During this period of ejection, the jet flow is entraining the surrounding flow field above the heat sink surface, while the cavity is entraining the flow field at the surface of the heat sink. The dual entrainment process is depicted in *figure 10 B)*. Both entrainment methods are assumed to increase the heat transfer by continually breaking down the thermal boundary layer at the heat sink surface.

Boundary layer entrainment and vortex induced turbulence are not the only methods of cooling which have been identified when analysing the multiple orifice heat sink. A differential pressure induced impingement was identified as a by-product from the propagating jet flow. A sub pressure zone is generated as both vortices become detached from the orifices. These propagating vortices entrain the surrounding flow field in the direction of ejection, which increases the volumetric flow rate of the jet. However, the entrainment of the fluid between the orifice pair creates a sub pressure zone. The counter rotating vortices at the top of the jet column entrain the fluid between both jet streams, this entrained fluid then detaches from the vortex by the differential pressure between the jet flow and sub pressure zone. This secondary entrainment forces the ambient flow in the opposing direction of ejection, and subsequently impinges on the surface of the heat sink. These jet flow characteristics are highlighted in *figure 10 C)* and *D)*. To identify the velocity and vorticity of the secondary entrainment, P.I.V. analysis must be conducted, this is planned for the second phase of this research.

5.3 Thermal Boundary Layer Analysis
The cooling method of the synthetic jet device was hypothesised to be a result of thermal boundary layer break-up. This results from the entrainment of the ambient flow field and local turbulence generation around orifice edge. The thermal boundary layer which is assumed to be diminished is analysed using this thermal boundary layer investigation. The results of the analysis are presented in *figures 11* and *12*.

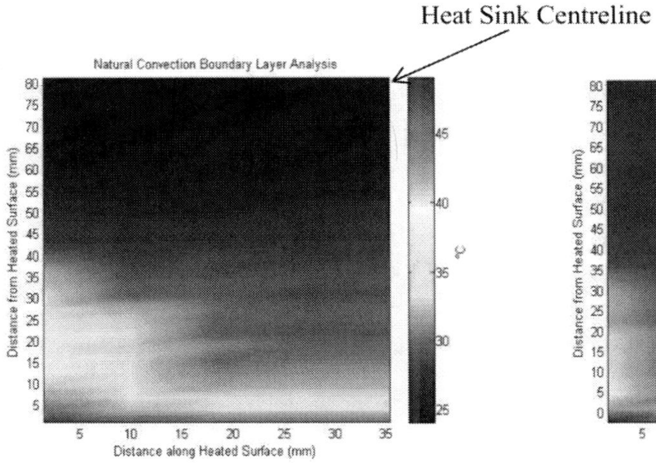

Figure 11. Thermal boundary layer analysis of the multiple orifice heat sink from natural convection

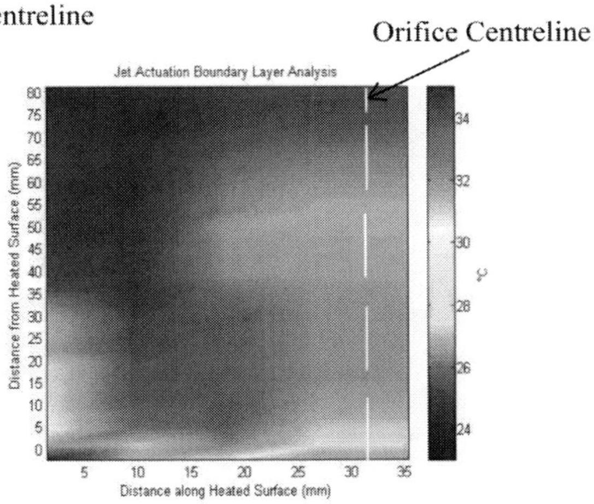

Figure 12. High resolution thermal boundary layer analysis of the multiple orifice heat sink from jet actuation

Figures 11 and *12* illustrate the temperature of the air occurring at the edge of the insulation, which is a result of the increased temperature of the heat sink underneath the insulation; this is also identified in the thermal imaging analysis. The boundary layer thickness for the natural convection analysis is highest at the edge of the insulation and is subsequently cooled along the length of the heat sink, which contrasts to conventional natural boundary layer development. This can be attributed to the high local temperatures at the heat sink/insulation interface. *Figure 12* illustrates the effect of synthetic jet actuation on the thermal boundary layer tempertaure. The figure indicates that the synthetic jet successfully disrupts the thermal boundary layer over the surface of the heat sink, satisfying the initial cooling process hypothesis.

6. Conclusion

An investigation into the heat transfer and flow structure developed by a synthetic jet emanating from a heat sink was undertaken. The heat transfer analysis provided sufficient evidence to indicate that a multiple orifice synthetic jet provided the greatest heat transfer enhancements, with the upper surface of the heat sink providing the greater heat transfer contribution. The flow structures responsible for the heat transfer enhancements were also identified using flow visualisation. The enhanced heat transfer characteristics resulted from boundary layer thinning from ambient flow field entrainment.

Acknowledgements
The authors acknowledge the technical staff of the Mechanical and Manufacturing Engineering Department in Trinity College Dublin.

References

[1] M. Baelmans, R. Gielen, T. Persoons, F. Rogiers, T. Saenen, T. Oevelen, Optimization of Convective Heat Transfer in Micro-Scale Electronics Cooling Applications, Houille Blanche-Revue Internationale De L'Eau, 66 (2011), 70 – 78

[2] S. V. Garimella, L-T. Yeh, T. Persoons, Thermal Management Challenges in Telecommunication Systems and Data Centres, IEEE Transactions on Components, Packaging and Manufacturing Technology, 2 (2012), 1302 – 1326

[3] M. Chaudhari, B. Puranik, A. Agrawal, Multiple Orifice Synthetic Jet for Improvement in Impingement Heat Transfer, International Journal of Heat and Mass Transfer, 54 (2011), 2056 – 2065

[4] T.L. Lupton, D.B. Murray, A.J. Robinson, The Effects of Varying Confinement Levels on the Heat Transfer to a Miniature Impinging Air Jet, 5th European Thermal-Sciences Conference, paper no. Jet 5, Eindhoven, 2008.

[5] M. Chaudari, B. Puranik, A. Agrawal, Effect of Orifice Shape in Synthetic Jet Based Impingement Cooling, Experimental Thermal and Fluid Science, 34 (2010), 246 – 256

[6] R. Farrelly, A. McGuinn, T. Persoons, D. B. Murray, Heat Transfer Behaviour and Flow Field Characterisation of Impinging Synthetic Jets for a Wide Range of Parameters, Proceedings of the International Heat Transfer Conference (2010), Washington D.C.

[7] M. Chaudari, B. Puranik, A. Agrawal, Heat Transfer Characteristics of Synthetic Jet Impingement Cooling, International Journal of Heat and Mass Transfer, 53 (2010), 107 – 1069

[8] T. McGuinn, A Study of the Heat Transfer Process and Related Flows of a Synehtic Jet Impinging Perpendicular to a Heated Surface, A thesis submitted to the University of Dublin (2011)

[9] A. Glezer, M. Amitay, Synthetic Jets, Annual Review of Fluid Mechanics, 34 (2002), 503 -529

[10] T. Persoons, General reduced-Order Model to Design & operate Synthetic Jet Actuators, AIAA Journal, 50 (2012), 916 - 927

[11] R. Holman, Y. Utturkar, R. Mittal, B. L. Smith, L. Cattafesta, Formation Criterion for Synthetic Jets, AIAA Journal, 43 (2005), 2110 - 2116

[12] T. Persoons, T. S. O'Donovan, A Pressure-Based Estimate of Synthetic Jet Velocity, Physics of Fluids, 19 (2007), 128104

[13] C. Carcasci, An Experimental Investigation on Air Impinging Jets using Visualisation Methods, International Journal of Thermal Science (1999) 38, 808 – 818

[14] R. Farrelly, A. McGuinn, T. Persoons, D. B. Murray, A Heat Transfer Comparison between a Synthetic Jet and a Steady Jet at Low Reynolds Number, Proceedings of 2008 ASME International Mechanical Engineering Congress (2008), 505 – 510

Copper-Carbon and Aluminum-Carbon Composites Fabricated by Powder Metallurgy Processes

Jean-François Silvain[1, 2], **Amélie Veillère**[1], **Yongfeng Lu**[2]
[1] Université Bordeaux 1, Institut de Chimie de la Matière Condensée de Bordeaux (ICMCB), CNRS, 33608 Pessac, France
[2] Department of Electrical Engineering, University of Nebraska – Lincoln, Lincoln, NE, 68588-0511, USA

E-mail: silvain@icmcb-bordeaux.cnrs.fr

Abstract. The increase in both power and packing densities in power electronic devices has led to an increase in the market demand for effective heat-dissipating materials, with high thermal conductivity and thermal- expansion coefficient compatible with chip materials still ensuring the reliability of the power modules. In this context, metal matrix composites: carbon fibers and diamond-reinforced copper and aluminum matrix composites among them are considered very promising as a next generation of thermal-management materials in power electronic packages. These composites exhibit enhanced thermal properties compared to pure copper combined with lower density. This article presents the fabrication techniques of copper/carbon fibers and copper/diamond and aluminum/carbon fibers composite films by powder metallurgy and hot pressing. The thermal analyses clearly indicate that interfacial treatments are required in these composites to achieve high thermomechanical properties. Interfaces (through novel chemical and processing methods), when selected carefully and processed properly will form the right chemical/mechanical link between metal and carbon, enhancing all the desired thermal properties while minimizing the deleterious effect.

1. Introduction

Electronic packaging refers to the protective features and safety devices that enable continued and efficient operation of electronic components, and does not refer to shipping containers. Electronic components typically refer to assemblage of resistors, capacitors, transistors, diodes, etc., in a complex well designed group as integrated circuits usually soldered to a printed circuit board. Since these components are very sensitive in all respects, the packaging is designed to prevent damage, avoid exposure to dirt and other adverse environment, dissipate heat very efficiently, enhance reliable service life and sustain efficient working conditions [1-3].

Composite materials show the highest potential for application in heat sinks for electronic applications. They can combine metal, polymer and ceramics in a broad range of different products and can be tailored to the final desired properties. It is critically important to note that final properties of composites depend not only on the basic constitutent materials but also on the methods of manufacture. Among the large number of different composites, the most efficient and the most commonly developed for heat dissipation purposes are metal matrix composites (MMCs). More specifically MMCs based on copper and aluminum matrices with a broad range of different reinforcements show great promise. Undoubtedly, the most efficient composites for heat sinks are those based on Cu and Al with a number of different reinforcements [4-9]. In general, MMCs show several improvements in comparison with currently used materials in electronic packaging, including:

Content from this work may be used under the terms of the Creative Commons Attribution 3.0 licence. Any further distribution of this work must maintain attribution to the author(s) and the title of the work, journal citation and DOI.
Published under licence by IOP Publishing Ltd

- Lower and tailorable CTE (the higher the volume fraction of reinforcement used, the lower the CTE),
- Higher heat dissipation capability,
- Light-weight, suitable for space applications,
- Higher stiffness at elevated temperatures (for dimensional stabilities at elevated temperatures).

In this paper, we mainly focus on understanding the thermal management requirements of electronic packaging, the key scientific and technological issues, advances in terms of materials, design and processing, and future prospects of the field. In specific, we will address how the competing physical and thermal requirements are optimized in heat sinks. Many of the issues addressed can be equally applicable to the design of any component with similar thermal and mechanical requirements including that of heat exchanger tubes. Though in this paper, we focus our attention on copper-carbon and aluminum-carbon system, concepts presented here can be easily extended to other material systems.

2. Experimental procedure: materials and processing

2.1 Copper matrix

Many techniques are used to manufacture powders. In the case of copper, generally either chemical or atomization methods are employed. Each technique plays a major part on final material properties, microstructure (grain size, distribution, morphology, and particle content), chemistry and finally, the cost of the powders. Chemical methods can be used to prepare metallic powders; the difference between each is linked to the initial components and the chemical reaction. It is possible to use oxides, sulphates or carbonyl compounds which are reduced, precipitated or decomposed in order to obtain copper powders. Among these chemical methods, electrolysis which is a method using an electric current to drive the metallic ion reduction is promising [10].

2.2 Heat sink fabrication

MMC fabrication processes are numerous; therefore, we will just discuss the process relevant to heat sink fabrication. It is necessary to sinter materials with a high density (porosity < 4%) in order to obtain high thermal conductivity. Indeed, the gas (air, argon…) in the porosities is a thermal insulator and can strongly hinder the thermal conductivity of the material. Fabrication processes of heat sinks can be classified in two groups: liquid route and solid route. The solid route process is a powder metallurgy process, i.e., the powder of metal matrix is mixed to the particle or short fiber reinforcement. This mix is then sintered in solid state. There are various powder metallurgy processes that can be distinguished by the method of use to produce the stress. The conventional powder metallurgy process consists in 1) shaping at ambient temperature and 2) strengthening the powder using at high temperature near the melting point. This last step is called sintering. This process does not allow obtaining composites with high density, which is why a pressure is usually applied during the sintering, referred to as hot sintering processes [11]. In the uniaxial hot pressing, a uniaxial pressure is applied at a temperature close to the melting point, which increases the density of the final composite (porosity < 5 vol.%). This technique allows producing composite materials with a high volume fraction of reinforcement (up to 50 vol. %). Nevertheless, the mechanical resistance of the tool limits the maximum stress that can be applied. This process is widely used to fabricate MMC with aluminum or copper matrices because these metals are ductile even at low temperature.

3. Results and discussions

3.1 Copper/Carbon fibers composites

3.1.1 Process

Copper and carbon fibers pitch type, named K6371T and K223HG, usually form the starting powders (Table 1). Typical Scanning Electron Microscopy (SEM) observations show that copper particles are present with dendrite shapes with sizes ranging from 25 to 30 μm. Chopped carbon fibers have diameters ranging from 9 to 10 μm and lengths between 100 to 300 μm. In the first step, copper powders and carbon fibers are mixed together. The mixed powder is then compacted in graphitic mould, which is heated by an induction system. The samples are pressed at 650 °C under 50 MPa during 20 minutes; the heating rate is 25 °C/min. The temperature is controlled with a thermocouple inserted inside the mould. A vacuum in the chamber of 0.66 Pa prevents the oxidation of copper powder during the heating cycle. The final density of the composite material is close to 97% and no specific degradation of the carbon fiber is observed. The optical micrographs of the samples show a strong anisotropic structure (Fig. 1), with an orientation of the carbon fiber perpendicular to the compaction direction.

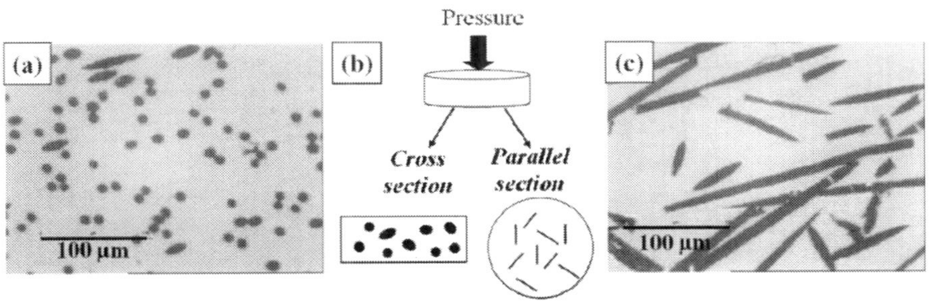

Figure 1: Optical micrographs of the composite material in (a) cross and (c) parallel section after densification and (b) schematic of the Carbon fiber orientation.

This anisotropy is due to the strains and stresses imposed by the processing-induced deformations and geometry, and will be reflected in the anisotropic thermal (k) and thermo-mechanical (CTE) properties of the composite.

3.1.2 Thermal properties

Thermal conductivity (k) and coefficient of thermal expansion (CTE) have been measured using a flash laser method and a horizontal dilatometer respectively. Due to the anisotropic structure of the composite, these properties are determined in two perpendicular directions namely parallel to the pressing direction ($k_{//}$ and $CTE_{//}$) and perpendicular to the pressing direction ($k_{\perp}$ and $CTE_{\perp}$). The $CTE_{//}$ of the composite are close to 17 $10^{-6}.K^{-1}$, whatever the type and the volume fraction of carbon fibers used. In contrast, the $CTE_{\perp}$ decreases when the carbon fibers volume fraction increases whatever the fiber type used (Table 2). These anisotropic properties are due to the strong anisotropy of carbon fiber properties (Table 1) and their orientation perpendicular to the pressing direction in the composite.

Table 1: Thermo-mechanical properties of matrices and reinforcements used

Material			Density	k ($W.m^{-1}.K^{-1}$)	CTE ($10^{-6} K^{-1}$)
Matrices	Cu		8.96	400	17
		K223HG	2.2	540 / 10	-1 / 12
		XN100	2.2	900 / 10	-1 / 12
	Diamond		3.5	800 - 2000	1

Same anisotropy can be observed for the thermal conductivity measured on the various composites. Table 2 shows the evolution of $k_{//}$ and $k_{\perp}$ for two types of carbon fibers (CF) with a volume fraction ranging from 30 to 40 vol. %. This table shows that $k_{//}$ decreases when the carbon fiber volume fraction increases; $k_{//}$ is just link to the CF volume fraction but not to the fiber type. However, the variation of $k_{\perp}$ with the volume fraction is different; $k_{\perp}$ is obviously link to the carbon fiber volume fraction but also to their intrinsic thermal conductivity along their axis. Indeed, when the fibers used have a thermal conductivity along their axis greater than that of copper, $k_{\perp}$ increase with the volume fraction (example of K223HG carbon fiber type). On the other hand, when the carbon fibers used have a thermal conductivity along their axis lower than that of copper, $k_{\perp}$ decrease with the volume fraction (example of K6371T carbon fiber type).

Table 2: Thermal conductivity parallel (//) and perpendicular (⊥) to the pressing direction and CTE perpendicular (⊥) to the pressing direction for copper/carbon fibers composites.

Cu/CF	*CF vol. %*	*Density*	K ($W.m^{-1}.K^{-1}$)	*CTE (10^{-6} K^{-1})*
Cu / K6371T	30	6.8	300 ⊥, 180 //	14
	35	6.5	330 ⊥, 160 //	12
	40	6.2	360 ⊥, 140 //	10
Cu / K223HG	30	6.8	240 ⊥, 180 //	14
	35	6.5	210 ⊥, 160 //	12
	40	6.2	180 ⊥, 140 //	10

3.2 Copper/Diamond composites

3.2.1 Material and process

Diamond has exceptional thermal properties with the highest thermal conductivity at room temperature of all known materials (λ = [800 – 2000] $W.m^{-1}.K^{-1}$). However, its very low CTE (α = 1.0 × 10^{-6} $°C^{-1}$) remains to be a problem. Thus, it is interesting to consider the use of diamond as reinforcements embedded in a thermal package substrate, such as a copper matrix (Cu). In this study, Cu/D composite films were fabricated by tape casting and hot pressing. Diamond powders from Henan Zhongxin Co., China, and dendritic copper powder from Ecka Poudmet, Germany, were used to process the Cu/D composites. After tape casting, the Cu/D composite tapes are first heat treated at 400°C under air during 2 hours (debinding treatment), then heat treated 60 minutes at 400°C under Ar/H_2 (reducing treatment). Finally, after stacking of 10 to 15 tapes, the Cu/D composites were hot pressed during 20 minutes at 650°C under 50 MPa to obtain dense materials.

3.2.2 Copper particles deposition onto diamond powders

Cu/D composites fabricated by powder metallurgy have shown weak interfacial bonding because pure liquid copper does not wet diamond particles. The non-reactive Cu/D interface is critically detrimental to the thermal properties of the Cu/D composites. Carbide forming materials, namely chromium or boron, are usually employed to bond diamond reinforcements to the copper matrix, either by being alloyed to the matrix [12, 13] or by being directly coated onto diamonds [14]. However, the carbide interphase also acts as a thermal barrier and affects the heat transfer between electronic conduction in copper and phonic conduction in diamond. We used here an innovative process that consists in depositing copper nanoparticles onto diamond reinforcements through C-O-P-O-Cu and/or C-P-O-Cu bonds prior to sintering. This new process creates strong chemical bonding between the diamonds and the copper matrix [15]. This process leads to dense and cohesive Cu/D composites without using carbide forming elements, thereby allowing high thermal performances with minimum diamond content making these Cu/D composites highly cost-effective.

Figure 2 shows diamond particle covered by Cu submicronic particles. One can see that the coverage of the diamonds surface by the copper particles is remarkably uniform. In addition, the size of the surface Cu particles can be adjusted depending on the reduction time and temperature, as shown on figure 8a and Figure 2b with Cu particle sizes ranging from 400 nm to 1 μm (after 60 minutes at 400°C under Ar/H_2), on Figure 2c with Cu particles sizes ranging from 300 to 800 nm (after 1 minute at 400°C under Ar/H_2) and Figure 2d with Cu particle sizes ranging from 100 nm to 400 nm (after 1 minute at 350°C under Ar/H_2). However, the influence of the size of the Cu nanoparticles coating on the final thermal properties of the composites is not yet clear. Regarding the chemical process responsible for the Cu particles deposition, XPS analysis has demonstrated that the mixing of the dispersant agent (phosphate ester) with the powders during formulation leads to the creation of phosphorus-functionalized nucleation sites on the diamond powders surface. During debinding, copper powders are oxidized and wire-shaped CuO nanostructures are observed to grow on the copper dendrites. Such CuO nanowires growth process has been reported in previous studies [25]. During the reduction treatment, copper sublimed from CuO nanowires and settles again onto the chemically active sites on the diamond particle surface through C-O-P-O-Cu and/or C-P-O-Cu chemical bridges.

During the hot pressing step, Cu-Cu metallic bonds are formed between the Cu nanoparticles onto the diamonds and the copper particles that will form the Cu matrix after densification. The Cu nanoparticles thus enable the formation of a cohesive interface between the copper matrix and the diamond reinforcements through strong chemical bonding without requiring any carbide forming materials.

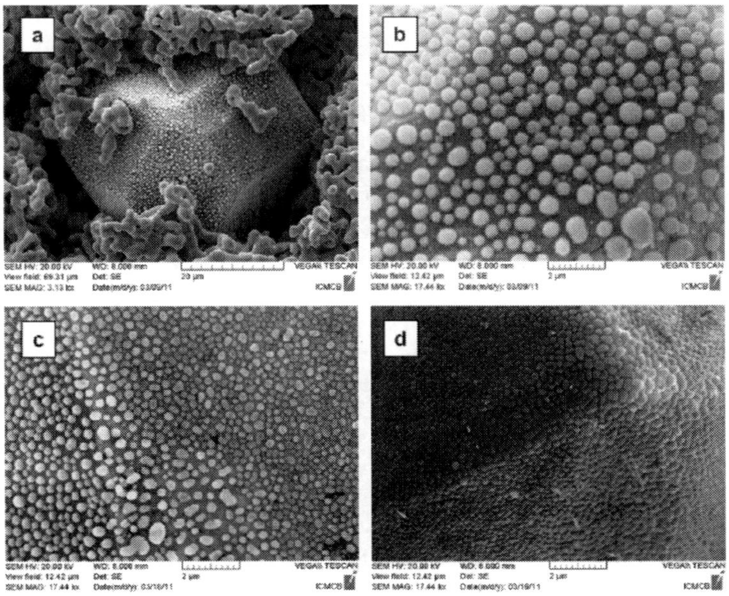

Figure 2: SEM micrographs of diamond reinforcements coated with copper particles: (a,b) after 60 minutes under Ar/H_2 at 400°C (c) after 1minute under Ar/H_2 at 400°C (d) after 1 minute under Ar/H_2 at 350°C

3.2.3 *Thermal properties*

As shown on Figure 3, the average thermal conductivity of multilayer composites materials, fabricated by tape casting process, increases from 320 to 460 $W.m^{-1}.K^{-1}$ with respect to diamond volume fractions ranging from 0 to 40 %. This confirms that the copper particles deposited onto the diamond reinforcements prior to sintering effectively act as bonding agents between diamond and the copper matrix, enabling efficient interfacial heat transfer between electronic and phonic conductions. However, we observe a drop of thermal conductivity to 313 $W.m^{-1}.K^{-1}$ with a diamond volume fraction of 50 % due to the low compaction of the $Cu_{50}D_{50}$ composite, which is detrimental to the thermal conductivity. This decrease in thermal conductivity means that the percolation threshold (i.e. the reinforcement volume fraction at which a continuous path between reinforcements exists in the composite) has been reached.

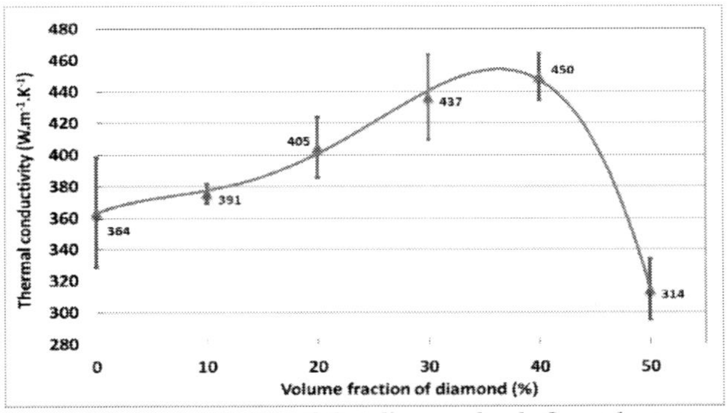

Figure 3: Experimental thermal conductivity of the diamond reinforced copper matrix composites as a function of diamond volume fraction

The percolation threshold occurs between 40 and 50 % in diamond volume fraction. We reach a maximum thermal conductivity of 460 $W.m^{-1}.K^{-1}$ with a diamond volume fraction of 40 %, which is a significant improvement compared to pure copper (λ_{Cu} = 400 $W.m^{-1}.K^{-1}$). We expect to improve this result in our ongoing effort using single tape systems. As shown on Figure 4, the averages measured CTEs of composites decrease linearly from 18.9×10^{-6} °C^{-1} to 11.8×10^{-6} °C^{-1} with diamond volume fractions ranging from 0 to 50 % and evolve with the same slope as the theoretical values. Unlike thermal conductivity, CTE is not affected by the low compaction exhibited by the $Cu_{50}D_{50}$ composites since it keeps decreasing between 40 and 50 %, although its decrease is less pronounced between these two volume fractions. The CTE of pure copper samples fabricated by tape casting and hot pressing is measured to be 18.9×10^{-6} °C^{-1}, which is higher than the theoretical value ($CTE_{Cu} = 17 \times 10^{-6}$ °C^{-1}). This also can be ascribed to the fact that our samples are multilayered systems, instead of single tapes. The interfaces between tapes may degrade the thermo-mechanical properties of stack. A minimum CTE of 11.8×10^{-6} °C^{-1} with a diamond volume fraction of 50 % is reached, which represents a strong enhancement relative to the CTE of pure copper ($CTE_{Cu} = 17 \times 10^{-6}$ °C^{-1}). These results demonstrate that the copper deposition process leads to a strong interfacial bonding enabling an efficient thermal expansion load transfer at the matrix/reinforcement interface.

3.3 Aluminum based composite

3.3.1 Materials and fabrication process

Spherical aluminum powder (F3731, Poudres Hermillon) with 0.88J/K of heat capacity and discontinuous pitch-based carbon fiber (Raheama® R-A301, Teijin Limited.) with 8mm of average diameter, 200mm of average length, 0.70J/K of heat capacity and 600 W/m.K of thermal conductivity in longitudinal fiber direction, were employed. Al-Si11.3at% alloy powder (F2071, Poudres Hermillon) was also prepared. And differential scanning calorimeter (DSC) revealed the melting point of this Al-Si alloy was 584.6°C. Discontinuous pitch-based carbon fibers was mixed with aluminum matrix powder, which was added 5vol% of Al-Si alloy powder, for 1 hour in air. Then, volume fractions of carbon fibers were controlled in 10, 20, 30, 40 or 50vol%. Mixed composite powder was consolidated by hot pressing with 60 MPa of uniaxial pressure, for 30 minutes of holding time at 600°C (*i.e.* between the melting points of aluminum and Al-Si alloy) of holding temperature. Temperature was monitored by a K-type thermocouple positioned in the hole spaced at the centre of carbon mold.

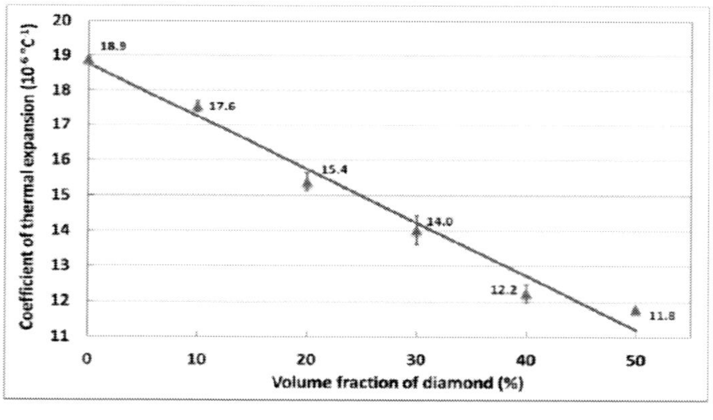

Figure 4: Experimental CTEs of the diamond reinforced copper matrix composites as a function of diamond volume fraction

3.3.2 Results

And as shown in Figure 5, carbon fibers tended to be aligned in plane due to uniaxial pressure in hot press process, while this alignment was less than perfect due to the high aspect ratio of carbon fiber.

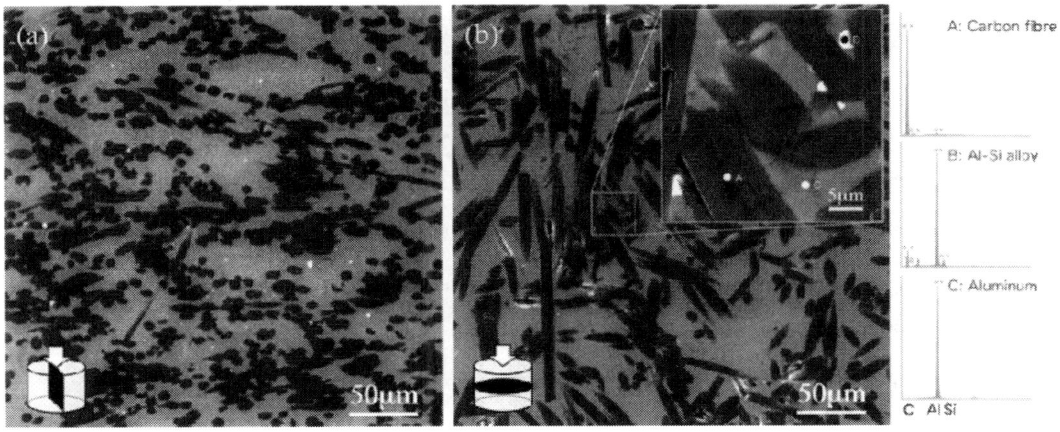

Figure 5: SEM micrographs of Al-CF50vol% composite with 5vol% of Al-Si alloy to (a) parallel and (b) vertical directions of applied pressure axis.

Figure 6(a) shows the thermal conductivity of Al-CF composite to transverse direction of carbon fibers. Thermal conductivities were decreased with carbon fiber addition, regardless of Al-Si alloy addition. To in-plane direction of carbon fiber, thermal conductivity of Al-CF composite, which was added 5vol% of Al-Si alloy in aluminum matrix, were increased with carbon fiber addition, and achieved to 258W/m.K in Al-CF50vol% composite as shown in Figure 6(b), although that was decreased without Al-Si alloy. Figure 7 shows the CTE of Al-CF composite to in-plane direction of carbon fibers. CTE of aluminum, 24.0×10^{-6}/K was slightly degraded to 25.3×10^{-6}/K whether Al-Si alloy was added or not. And CTEs of Al-CF composite, with/without 5vol% of Al-Si alloy, were improved with less than 20vol% of carbon fiber. However, CTE of Al-CF30vol% without Al-Si alloy was slightly higher than that of Al-CF20vol%, and not stable. Furthermore, CTE of Al-CF composite without Al-Si alloy was immeasurable with more than 40vol% of carbon fiber due to the deterioration of the composite attributed to low relative density. On the other hand, CTE of Al-CF composite with 5vol% of Al-Si alloy was further improved and achieved to 7.0×10^{-6}/K in Al-CF50vol% composite.

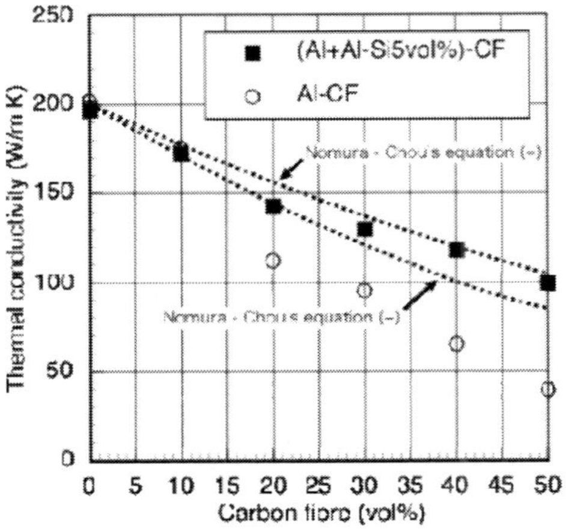

a)

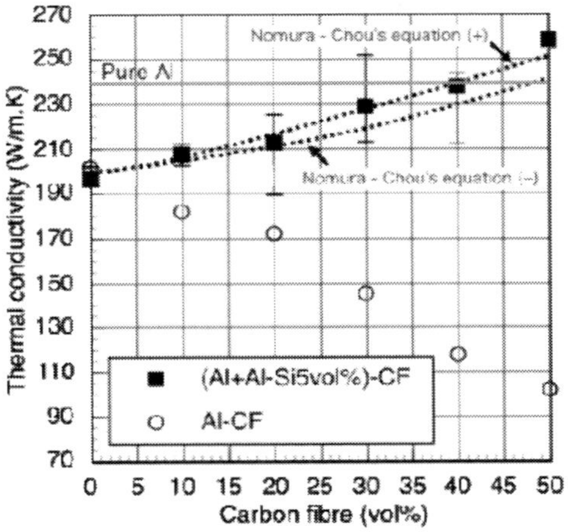

b)

Figure 6: Thermal conductivity of Al-CF composite to (a) transverse direction and (b) in-plane direction of carbon fibers.

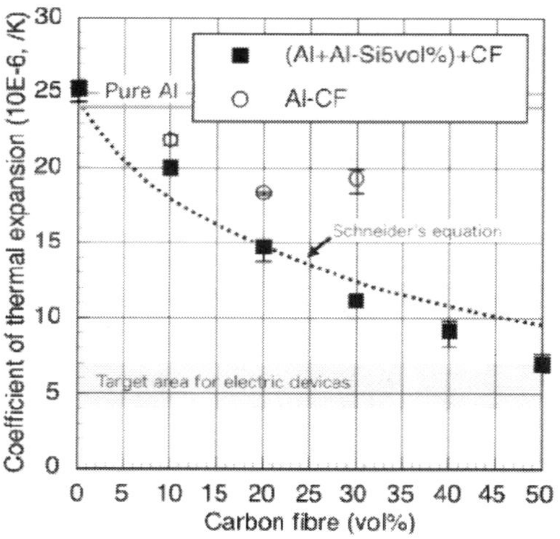

Figure 7: CTE of Al-CF composite to in-plane direction of carbon fibers.

Figure 8(a) shows TEM micrographs of Al/CF interface in Al-CF50vol% composite. It was observed needle like aluminum carbide (Al_4C_3) crystals, and these crystals were oriented to same direction. And the fringe indicated the strain in aluminum matrix. Figure 8(b) shows TEM micrographs of tip of carbon fiber in Al-CF50vol% composite. Al_4C_3 crystals were observed as clusters, carbon fiber tip was covered with Al_4C_3.

3.3.3 Discussion

In conventional hot pressing at 600°C (i.e. between the melting points of Al (660°C) and Al-Si (584.6°C)), Al-Si alloy changes to liquid phase. And this liquid phase of Al-Si alloy infiltrates into slight spaces such as aluminum particle boundaries, between carbon fibers and Al/CF interface. Consequently few quantity of Al-Si alloy allows to fabricate fully-dense composite without particular or complicate process, although thermal conductivity and CTE of aluminum are slightly decreased due to 5vol% of Al-Si alloy addition. Not only the pitch-based carbon fiber which we have employed, but also other several pitch-based carbon fibers, which have excellent thermal conductivity to longitudinal direction have low thermal conductivity to transverse direction. Therefore, thermal conductivity of Al-CF composite is decreased by carbon fiber addition, to transverse direction of carbon fiber. On the other hand, the thermal conductivity of Al-CF composite is improved to in-plane direction of carbon fibers although that is decreased in Al-CF composite, which is fabricated without Al-Si alloy, due to low relative density. Thermal conductivity of Al-CF50vol%, 258W/m.K is higher than the value of pure aluminum, 240W/m.K.

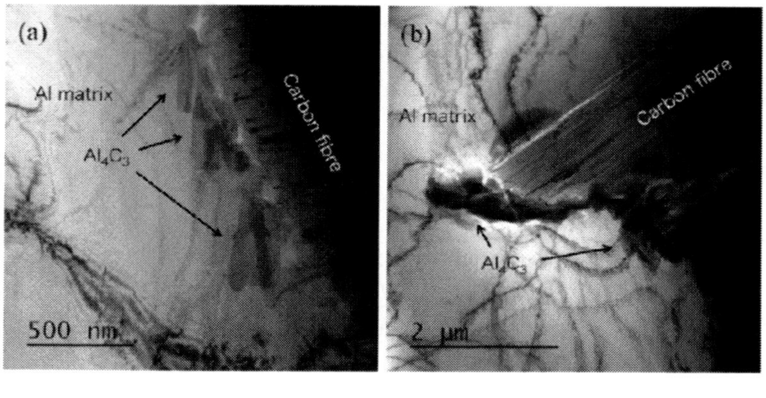

Figure 8: TEM micrographs of Al-CF50vol% composite with 5vol% of Al-Si alloy: (a) Al/CF interface and (b) tip of carbon fiber.

The assemblies for computer are generally constructed by silicon (4.1×10^{-6}/K) or alumina (6.7×10^{-6}/K). To avoid the fracture by thermal stress, the heat sink material also should have a similar CTE to them, around $4.0\sim7.0\times10^{-6}$/K to parallel direction to assembly/heat sink contact surface. Here, it seems that our Al-CF50vol% composite with 7.0×10^{-6}/K of CTE is suitable as a heat sink material. As shown in Figure 7, it is revealed CTEs of our Al-CF composites tend to be amenable with the values estimated by Schneider's equation while it should be taken account this equation is for continuous fiber reinforced composite. Here, it is required a strong interfacial link to bring out remarkable CTE of carbon fiber in aluminum matrix because it is assumed that matrix/fiber interface is perfectly linked. It has been reported interfacial Al_4C_3 is valid for effective load transfer due to strong chemical bond at Al/Carbon interface. In our Al-CF composite, it seems Al/CF interface is not delaminated due to reinforce by interfacial Al_4C_3 formation. Indeed, the thermal strain around Al/CF interface is shown as the fringe in TEM micrograph (see Figure 8). Therefore, it seems aluminum carbide at Al/CF interface (*i.e.* Al-C chemical bond) contributes to effective CTE improvement.

4. Conclusions

In the case of copper/carbon fibers composites, the orientation of the carbon fibers in the matrix induces anisotropic properties of the composite. Due to the processing method, we have measured the thermal conductivity parallel to the loading direction, which shows lower thermal conductivity of the composite. The coefficient of thermal expansion (CTE) has been measured perpendicular to the loading direction. The optimized interfaces obtained with this method give a higher thermal conductivity of the final composites compare to the Cu-X/D_{30}. Indeed, this technique allows to create a chemical link between the copper and the diamond through C-O-P-O-Cu and/or C-P-O-Cu bridges which have a thermal conductivity nearer to copper than carbides. The thermal conductivities of composites prepared with an interfacial treatment, Cu/$D_{30\text{-Cu dots}}$ are greater than pure copper (400 $W.m^{-1}.K^{-1}$) and twice that of the Cu/D_{30} reference composite.

Al-CF composite was fabricated by conventional powder metallurgy process, hot press with 5vol% of Al-Si alloy between the melting points of Al and Al-Si alloy. Liquid phase of Al-Si alloy allows to fabricate fully-dense composite without particular or complicate process even if thermal conductivity and CTE of aluminum are slightly decreased due to 5vol% of Al-Si alloy addition. With 50vol% of carbon fiber, thermal conductivity and CTE of Al-CF composite were achieved to 258W/m.K and 7.0×10^{-6}/K to in-plane direction of carbon fiber. 258W/m.K is slightly higher than pure aluminum (240W/m.K), thermal conductivity was improved by carbon fiber addition. And 7.0×10^{-6}/K of CTE satisfies the value as the heat sink materials for the computer assemblies, around $4.0\sim7.0\times10^{-6}$/K to parallel direction to assembly/heat sink contact surface. Then, it seems interfacial Al_4C_3 formation, *i.e.* Al-C chemical link can effectively contribute to heat transfer and Al/CF interface strength.

5. Acknowledgments

The authors would like to thank the French "Délégation Générale pour l'Armement" and the Region Aquitaine for its financial support.

6. References

1. A. Luedtke, Thermal management materials for high-performance applications. Advanced Engineering Materials, 6(3), 142–144, 2004.
2. C. Zweben, Advances in composite materials for thermal management in electronic packaging, 1998.
3. C. Zweben, High-performance thermal management material, Advanced packaging, 2006.
4. P.M. Geffroy and J.F. Silvain, Structural and thermal properties of hot pressed Cu/C matrix composites materials used for the thermal management of high power electronic devices, Materials Science Forum, 534-536:1505-1508, 2007.

5. J.D. Mathias, P.M. Geffroy and J.F. Silvain, Architectural optimization for microelectronic packaging. Applied Thermal Engineering, 29(11-12), 2391–2395, August 2009.
6. A. Veillère, PhD thesis, "Drain thermique adaptatifs : Cuivre allié / Fibres de carbone", Université de Bordeaux 1, France, 2009.
7. J.F. Silvain, V. Denis-Lutard, P.M. Geoffroy, J.M. Heintz, Adaptive composite materials with novel architectures, Materials Science Forum, 631-632:149-154, 2010.
8. J.F. Silvain, C. Vincent, J.M. Heintz, N. Chandra, Novel processing and characterization of Cu/CNF nanocomposite for high thermal conductivity applications, Composites Science and Technology 69, 2474–2484, 2009.
9. A. Veillere, A. Sundaramurthy, J.M. Heintz, J. Douin, M. Lahaye, N. Chandra, S. Enders, J.F. Silvain, Relationship between interphase chemistry and mechanical properties at the scale of micron in Cu–Cr/CF composite, Acta Materialia 59:1445–1455, 2011.
10. D.L. Saums, Developments in selective high thermal conductivity orientation in CTE-compatible substrate and package component materials, 20th IEEE Semi-Therm Symposium, 2004.
11. D. Bouvard, "Powder Metallurgy", ISTE Publishing Company, 384 pages, 2007.
12. T. Schubert, L. Ciupinski, W. Zielinski, A. Michalski, T. Weißgarber, and B. Kieback, Interfacial characterization of Cu/diamond composites prepared by powder metallurgy for heat sink applications, Scripta Materialia, 58(4), 263–266, 2008.
13. L. Weber and R. Tavangar, On the influence of active element content on the thermal conductivity and thermal expansion of Cu-X (X = Cr, B) diamond composites, Scripta Materialia, 57(11), 988–991, 2007.
14. Y. Xia, Y.Q. Song, C.G. Lin, S. Cui, Z.Z. Fang, Effect of carbide formers on the microstructure and thermal conductivity of diamond-Cu composites for heat sink materials, Transactions of Nonferrous Metals Society of China, 19: 1161-6, 2009.
15. J.F. Silvain, J.M. Heintz, C. Vincent, Procédé de formation d'un dépôt métallique à la surface d'un substrat et applications, French patent FR1055422, 5th of July 2010.

In Situ Testing of Metal Micro-Textured Thermal Interface Materials in Telecommunications Applications

R Kempers[1,3] and S Kerslake[2]

[1]Alcatel-Lucent Ireland, Wireless Advanced Technologies, Blanchardstown Business & Technology Park, Dublin 15, Ireland
[2]Alcatel-Lucent Telecom, Swindon, United Kingdom
[3]Department of Mechanical and Manufacturing Engineering, Parsons Building, Trinity College Dublin

Email: kempersr@tcd.ie

Abstract. A metal micro-textured thermal interface material (MMT-TIM) has been developed to address the shortcomings of conventional TIMs for Remote Radio Heat (RRH) applications. The performance of the MMT-TIM was characterized in-situ by monitoring the temperatures of the dominant heat generating devices in an RRH Power Amplifier for a fixed input power. Measurements show that the use of the MMT-TIM results in significantly lower devices temperatures than achieved with the conventionally used graphite pads with a maximum temperature drop of 14.9 °C observed. The effect of power cycling on the long term performance trends is also examined.

1. Introduction

The mitigation of thermal contact resistance is essential to the performance of conduction-based electronic thermal management solutions which are employed in numerous ICT applications. Typically the most feasible strategy to reduce thermal contact resistance is to insert a thermal interface material (TIM) of higher thermal conductivity between the mating surfaces to conform to the contacting surface asperities and displace any micro and macroscopic air voids, thereby providing an improved path for heat transfer between the two components.

To work effectively, TIMs must physically conform to the mating surfaces under reasonable assembly pressures and exhibit low contact resistance with adequate bulk thermal conductivity. The bond-line thickness values are kept as low as possible to help reduce bulk thermal resistance; however the thickness must be sufficiently large to enable the TIM to comply with surface irregularities and non-planarities.

Many different TIMs are commercially available that attempt to meet these requirements in different ways. These include a range of adhesives, greases, elastomeric pads, phase-change, carbon and nano-structured materials [1-3]. However, the main weakness of many commercially available TIMs remains their relatively poor thermal performance. Many of these technologies typically consist of a low-conductivity organic phase, such as silicone grease, interspersed with higher conductivity metal (e.g. silver, copper) or ceramic particles (e.g. aluminium oxide, zinc oxide or boron nitride) to enhance the bulk thermal conductivity of the material.

Thermal greases and adhesives are widely used today and are composed of thermally conductive particles dispersed in a fluid matrix. The matrix can be a grease, an adhesive, or an elastomer.

Content from this work may be used under the terms of the Creative Commons Attribution 3.0 licence. Any further distribution of this work must maintain attribution to the author(s) and the title of the work, journal citation and DOI.
Published under licence by IOP Publishing Ltd

Greases have the lowest viscosity and thus form the thinnest interfaces at a given particle filler loading. Disadvantages of greases include the need for mechanical clamps to maintain contact pressure, creep of material away from the interface over time and/or temperature cycles and poor shelf life due to settling of particles. Adhesives offer permanent assembly, but may more readily trap voids due to their higher viscosity. Elastomers reduce the mess from material creeping away from the bond line, however mechanical clamping is again required. And assembly pressures are higher, especially at large displacements.

In most of these materials (i.e. greases, adhesives and elastomers), the thermal conductivities are relatively low—typically just one order of magnitude better than the organic matrix (100 times lower than pure metals). This is because the organic matrix forms the continuous phase within the materials whereas the thermally conductive particles are discrete and heat transfer is limited by multiple point-to-point contacts between adjacent particles. The limited number of contact points limits the thermal conductivity. For example, despite using extremely high conductivity filler materials, such as silver ($k \approx 420$ W/m·K), the effective thermal conductivity of the best commercially available TIMs is on the order of 3 to 10 W/m·K, which is considerably lower than the thermal conductivities of typical mating components. Significant work has been done in the past to maximize the particle loading and the total particle-to-particle contact area, however additional improvements have been limited due to the maturity of the systems. In addition, dispensing and flow of the particle-matrix composite can result in voids being trapped within the bond.

Pure, soft metal thermal interface materials such as indium are also used. Although the thermal conductivities of these materials are significantly higher than those of metal-organic composites, the ability of these materials to comply with non-flat and high rough surfaces is limited. For some materials, especially indium, fretting corrosion may become a reliability problem, especially where numerous thermal cycles are experienced. Also the relatively low melting point of indium would limit applicability and assembly options

Finally, graphite pads are used extensively in Alcatel-Lucent hardware, particularly in radio frequency power amplifier (RF PA) assemblies off Remote Radio Heads (RRH). These TIMs consist of graphite flakes that are exfoliated by thermally vaporizing an intercalant ion inserted between its layers, generating an internal pressure that causes the intercalated graphite to expand in the direction perpendicular to the layers. The expanded graphite flakes are then mechanically consolidated together with a matrix into a cohesive sheet of flexible graphitic material [4, 5]. The results is a flexible sheet that has a reasonable through-plane thermal conductivity (3-5 W/m·K) [4]. However, due to the highly anisotropic nature of the structure, the in-plane conductivity is significantly higher, ranging from 140 to 500 W/m·K according to Smalc et al. [5]. As a result, they are a viable option for heat spreading applications [5]. However, despite being a flexible sheet and possessing a relatively high thermal conductivity, their thermal performance is limited by the lack of mechanical compliance as they are compressed between surfaces. Even at compressive pressures of 1 MPa, graphite pads exhibit only approximately 5% bulk strain [4]; compliance is more limited when high aspect ratio surface features are present. Furthermore, graphite pads typically lead to an increased electrical contact resistance when assembled between components. For example, when used in certain RF PA applications, poor grounding through these joint results in lowered PA efficiency.

Generally, the bulk thermal conductivities of the thermal interface materials currently used are very low compared to the thermal conductivities of the components being assembled. As a result, the thermal interface between the two components remains the dominant resistance to heat transfer. In many high thermal energy dissipating systems, the TIMs can account for up to 50% of the available thermal budget of the package [6]. With the implementation of high-performance liquid and phase-change cooling strategies, this percentage becomes even greater and thermal interfaces become a critical "bottleneck" to effective cooling. If the thermal management of an electronic device is inadequate, unacceptable temperature levels may be reached which can adversely affect device performance, reliability and lifespan [7]. These thermal issues have spawned a global effort towards the development of exotic TIMs with complex formulations [1-3, 6-9].

2. Metal Micro-Textured Thermal Interface Materials

To address the performance limitations of conventional thermal interface materials, particularly graphite pads used in RRH PA applications, a metal micro-textured thermal interface material (MMT-TIM) has been developed [10, 11]. The concept is illustrated in Fig. 1. These materials consist an array of small-scale (≈0.1 mm to 1 mm) raised features. When this structure is compressed between two mating surfaces, the features plastically deform and conform to the contacting bodies as illustrated in Fig. 1.

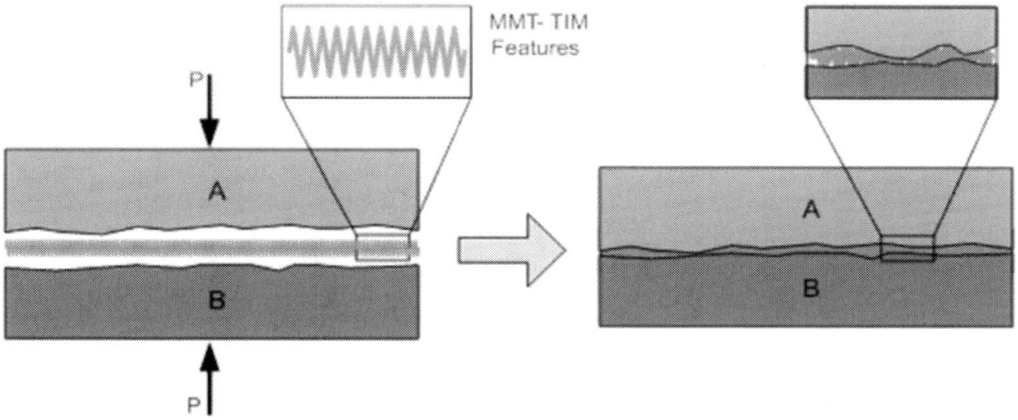

Fig. 1: Metal Micro-Textured Thermal Interface Material Concept

Plastic deformation of the features into the asperities of the contacting surfaces ensures intimate contact with the contacting bodies. The result is an array of conformable, yet continuous solid metal features of low thermal resistance (or high effective thermal conductivity) that are in intimate contact with the mating surfaces due to the plastic deformation of the raised features.

Given the nature of this concept, there is clearly a large design space within which the materials and geometries of the MMT-TIMs could be optimized to minimize thermal resistance and compressive pressure while maximizing mechanical compliance and effective thermal conductivity. Initial investigations into MMT-TIMs explored the effect of size and shape of the raised features on their thermal-mechanical response during compression [10, 11].

Complete thermal-mechanical models which simultaneously characterize the complex feature deformation and effective thermal resistance of MMT-TIMs have been developed to serve as design tools for these materials [12, 13].

The present embodiment of this technology is shown in Fig. 2. In this embodiment, the TIM geometry consists of continuously formed truncated domes approximately 0.5 mm in size extending in both directions from a mid-plane however. This geometry is desirable because it readily allows for buckling and compression when assembled between two components. The open "holes" in the tops and bottoms of the raised features reduce the stiffness of these structures and allow the compression and flow of metal as the entire structure is compressed. These MMT-TIMs were fabricated from flat, 25 μm tin foil using a proprietary micro-stamping operation.

The compressed geometries this MMT-TIM design is shown in Fig. 3. Here the truncated domes have undergone large-scale compressive deformation including bending and buckling modes that results in flattened structures that form a continues path between the assembled bodies including areas that are in intimate contact with the compressing surfaces. This creates highly effective thermal and electrical paths between the mating components.

The objective of the present study is to characterize the thermal performance of a remote radio head (RRH) using the MMT-TIM shown in Fig. 2 installed in the critical interface between the power amplifier (PA) board and the main natural convection heat sink.

(a) (b)

Fig. 2: a) Photograph and, b) SEM image of a MMT-TIM prior to compression

(a) (b)

Fig. 3: a) Photograph and, b) SEM image of a MMT-TIM subsequent to compression

2. Experiment Characterization

A drawing of the RRH in-situ test setup employed in the present study is shown in Fig. 4. The upper part of the RRH consists of a PA heat sink on which the PA board is mounted. Here The PA transistors (shown in red) are the dominant heat-generating components on the PA board (blue). This PA is configured of two main and peak pairs (Doherty configuration). The PA transistors are soldered to a flush mounted copper coin embedded and flush mounted with the bottom of the PCB. These coins serve primary RF ground purpose but do afford some heat spreading. The TIM is between the PA board and the heat sink and an RF shield is mounted over the PA board. Forty M3 screws torqued to 1.0 N·m bolt the RF shield to the heat sink, sandwiching the PCB tightly against the heat sink. Standard screw fasteners and torque patterns were maintained.

The aluminum heat sink is approximately 500 mm tall and 270 mm wide and contains several heat pipes embedded in the base to improve heat spreading. The fins are optimized for natural convection.

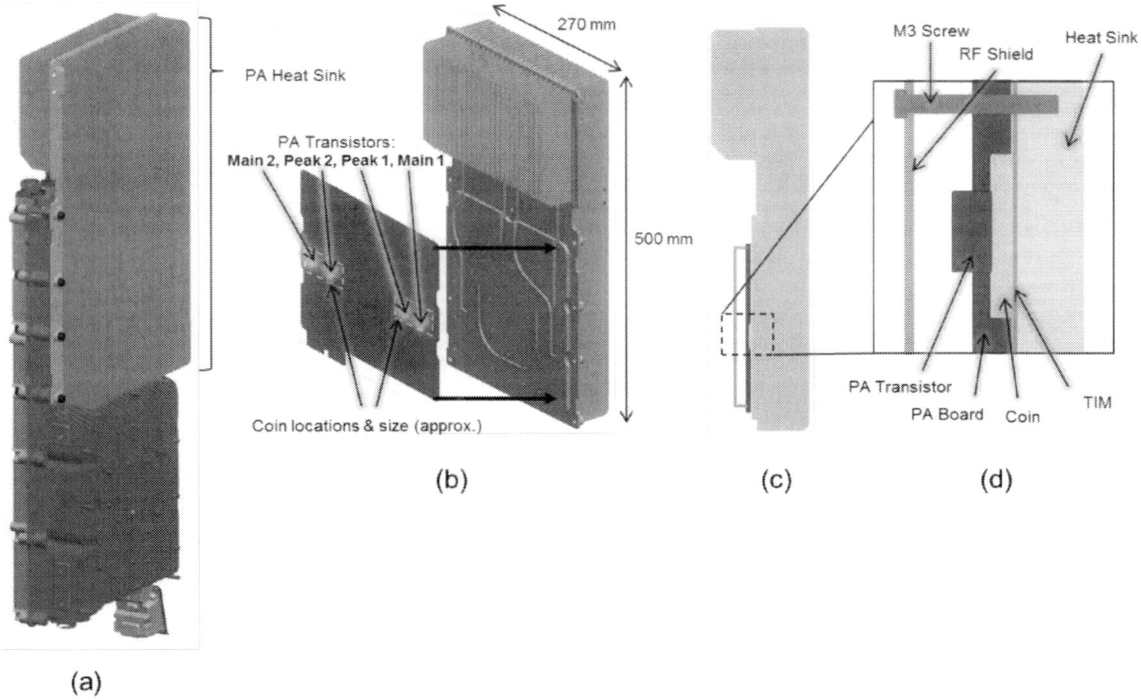

Fig. 4: Drawing of the RRH PA assembly used in the present work a) overall RRH with PA heat sink indicated, b) PA board (transistors in red) and PA contact interface with heat sink, c) side view of assembly showing d) cross-section of assembled component stack-up

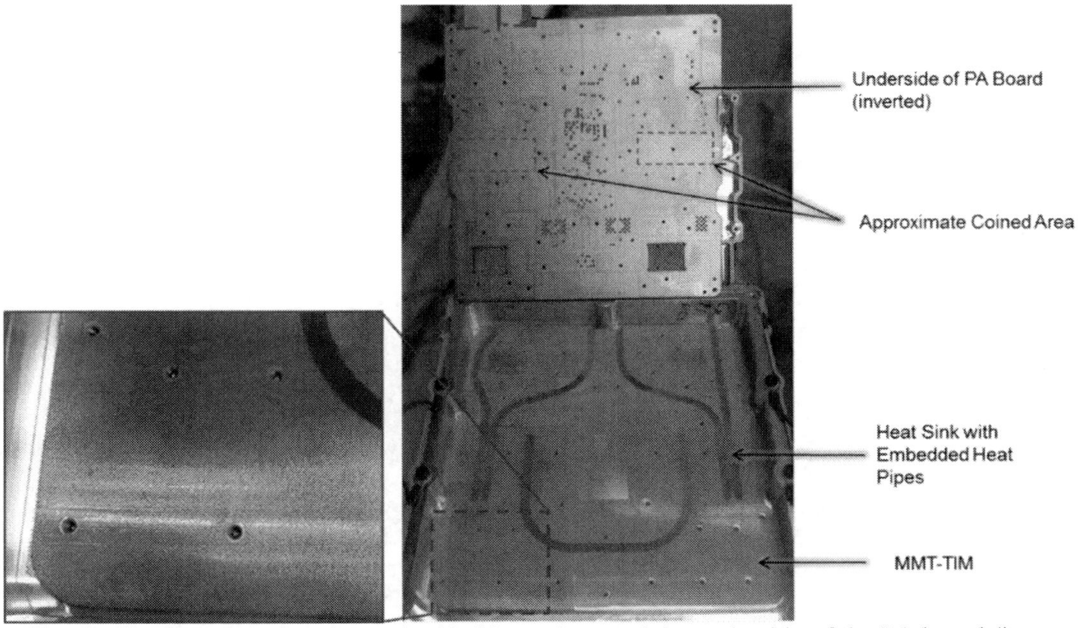

Fig. 5: The heat sink showing embedded heat pipes and the underside of the PA board (inverted relative to the heat sink) with a close up of the MMT-TIM prior to installation

A photograph of the contact area of the heat sink and the underside of the PA board (inverted relative to the heat sink) is shown in Fig. 5. Here the approximate size and shape of the flush-mounted coins in the PA board are indicated. Also visible are the embedded heat pipes in the heat sink base and the MMT-TIM prior to installation.

The outer case of each PA transistor was instrumented with a T-type thermocouple using a high-temperature epoxy. An additional thermocouple was used to measure the ambient air temperature. The required heat load was generated by the transistors themselves by wiring them in a bias mode. Four TTi CPX4000DP DC power supplies provided the operating and bias voltages. The power dissipated by each device was controlled by varying the bias voltage of the circuit and measured by computing the product of the current draw and operating voltage. A fixed power of 81.5 W on the two main devices and 23.6 W on the two peak devices was used. This corresponds to the maximum thermal output of these devices in the field. Ambient temperature was approximately 20 °C.

3. Results & Discussion

Four different interface conditions were tested and compared: i) the PA assembled with no TIM, ii) using a flat, 25 tin foil, iii) the conventionally specified graphite pad , iv) the tin MMT-TIM presented in the previous section. The thermal resistance of this interface (and indeed the system) is difficult to characterize due to the presence of multiple heat sources with thermal communication between them. Consequently, results here are presented using the heat source temperatures rather than thermal resistance.

The steady-state temperature of each device above ambient is shown in Fig. 6 for each of the four interface conditions. As could be expected, using no TIM resulted in the highest observed device temperatures with outer case temperature of Main 1 at 88.5 °C. The conventionally used graphite pad lowered this temperature somewhat while the MMT-TIM resulted in significantly lower device temperatures across the board. For comparison, an un-textured flat Sn foil was also tested as thermal interface material however, only minimal temperature reductions were observed.

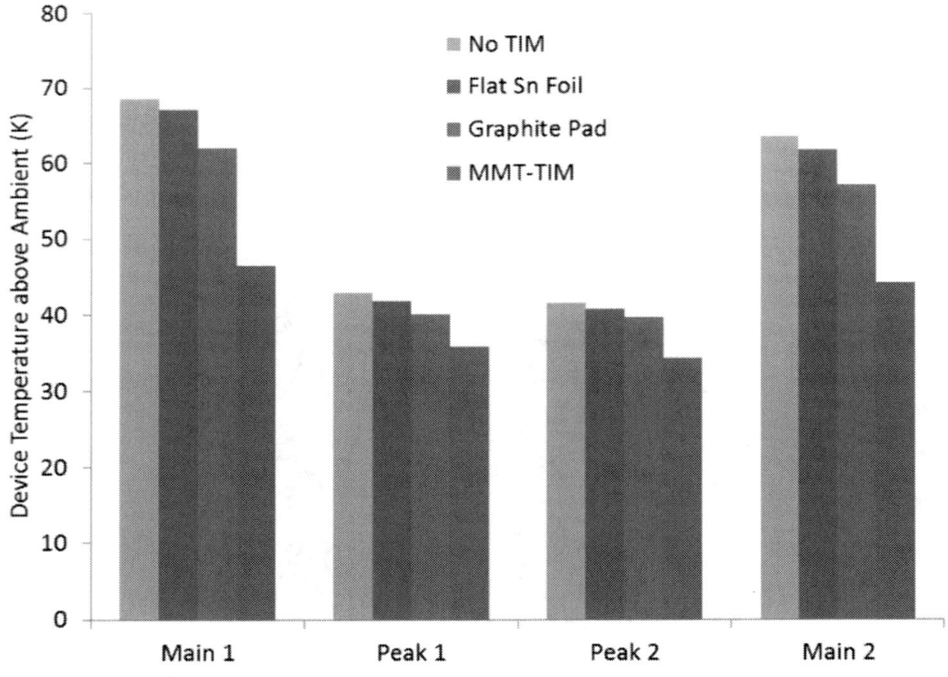

Fig. 6: Device operating temperatures for different TIMs

These tests indicate the interface resistance between the PA board and the heat sink is of crucial importance to the performance of the RRH. For the example of transistor Main 1, the temperature difference between using no TIM and a MMT-TIM is 22 °C—nearly one third of the 68.5 °C temperature difference between of Main 1 and ambient with no TIM.

A more detailed comparison between the conventionally specified graphite pad and the MMT-TIM is shown in Fig. 7. The results indicate clearly that use of the MMT-TIM in place of the graphite pad resulted in a significant drop in terms of absolute device temperatures: The "Main 1" device operating temperature dropped from 81.1 °C to 66.2 °C; a 14.9 °C temperature reduction. The "Main 2" device saw a temperature reduction of 12.2 °C. The temperature reductions in the peak devices were commensurate with their somewhat lower power dissipations: approximately 4.3 and 3.3 °C in "Peak 1" and "Peak 2", respectively.

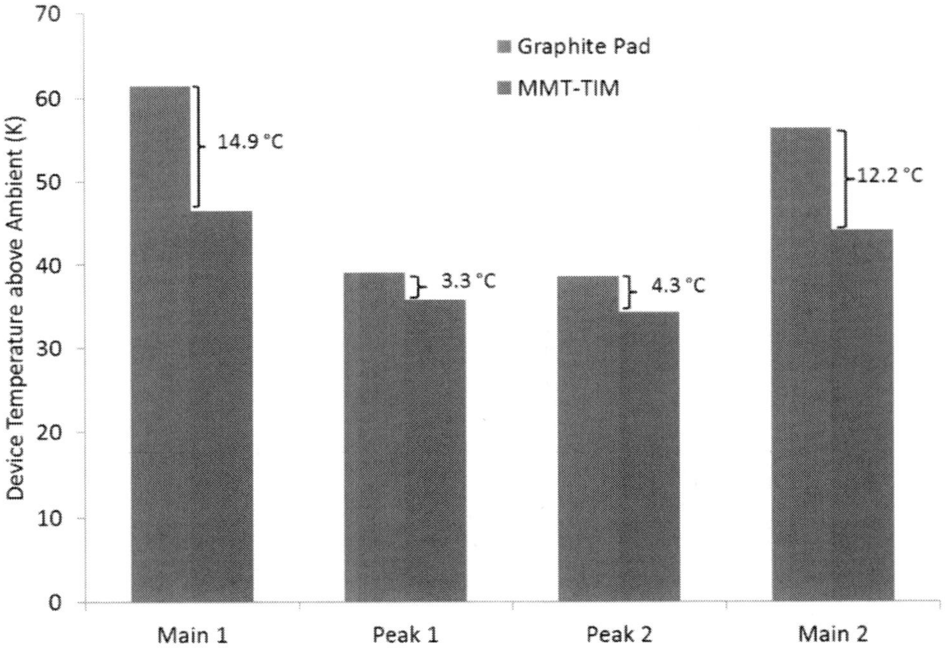

Fig. 7: Detailed comparison between conventional graphite pad and MMT-TIM in terms of resultant device temperatures

The results further illustrate the thermal benefit from using a MMT-TIM in this assembly compared to a conventional graphite pad. This level of temperature reduction has significant implications on overall system design. For example, this level of temperature reduction can have significant positive implications upon device and system reliability. Alternatively, the decreased thermal resistance of this interface can be offset by employing a smaller and lighter heat sink in order to minimize the overall size and weight of the RRH.

The results presented so far characterize the MMT-TIM at shortly after installation at beginning of life. Another crucial aspect is how the TIM performs over time as the assembled interfaces heat up and cool down, expanding and contracting over time. Power cycling tests are recommended for assessing TIM reliability since these tests resemble most closely the actual use of the material [14].

An initial investigation to asses the reliability of the MMT-TIMs in the present application was performed using this setup by alternately powering the four PA transistors between zero and full (rated) power. At each setting the system was allowed to come to steady-state and device temperatures were monitored. The device temperatures at full power (steady-state) are shown in Fig. 8 over 100 power cycles (which took approximately one week). Here no significant changes in device temperature occurred.

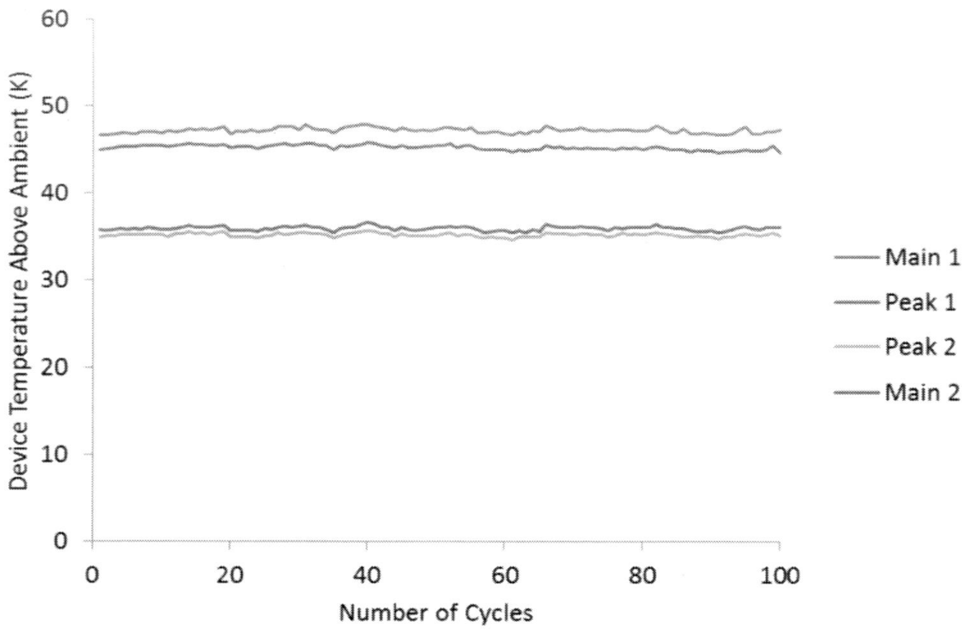

Fig. 8: PA transistor steady-state temperatures during power cycling

4. Conclusions

In-situ tests of a metal micro-textured textured thermal interface material (MMT-TIM) in an RRH PA application have demonstrated a significant reduction in system thermal resistance and therefore device operating temperatures. These temperature reductions will allow for improved system reliability or could be traded off against a smaller heat sink to decrease overall system size and weight.

System power cycling tests were employed to make an initial assessment into the reliability of MMT-TIMs in this application and it was shown that the device temperature remained constant over 100 cycles from zero to full power.

This experimental data will be combined with system-level numerical simulations in order to establish the fractional contribution played by MMT-TIM to the total thermal resistance of the RRH and evaluate possible reductions in heat sink size and weight as a result of lowering the interface resistance.

References

1. Liu, J., Michel, B., Rencz, M., Tantolin, C, Sarno, C., Miessner, R., Schuett, K-V., Tang, X., Demoustier, S. & Ziaei, A., 2008, "Recent Progress of Thermal Interface Material Research—An Overview", Proceedings of the 14th Workshop on Thermal Issues in ICs and Systems (THERMINIC), Rome, Italy, September 24-26, 2008

2. Chung, D.D.L., 2012, "Carbon materials for structural self-sensing, electromagnetic shielding and thermal interfacing", Carbon, 50, pp. 3342 – 3353.

3. McNamara, A. J., Joshi, Y., & Zhang, Z.M., 2012, "Characterization of nanostructured thermal interface materials - A review", International Journal of Thermal Sciences, 62, pp. 2-11, doi:10.1016/j.ijthermalsci.2011.10.014

4. Smalc, M., Norley, J., Reynolds, R.A., Pachuta, R. & Krassowski, D.W., 2013, "Advanced Thermal Interface Materials Using Natural Graphite". Proceedings of the International Electronic Packaging Technical Conference and Exhibition (InterPACK), Maui, USA, July 6-11, 2003.

5. Smalc, M., Shives, G., Chen, G., Guggari, S., Norley, J. & Reynolds, R.A., 2005 "Thermal Performance of Natural Graphite Heat Spreaders". Proceedings of the International Electronic Packaging Technical Conference and Exhibition (InterPACK), San Francisco, USA, July 17-22, 2005.

6. Smith B., Brunschwiler, T., & Michel, B. 2008, "Comparison of transient and static test methods for chip-to-sink thermal interface characterization". Microelectron. J., doi:10.1016/j.mejo.2008.06.079

7. Linderman, R., Brunschwiler, T., Smith B. & Michel, B., 2007, "High performance thermal interface technology overview". Proceedings of the 13th Workshop on Thermal Issues in ICs and Systems (THERMINIC), pp. 129-134.

8. Savija, I., Culham, J.R., Yovanovich, M.M. & Marotta, E.E., 2003, "Review of thermal conductance models for joints incorporating enhancement materials" Journal of Thermophysics and Heat Transfer, 17, pp. 43-52.

9. Nguyen, J. J., Bougher, T. L., Pour Shahid Saeed Abadi, P., Sharma, A., Graham, S. & Cola, B. A., 2013, "Postgrowth Microwave Treatment to Align Carbon Nanotubes" Journal of Micro and Nano-Manufacturing. 1(1), 014501; doi: 10.1115/1.4023162

10. Kempers R., Robinson, A.J. & Lyons, A 2009, "Characterization of Metal Micro-Textured Thermal Interface Materials" Therminic 2009, Leuven, Belgium

11. Kempers, R., Robinson, A. J. & Lyons, A "Characterization of Thermal Contact Resistance in Metal Micro-Textured Thermal Interface Materials using Electrical Contact Resistance Measurements" Defect and Diffusion Forum Vols. 297-301 (2010) pp. 1190-1198. doi:10.4028/www.scientific.net/DDF.297-301.1190

12. Kempers, R., Frizzell, R., Lyons, A. & Robinson, A.J. 2009, "Development of a Metal Micro-Textured Thermal Interface Material" ASME InterPACK 2009, San Francisco.

13. Kempers, R., Lyons, A. & Robinson, A. J. 2014, "Modelling and Experimental Characterization of Metal Micro-Textured Thermal Interface Materials" Journal of Heat Transfer, 136, pp. 011301-1-011301-11DOI: 10.1115/1.4024737

14. Due, J. & Robinson, A.J., 2013, "Reliability of thermal interface materials: A review" Applied Thermal Engineering, **50**, pp.455-463.

Micro-structured rough surfaces by laser etching for heat transfer enhancement on flush mounted heat sinks

L Ventola[1], L Scaltrito[1,2], S Ferrero[1,2], G Maccioni[2], E Chiavazzo[1], P Asinari[1]

[1] multi-Scale ModeLing Laboratory (SMaLL), Energy Department, Politecnico di Torino, Corso Duca degli Abruzzi 24, 10129, Turin, Italy.
[2] Microla Optoelectronics S.r.l., Localita Baraggino 10034 Chivasso, Turin, Italy.

E-mail: `pietro.asinari@polito.it`

Abstract. The aim of this work is to improve heat transfer performances of flush mounted heat sinks used in electronic cooling. To do this we patterned 1.23 cm^2 heat sinks surfaces by micro-structured roughnesses built by laser etching manufacturing technique, and experimentally measured the convective heat transfer enhancements due to different patterns. Each roughness differs from the others with regards to the number and the size of the micro-fins (e.g. the micro-fin length ranges from 200 to 1100 μm). Experimental tests were carried out in forced air cooling regime. In particular fully turbulent flows (heating edge based Reynolds number ranging from 3000 to 17000) were explored. Convective heat transfer coefficient of the best micro-structured heat sink is found to be roughly two times compared to the smooth heat sinks one. In addition, surface area roughly doubles with regard to smooth heat sinks, due to the presence of micro-fins. Consequently, patterned heat sinks thermal transmittance [W/K] is found to be roughly four times the smooth heat sinks one. We hope this work may open the way for huge boost in the technology of electronic cooling by innovative manufacturing techniques.

1. Introduction

The heat fluxes electronic devices require to be dissipated continuously increase. The need of solutions able to deal with these high heat fluxes has led to explore a huge variety of methods and technologies. Liquid and two-phase based solutions have already been developed to properly cool down devices with high power density (e.g. high performance supercomputers). Nevertheless, forced air cooling techniques will remain the most suitable solutions in the foreseeable future for a large class of electronic devices, i.e. notebook and portable devices to name a few, because of their cheapness and reliability.

In the contest of air forced cooling regime, a variety of convective heat transfer enhancement techniques have been proposed in literature, such as plane fins [1, 2], pin fins [3, 4, 5], dimpled surfaces [6, 7, 8], surfaces with arrays of protrusions [9, 10], metal foams [11] and artificial surface roughness [12]. The term *artificial roughness* means any regular pattern build on the surface and purposely designed to enhance convective heat transfer. For instance, in such a category, we may include ribs [13, 14, 15] and, more recently, (shark-skin-like) scale roughened surfaces [16, 17]. The resulting heat transfer enhancement of the scale roughened surface is surprisingly

Content from this work may be used under the terms of the Creative Commons Attribution 3.0 licence. Any further distribution of this work must maintain attribution to the author(s) and the title of the work, journal citation and DOI.
Published under licence by IOP Publishing Ltd

Table 1. Optimized laser parameters for copper processing.

Power $[W]$	frequency $[kHz]$	scan speed $[mm/s]$
7	10	100

good compared to rib roughened and dimpled surfaces [18]. The latter result proves that there is still room for improving the optimal design of artificial surface roughness. On the other hand innovative manufacturing techniques, e.g. additive manufacturing, are able to control process roughness of built surfaces. Recent works show the possibility to exploit the latter feature to produce flush mounted and finned heat sinks whose rough surfaces are able to enhance convective heat transfer[19].

In this work we focus on the possibility to enhance convective heat transfer of 1.23 cm^2 heat sinks surfaces patterned by artificial micro-structured roughnesses.

2. Laser etching manufacturing technique and samples characterization

In this paper, a novel technique, using nanoseconds DPSS laser direct writing on Copper substrate is studied. This class of laser is a flexible and effective tool characterized by low operating cost [20]. Laser radiation can be focused to micro-sized spots. A galvanometric head is used to control the laser beam position, thus a wide variety of low cost geometries can be obtained [21, 22]. Limited heat transport to minimize thermal degradation of the material outside the production zone is achieved by employing nanosecond pulse width. Short pulse widths, under the right conditions, can promote the spallation process that minimizes the volume of molten material produced and the subsequent no loss of geometry control because of its displacement. Moreover, the short pulse duration leads to high peak power, and therefore a very small heat-affected zone (HAZ) can be obtained.[23] Different lasers, such as picoseconds and femtoseconds allow a better control of the process, but the use of such classes of laser would increase dramatically the production costs, for instance the picoseconds and the femtoseconds laser can be up to five and ten times more cost intensive, respectively [24, 25]. On the contrary, nanosecond laser used in this study can be employed for a large scale of heat sinks production due to its low production costs. Besides, the nanoseconds laser are maintenance free at least up to three years per 24hr/7days working time.

We used laser etching manufacturing technique to build artificial micro-structured roughnesses on four copper samples. A DPPS laser system with emission at 532 nm has been employed. The laser parameters used for the machining are shown in Table1. Each sample is a cuboid of dimensions 11.1 mm x 11.1 mm x 5 mm, whom 11.1 mm x 11.1 mm surface has been manufactured in order to obtain a specific artificial roughness. Each roughness is made of micro-fins, namely cones or truncated cones, of uniform size arranged in staggered layout. Pictures of the samples are shown in Fig.1. Morphology of each samples has been characterized by a 3D optical scanner ATOS Compact Scan 2M (GOM GmbH). Field Emission Scanning Electron Microscope (FESEM) images of micro-fins built on each sample are shown in Fig.2. Geometrical parameters of tested samples are summarized in Table 2.

3. Experimental procedure

To measure the convective heat transfer coefficient a sensor purposely developed in a previous work[26] was used. This sensor is based on the key idea of using a thermal guard. A heater in the bottom of the sample generates a controlled heat flux. The sample is surrounded by a temperature controlled "guard" made of copper. This guard is kept at the same temperature of the sample, to force the heat to flows into the sample only. This ensures that the heat flux

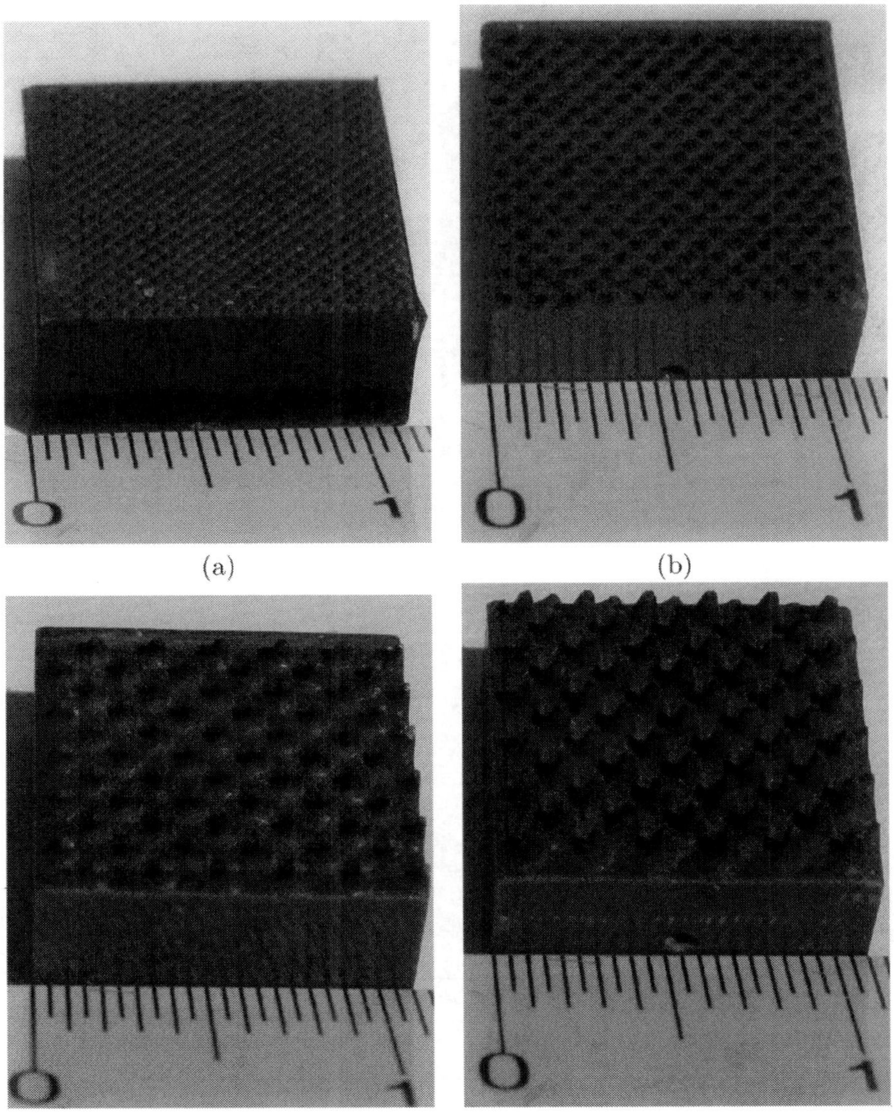

(a) (b) (c) (d)

Figure 1. Picture of the four tested samples.

through the sample is essentially one-dimensional, going from the heater to the air flow passing through the sample itself. Temperature are measured by thermocouples in the sample and in the upstream and downstream (referring to fluid flow) part of the guard. The upstream and downstream part of the guard experience different values of convective heat transfer coefficient, due to leading edge effect. Hence, the guard will be slightly non isothermal, and some (undesired) temperature gradient between the sample and the guard (of the order of 0.2 K) could arise. For this reason a thermal insulating shield made of Teflon is placed between sample and guard. Its function is to minimize parasite conductive heat flux Q_k due to residual temperature gradients between sample and guard. In Fig 3 a view of the sensor is shown.

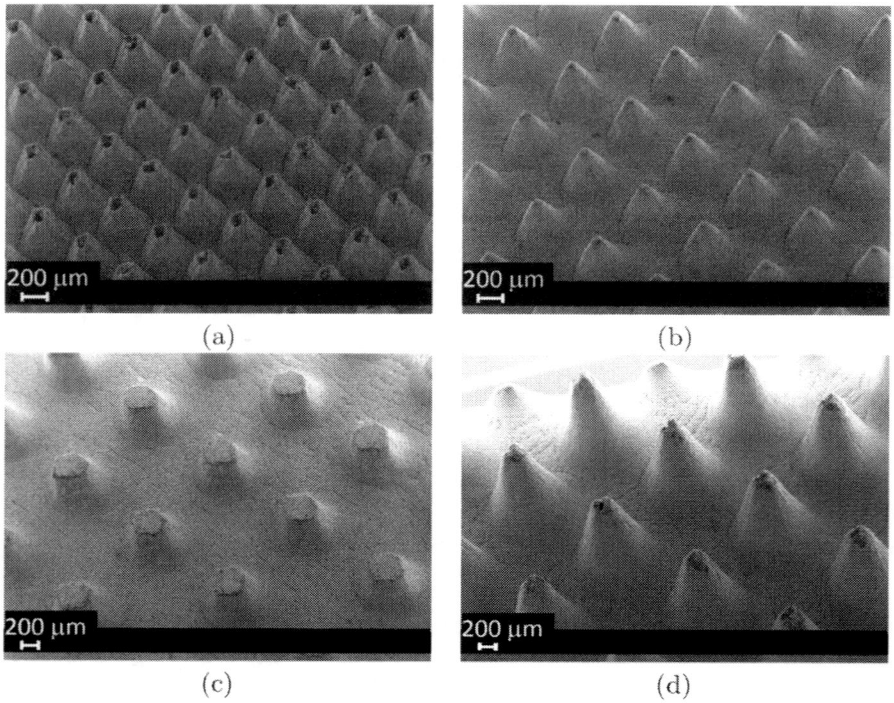

Figure 2. Morphological characterization of the four tested samples by Field Emission Scanning Electron Microscope (FESEM).

Table 2. Geometrical parameter of tested samples (see Fig.1): H, d_b, d_u, p and N are the average height, base diameter, upper diameter, distance (pinch) and number of micro-fins built on each sample, respectively (d_u is equal to 0 for cone-shaped fins and higher than 0 for truncated-cone-shaped fins, see Fig. 2). A is the sample surface area and $A_s = 123\ mm^2$ is the smooth sample surface area.

Sample	H $[mm]$	d_b $[mm]$	d_u $[mm]$	p $[mm]$	N	A $[mm^2]$	A/A_s
(a)	0.20	0.48	0.00	0.49	465	154	1.25
(b)	0.35	0.52	0.00	0.65	253	185	1.50
(c)	0.65	0.85	0.35	1.35	72	177	1.44
(d)	1.10	1.00	0.00	1.35	61	240	1.95

The main portion of the power P provided by the heater to the sample is transferred into the air flow by convective heat transfer, while a minor, but far from negligible, portion of P is transferred to the wind channel walls through radiative heat transfer. Denoting by Q_c and Q_r the convective heat flux and the radiative heat flux respectively, we can calculate them by the following:

$$Q_r = \sigma_B \epsilon A(T_s^4 - T_w^4), \tag{1}$$

$$Q_k = k[T_s - (T_{g1} + T_{g2})/2], \tag{2}$$

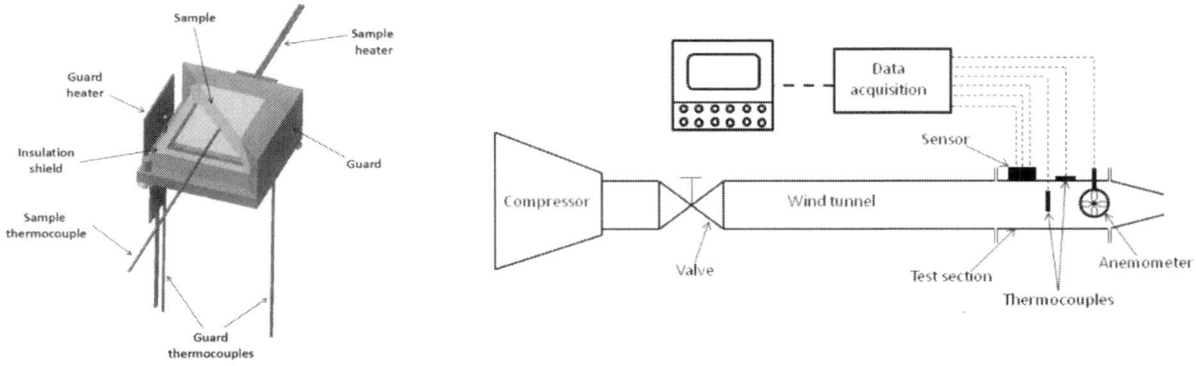

(a) (b)

Figure 3. Isometric view of the convective heat transfer sensor (a), schematic diagram of the experimental facility (b) (see Ref. [26]).

$$Q_c = P - Q_r - Q_k = P - \sigma_B \epsilon A(T_s^4 - T_w^4) - k[T_s - (T_{g1} + T_{g2})/2], \tag{3}$$

where $\sigma_B = 5.67 \times 10^{-8}\ W/m^2/K^4$ is the Stefan-Boltzmann constant, $\epsilon = 0.23$ is the emissivity of the sample surface, T_s is the sample temperature measured by the thermocouple inserted in the center of the sample, T_w is the temperature of the channel wall, $k = 0.01\ W/K$ is the sample-to-guard coupling conductive thermal transmittance [26], T_{g1} and T_{g2} are the temperatures measured by the (upstream and downstream) thermocouples installed into the thermal guard and A is the sample surface (see Table2). It is worth to note that Q_r and Q_k are less than 10% and 1% of P respectively, among all the tests. For each sample thermal transmittance Tr $[W/K]$ and convective heat transfer h $[W/m/K]$ can be calculated as:

$$Tr = \frac{Q_c}{T_s - T_a} = \frac{P - \sigma_B \epsilon A(T_s^4 - T_w^4) - k[T_s - (T_{g1} + T_{g2})/2]}{T_s - T_a}, \tag{4}$$

$$h = \frac{Q_c}{A(T_s - T_a)} = \frac{P - \sigma_B \epsilon A(T_s^4 - T_w^4) - k[T_s - (T_{g1} + T_{g2})/2]}{A(T_s - T_a)}, \tag{5}$$

where T_a is the air temperature.

The flow loop of the experimental system used in this work is the same used in previous works [19][26] and it is schematically shown in Fig. 3. An horizontal rectangular wind channel with cross section of $228\,mm \times 158\ mm$ (hydraulic diameter 187 mm) and an entrance length of $5\,m$ (corresponding roughly to 26 hydraulic diameters) is used. The air is blown by a Savio s.r.l. centrifugal fan type SFL 25-A, with a throttling valve for regulating the mass flow rate. At the end of the channel, downstream from the test section, a vane anemometer Testo 450 by Testo AG (sensitivity $\pm 0.1\ m/s$) was used for measuring the axial velocity. The air temperature is measured at the same location where the anemometer is installed. Similarly, the channel wall temperature is measured by a thermocouple installed on the inner surface of the channel. Chromel-Alumel (type K) thermocouples were used for all temperature measurements. Two PS3003 variable power suppliers are used to feed both the sample and the guard heater. Finally, a six-digits, electronic multimeter (Agilent 34401A) is used to measure the voltage to the sample heater.

Journal of Physics: Conference Series **525** (2014) 012017 doi:10.1088/1742-6596/525/1/012017

4. Results

Convective heat transfer coefficient h $[W/m^2/K]$ and thermal transmittance Tr $[W/K]$ of each sample have been measured for different velocities, in order to investigate the heat transfer phenomenon in a wide range of fluid flows. In addition h and Tr have been measured also for a smooth (without micro-fins) surface sample, obtained by traditional milling.

To each value of mean velocity u inside the wind tunnel we associated the dimensionless Reynolds number Re_L, calculated according to the formula:

$$Re_L = uL/\nu \tag{6}$$

with $\nu = 1.493\ 10^{-5}\ m^2/s$ being kinematic viscosity of air and $L = 20\ mm$ being the characteristic heated edge, namely the guard. Conversely, the average convective heat transfer coefficient h was expressed in terms of dimensionless number $Nu/Pr^{\frac{1}{3}}$, whereas:

$$Nu = hL/k \tag{7}$$

is the dimensionless Nusselt number and

$$Pr = \nu/\alpha \tag{8}$$

is the dimensionless Prandtl number with $\alpha = 2.224\ 10^{-5}\ m^2/s$ and $k = 2.83\ 10^{-2}\ W/m/K$ being the thermal diffusivity and thermal conductivity of air respectively.

It is worth to note that temperature gradient along the micro-fins length is negligible among all the samples, i.e. fin efficiency of micro-fins could be considered 100% among all the samples. To prove the latter claim we calculated the dimensionless fin parameter mL for each conical fin, where L is the fin height H defined in table 2, and m is defined as:

$$m = (\frac{4h_{max}}{k_{Cu}d_b})^{0.5} \tag{9}$$

where $k_{Cu} = 390\ W/m/K$ is the thermal conductivity of copper, and h_{max} is the maximum convective heat transfer coefficient experimentally measured. the mL parameter is less then 0.05 for all the fin configurations, consequently fin efficiency of micro-fins could be considered 100% among all the samples according [27].

In Fig.4 the experimental results in term of thermal transmittance Tr $[W/K]$ as a function of Reynolds number Re_L are shown. It could be noticed in Fig.4 that micro-structured sample (d) achieve in the high Reynolds number regime a thermal transmittance which is roughly four times the reference smooth sample one. In addition we note that, in the air velocity range under consideration in this study, samples with higher micro-fin height H (see table 2) are characterized by higher thermal transmittance than the other samples. This is due to two main reasons: The first is that height of the fins strongly affects the fluid dynamics interactions between the flow field and the fins themselves, and consequently the convective heat transfer phenomena. The second is that, due to peculiar geometrical dimensions of the samples under consideration in this study, higher fins belong to samples with higher surface area, except for sample (c) which has higher fins but a slightly lower surface area than sample (b) (see table 2). Obviously samples with higher surface area available for convective heat transfer are supposed to have higher thermal transmittance Tr than the other samples.

In order to split the contribution of the increase in surface area and of fluid dynamic interactions on heat transfer enhancement, it is worth to analyze the dimensionless Nusselt number Nu_L. This quantity allow us to remove the contribution of the increase in surface area on heat transfer enhancement, and consequently to study the portion of heat transfer enhancement due to fluid dynamics interactions only. In Fig.5 the experimental results in term

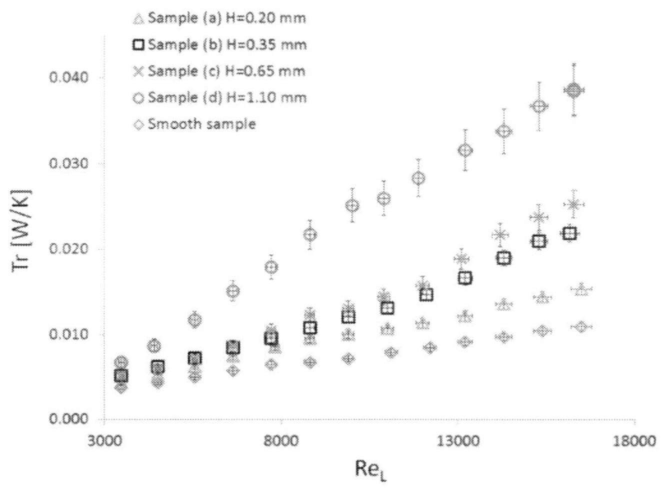

Figure 4. Experimental results for micro-structured artificial rough samples and reference smooth sample, Tr versus Re_L.

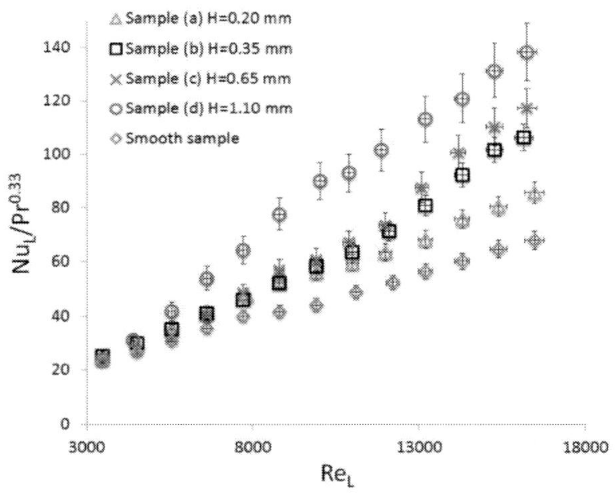

Figure 5. Experimental results for micro-structured artificial rough samples and reference smooth sample, Nu_L versus Re_L.

of Nusselt number Nu_L as a function of Reynolds number Re_L are shown. It could be noticed that micro-structured sample (d) achieve in the high Reynolds number regime a Nu_L value which is roughly two times the reference smooth sample one. In addition we note that, in the air velocity range under consideration in this study, samples with higher micro-fin height H (see table 2) are characterized by higher value of Nu_L than the other samples. The reason behind this phenomenon could be found analyzing the near-wall velocity profile.

The pressure gradient inside a pipe can be expressed as a function of the dimensionless friction factor f, namely $\nabla p = (f/D)\rho v^2/2$. In case of smooth pipe turbulent flow, the friction factor can expressed by the phenomenological correlation proposed by Blasius [28],

$f_B = f_B(Re_D) = 0.3164\, Re_D^{-1/4}$, where Re_D is the Reynolds number based on the pipe diameter and on the average flow speed. The shear stress at the wall of a pipe τ is related to the pressure gradient inside the pipe ∇p by the linear relation $\tau = \nabla p\, D/4$, where D is the pipe diameter. Combining the previous relations yields $\tau = (f/8)\,\rho\, v^2/2$. The shear stress can be used to define the friction velocity v_*, namely $v_* = \sqrt{\tau/\rho} = v\,\sqrt{f/8}$, and, consequently, the friction length $y_0 = \nu/v_*$, where ν is the kinematic viscosity of the fluid. The friction length y_0 is useful to define a dimensionless distance from the wall, $y^+ = y/y_0$ (with y being the distance from the wall) and to formulate the following logarithmic law due to Von Kármán, $v^+ = v/v_* = \kappa^{-1} \ln y^+ + A$, where κ is the Von Kármán's constant and A is a constant for smooth walls. The friction length can be expressed as

$$y_0 = \frac{D}{Re_D\,\sqrt{f_B/8}}, \tag{10}$$

where

$$Re_D = uD/\nu \tag{11}$$

is the Reynolds number based on the wind channel hydraulic diameter $D = 0.187\ m$. In our experiments Re_D and y_0 lay in the ranges $[3\ 10^4 \div 1.6\ 10^5]$ and $[0.026 \div 0.11\ mm]$, respectively. Consequently, even in case of the highest Reynolds number $Re_D = 1.6\ 10^5$, i.e. thinnest boundary layer ($y_0 = 0.026\ mm$), we find $H/y_0 < 45$ among all the tested samples. That means micro-fins of all the tested samples lies in the buffer layer region or in the early log-law region [29], depending on the sample tested, i.e. fin height H, and on the flow field conditions, i.e. value of y_0. Hence fins lies in a portion of the boundary layer in which the velocity is still very sensitive to the distance from the wall, namely the fin root surface. Consequently we can argue higher fins experience an higher heat transfer because they are flushed by a stronger flow field than the shorter ones.

On the other hand, it could be noticed in Fig. 5 that in the low Reynolds number region there is a superposition of experimental points belonging to different samples. Also in this case we can provide an explanation of the phenomenon analyzing the nature of the near wall flow field. Data in Fig. 5 show different samples experience the same curve $Nu_L = f(Re_L)$ if the corresponding fin height is less than $l = 12y_0$, namely roughly two times the viscous sub layer thickness $y_{VSL} = 5y_0$. In particular $y_0 = 0.093mm$ for very low Reynolds number, i.e. $Re_D = 3.7\ 10^4$ corresponding to $Re_L = 4\ 10^3$, (see Eq.10) leading to $l = 1.1mm$ which is exactly the size of the higher fin height (see Tab.2, sample (d)). Consequently all the samples experience the same value of Nu_L for $Re_L \leq 4\ 10^3$. As Reynolds number increase, y_0 decrease (i.e. thinner boundary layer), leading size of fin height of sample (d) to exceed the threshold level $l = 12y_0$ and consequently to experience higher values of Nu_L than the other samples. For $Re_D = 7.5\ 10^4$ corresponding to $Re_L = 8\ 10^3$, $y_0 = 0.054mm$ leading to $l = 0.65mm$ which is exactly the fin height of sample (c). As shown in Fig.5 for $Re_L \leq 8\ 10^3$ all finned samples but sample (d) experience the same curve $Nu_L = f(Re_L)$. On the contrary, when Reynolds number increase further each sample experience a different trend, due to the different sample geometrical features an consequently to the different nature of interactions between the fluid flow and the fin walls, as discussed above.

Uncertainty values shown in Figs.4, 5 have been calculated with regard to a significant level $\alpha = 0.05$ (5 %). Both type A and type B uncertainties have been considered. Concerning type B uncertainties, Eqs. (4, 5) allows one to compute Tr and h respectively, as a function of other measurements $(P, T_s, T_{g1}, T_{g2}, T_a, T_w, \epsilon)$. These independent quantities can be organized in a vector, namely $\boldsymbol{q} := \{P, T_s, T_{g1}, T_{g2}, T_a, T_w, \epsilon\}$, and $q_i \in \boldsymbol{q}$ is the generic i-th quantity. The standard uncertainty $\Sigma_{h,B}$ can be computed by the uncertainty estimation method [30] as

$$\Sigma_{h,B} = \sqrt{\sum_{i=1}^{7}\left(\Sigma_{q_i}\frac{\partial h}{\partial q_i}\right)^2}, \tag{12}$$

where Σ_{q_i} is the standard uncertainty for the quantity q_i, reported in [26], Table 1. On the other hand, type A uncertainties have been calculated using both t-student and chi-squared distributions by mean of a novel method proposed in [26]. Combining type A and type B uncertainties [26] is possible to calculate the overall uncertainty values, as shown in Figs.4, 5. It is worth to highlight that the previous procedure allows one to compute tolerance intervals, which are wider than confidence intervals and more robust estimates of the experimental uncertainty. The maximum and mean relative uncertainty for the convective heat transfer coefficient is $\pm 9.4\%$ and $\pm 6.4\%$, respectively.

5. Conclusions

In this work heat transfer performances of forced air cooled flush mounted heat sinks have been improved by pattering heat sinks surfaces with micro-structured artificial roughnesses obtained by laser etching manufacturing technique. Each structured roughness is made of micro-metric structures, namely truncated cones, of uniform size arranged in staggered layout. In particular each sample differs from the others with regards to the number and the size of the micro-fins (e.g. the micro-fin length ranges from 200 to 1100 μm). Experimental tests were carried out in forced air cooling regime. In particular fully turbulent flows (heating edge based Reynolds number ranging from 3000 to 17000) have been explored. Convective heat transfer coefficient of the best micro-structured heat sink is found to be roughly two times compared to the smooth heat sinks one. In addition, surface area roughly doubles with regard to smooth heat sinks, due to the presence of micro-fins. Consequently, patterned heat sinks thermal transmittance [W/K] is found to be roughly four times the smooth heat sinks one. An effective physical interpretation of the enhancement and of its dependency on the air velocity is provided in this work. In particular, the dependency of the enhancement on both geometrical and fluid dynamics characteristic length scales is demonstrated. In particular this study demonstrate that micro-fin whose height lays between the buffer layer and the early log-law region, i.e. $30 \leq y_0 \leq 45$, can lead to a huge enhancement in thermal transmittance due to interactions between the fluid flow and the fin walls, besides the augmentation in surface area available for heat transfer. The enhancement in thermal transmittance due to fluid-wall interaction has been proven to be effective when micro-fin height exceed $12y_0$, i.e. roughly two times the viscous sub layer thickness, for the samples under consideration. For this reason, this works provides preliminary guidelines to the design of micro-structured artificial rough surface for enhanced heat transfer.

Beyond laser etching, there are other interesting techniques to realize micro-structured rough geometries, e.g. additive manufacturing [19]. As an outlook, based on the present results, we plan further studies focusing on investigation and optimization of micro-structured surfaces and design of hierarchical solutions able to coupling micro- and macro- structures.

Acknowledgments

Authors would like to acknowledge the THERMALSKIN project: Revolutionary surface coatings by carbon nanotubes for high heat transfer efficiency (FIRB 2010 Futuro in Ricerca", grant RBFR10VZUG).

References

[1] J.R. Culham, Y.S. Muzychka, Optimization of plate fin heat sinks using entropy generation minimization, IEEE Transactions on Components and Packaging Technologies 24:2 (2001) 159-165.

[2] C.J. Shih, G.C. Liu, Optimal design methodology of plate-fin heat sinks for electronic cooling using entropy generation strategy, IEEE Transactions on Components and Packaging Technologies 27:3 (2004) 551-559.
[3] W.A. Khan, J.R. Culham, M.M. Yovanovich, The Role of Fin Geometry in Heat Sink Performance, Journal of Electronic Packaging 128:4 (2006) 324-330.
[4] S.W. Chang, T.L. Yang, C.C. Huang, K.F. Chiang, Endwall heat transfer and pressure drop in rectangular channels with attached and detached circular pin-fin array, International Journal of Heat and Mass Transfer 51:21-22 (2008) 5247-5259.
[5] Y.-T. Yang, H.-S. Peng, Numerical study of pin-fin heat sink with un-uniform fin height design, International Journal of Heat and Mass Transfer 51:19-20 (2008) 4788-4796.
[6] P.M. Ligrani, G.I. Mahmood, J.L. Harrison, C.M. Clayton, D.L. Nelson, Flow structure and local Nusselt number variations in a channel with dimples and protrusions on opposite walls, International Journal of Heat and Mass Transfer 44:23 (2001) 4413-4425.
[7] M.A. Elyyan, A. Rozati, D.K. Tafti, Investigation of dimpled fins for heat transfer enhancement in compact heat exchangers, International Journal of Heat and Mass Transfer 51:11-12 (2008) 2950-2966.
[8] Y. Chen, Y.T. Chew, B.C. Khoo, Enhancement of heat transfer in turbulent channel flow over dimpled surface, International Journal of Heat and Mass Transfer 55:25-26 (2012) 8100-8121.
[9] S.V. Garimella and P.A. Eibeck, Heat transfer characteristics of an array of protruding elements in single phase forced convection, Internal Journal of Heat and Mass Transfer 33:12 (1990) 2659-2669.
[10] G.I. Mahmood, M.Z. Sabbagh, P.M. Ligrani, Heat Transfer in a Channel with Dimples and Protrusions on Opposite Walls, Journal of Thermophysics and Heat Transfer 15:3 (2001) 275-283.
[11] S. Mancin, C. Zilio, A. Diani, L. Rossetto, Air forced convection through metal foams: Experimental results and modeling, International Journal of Heat and Mass Transfer 62:1 (2013) 112-123.
[12] E.K. Kalinin, G.A. Dreitser, Heat Transfer Enhancement in Heat Exchangers, Advances in Heat Transfer 31:C (1998) 159-332.
[13] J.S. Park, J.C. Han, Y. Huang, S. Ou, R.J. Boyle, Heat transfer performance comparisons of five different rectangular channels with parallel angled ribs, International Journal of Heat and Mass Transfer 35:11 (1992) 2891-2903.
[14] X. Gao, B. Sunden, Heat transfer and pressure drop measurements in rib-roughened rectangular ducts, Experimental Thermal and Fluid Science 24:1-2 (2001) 25-34.
[15] G. Tanda, Heat transfer in rectangular channels with transverse and V-shaped broken ribs, International Journal of Heat and Mass Transfer 47:2 (2004) 229-243.
[16] S.W. Chang, T.-M. Liou, M.H. Lu, Heat transfer of rectangular narrow channel with two opposite scale roughened walls, International Journal of Heat and Mass Transfer 48:19-20 (2005) 3921-3931.
[17] S.W. Chang, T.M. Liou, K.F. Chiang, G.F. Hong, Heat transfer and pressure drop in rectangular channel with compound roughness of V-shaped ribs and deepened scales, International Journal of Heat and Mass Transfer 51:3-4 (2008) 457-468.
[18] F. Zhou, G.W. Demoulin, D.J. Geb, I. Catton, Closure for a plane fin heat sink with scale-roughened surfaces for volume averaging theory (VAT) based modeling, International Journal of Heat and Mass Transfer 55:25-26 (2012) 7677-7685.
[19] L. Ventola, F. Robotti, M. Dialameh, F. Calignano, D. Manfredi, E. Chiavazzo and P. Asinari, Rough surfaces with enhanced heat transfer for electronics cooling by direct metal laser sintering, International Journal of Heat and Mass Transfer 75 (2014) 5874.
[20] M. Mullenborn, H. Dirac, J. Petersen, S. Bouwstra, Fast three dimensional laser micromachining of silicon for microsystem, Sensors and Actuators 52 (1996) 121-125.
[21] J. Chang, B. Warner, E. Dragon, M. Martinez, Precision micromachining with pulsed green laser, J. Laser Appl. 10:6 (1998) 285-290.
[22] M.R.H. Knowles, G. Rutterford, A.I. Bell, A.J. Andrews, G. Foster-Turner, A.J. Kearsley, Sub-micron and high precision micromachining using nanosecond lasers, Proceedings of ICALEO 112-120 (1998).
[23] T. Otani, L. Herbst, M. Heglin, S.V. Govorkov, A.O. Wiessner, Microdrilling and micromachining with diode-pumped solid state laser, Appl. Phys. 79 (2004) 1335-1339.
[24] www.rofin.it/
[25] www.lightcon.com/
[26] E. Chiavazzo, L. Ventola, F. Calignano, D. Manfredi, P. Asinari, A sensor for direct measurement of small convective heat fluxes: Validation and application to micro-structured surfaces, Experimental Thermal and Fluid Science 55 (2014) 42-53.
[27] A. Bejan and A.D. Kraus (Editors), Heat Transfer Handbook, John Wiley, Hoboken (2003).
[28] H. Schlichting, K. Gersten, Boundary-layer theory, Springer, 8th ed., Berlin, 2000.
[29] S.B. Pope, Turbulent Flows, Cambridge University Press, 2000.
[30] J. Mandel, The Statistical Analysis of Experimental Data, Dover, New York, (1984).

Separation & free-stream turbulence: implications for surface aerodynamics & heat transfer

J P J Stevenson[1], E J Walsh[2], K P Nolan[3] and M R D Davies[1]

[1] Stokes Institute, University of Limerick, Limerick, Ireland
[2] Department of Engineering Science, University of Oxford, Oxford, UK
[3] Bell Labs Ireland, Thermal Management Research Group, Alcatel-Lucent, Dublin, Ireland

E-mail: jonathan.stevenson@ul.ie

Abstract. Preliminary results from a Particle Image Velocimetry (PIV) investigation of the separation-reattachment flow over a flat plate are presented. The experiments address the effects of two key variables: flow approach angle (manipulated indirectly with a trailing edge flap) and free-stream turbulence level (introduced upstream with grids). The plate thickness Reynolds number is fixed throughout and lies within the transitional regime.

In the first test series (I), it is shown that increasing the turbulence level and reducing the approach angle cause the mean leading-edge separation bubble to shrink. The effect of free-stream turbulence, in particular, diminishes progressively as its level is raised. In the second series (II), downstream development of the reattached boundary layer is found to unfold rapidly at first but plateau after approximately three bubble-lengths. Momentum thickness Reynolds and Stanton numbers develop independently of the free-stream turbulence thereafter, and are well described by shifted turbulent correlations. Heat transfer potential ultimately depends upon the balance between frictional loss, bubble size and downstream mixing.

1. Introduction

Many industrial heat-exchanging surfaces operate at Reynolds numbers and angles of attack high enough to produce regions of transitional separation at their upstream side. In these cases, the flow detaches at the leading edge of the surface, undergoes transition and reattaches as a developing turbulent boundary layer [1].

The region enclosed by the separated layer is known as a *separation bubble.* The bubble recirculates under shear and inhibits local heat transfer by its own thermal inertia. Downstream, recovery of the reattached wall layer to turbulent equilibrium is notably slow [2] and depends, amongst other factors, on surface geometry, upstream approach incidence and streamwise pressure gradient [3]. But under suitable conditions, the enhanced diffusion *might* provide a workable means for increasing heat transfer.

In the present work, Particle Image Velocimetry (PIV) is used to study the effects of upstream approach angle and free-stream turbulence on such a flow for a nominal thickness Reynolds number Re_I ($= U_I D/\nu$) of 480. Two studies are conducted: the first, denoted 'series I', provides time-mean and instantaneous views of the bubble zone. The second, denoted 'series II', focuses on the mean development and heat transfer characteristics of the reattached wall layer.

Content from this work may be used under the terms of the Creative Commons Attribution 3.0 licence. Any further distribution of this work must maintain attribution to the author(s) and the title of the work, journal citation and DOI.
Published under licence by IOP Publishing Ltd

2. Experimental facility

All measurements were performed in a subsonic open-return wind tunnel facility. The tunnel has a honeycomb-mesh screen cascade at its inlet and a four-square bank of axial fans (in suction mode) at its outlet. The background turbulence level in the test section is 0.7 %. The PIV system consists of a 200 mJ pulsed laser and full-frame 11 MP CCD camera (with a 105 mm f/2.8 macro lens) mounted separately on three-axis traverse platforms. Seeding is provided by a commercial fog generator. Figure 1 shows the arrangement.

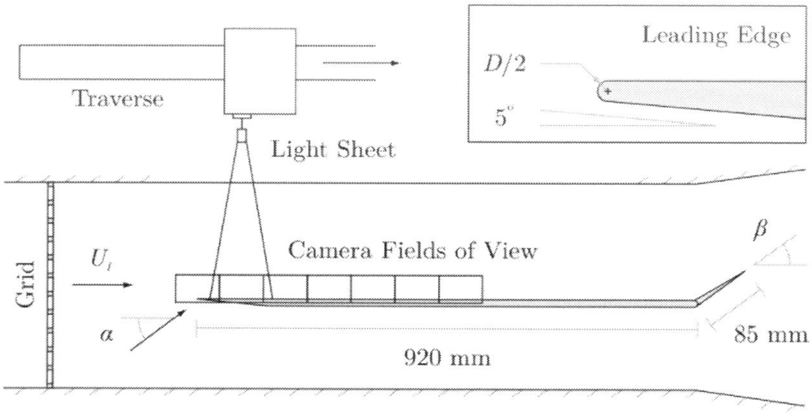

Figure 1: Schematic of the PIV test setup (not to scale).

The round-nosed plate model (295 mm wide × 920 mm long) has a hinged flap at its trailing edge. This allows the circulation, and therefore the upstream approach incidence α of the flow, to be varied indirectly. The flow separates at the leading edge until the flap is positioned at an upward setting β of about 45°, for which an attached *Blasius* boundary layer is produced downstream (figure 2). Note that coordinate x is measured from the leading edge of the plate.

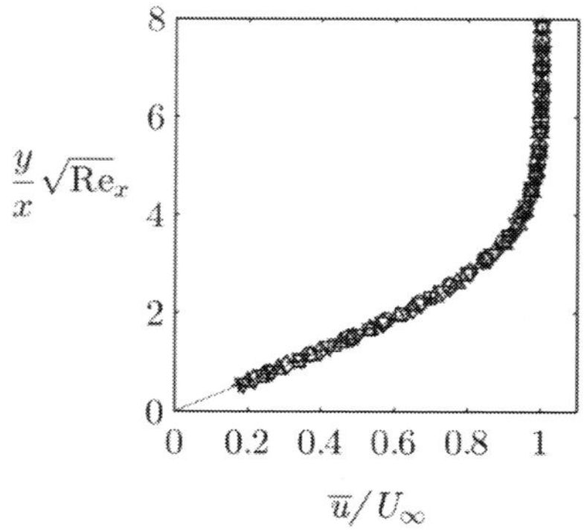

Figure 2: Boundary layer velocity profiles at the 45° flap setting. —— Blasius; ○ $\mathrm{Re}_x = 50100$; □ $\mathrm{Re}_x = 62700$; ◇ $\mathrm{Re}_x = 74900$; △ $\mathrm{Re}_x = 87300$; ▽ $\mathrm{Re}_x = 100000$; + $\mathrm{Re}_x = 111900$.

Free-stream turbulence may be introduced upstream of the plate using grids (with square bars of width d). Figures 3 and 4 show the development of streamwise turbulence intensity T_u, micro-scale λ_x and integral scale Λ_x for the grids of the present study. Roach's correlations [4] are also shown. Power spectra were further computed at each station and it was confirmed, by comparison with von Kármán's spectrum, that the turbulence was sufficiently isotropic.

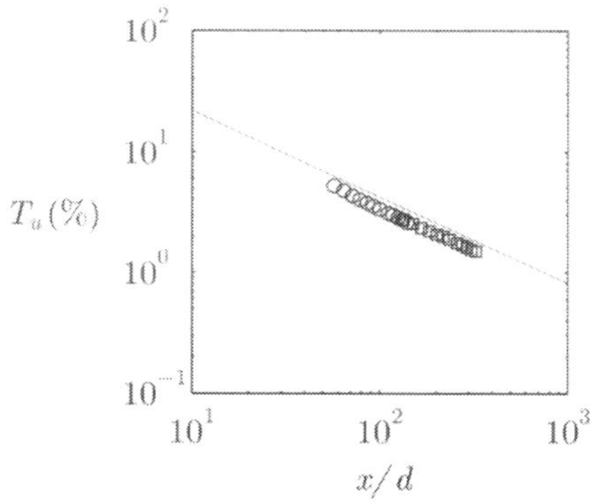

Figure 3: Decay of streamwise intensity T_u. ○ Coarse Grid; □ Fine Grid; —— Roach's [4] square-bar grid correlation.

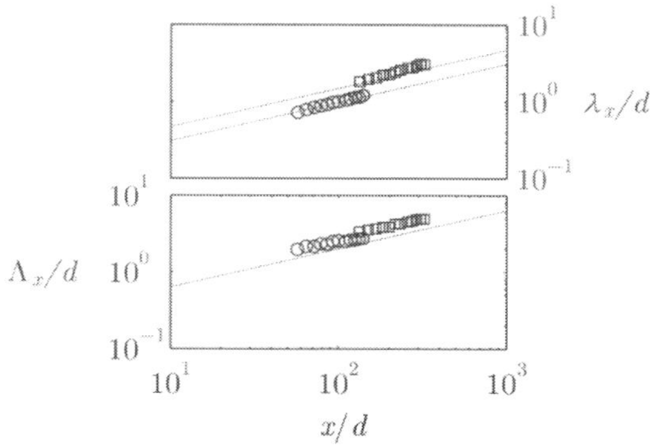

Figure 4: Growth of length scales Λ_x and λ_x. ○ Coarse Grid; □ Fine Grid; —— Roach's [4] length-scale correlations.

Table 1: Series I test cases.

T_u (%)	[λ_x Λ_x] (mm)	β (α) (°)	Re_I
0.7	—	0 (30)	480
0.7	—	15 (26)	480
0.7	—	30 (20)	480
3.1	[4.8 9.1]	0 (28)	460
3.1	[4.8 9.1]	15 (24)	500
3.1	[4.8 9.1]	30 (19)	480
6	[4.7 13.2]	0 (29)	460
6	[4.7 13.2]	15 (25)	480
6	[4.7 13.2]	30 (20)	490

Table 2: Series II test cases.

T_u (%)	[λ_x Λ_x] (mm)	β (α) (°)	Re_I
0.7	—	0 (33)	480
0.7	—	22.5 (25)	480
5.7	[4.8 13.5]	0 (33)	460
5.7	[4.8 13.5]	22.5 (24)	480

3. Results

Details of the respective experiment series are provided in tables 1 and 2. Note that the values of T_u, λ_x and Λ_x refer to those at the plate leading-edge station.

3.1. Bubble structure (series I)

Figure 5 shows a vector-contour map of mean absolute velocity $\overline{\mathbf{V}}$ within the leading edge bubble for the 0° flap background-turbulence case. This flow field was representative of all separation bubbles from series I. The coordinate origin was placed at the nose-plate junction and the downstream terminus of the zero-contour for mean streamwise velocity $\overline{u}$, relative to this origin, was used to define the bubble length l_R. Spatial coordinates were normalised accordingly.

The vector detail reveals finely the internal structure of the bubble. The flow entrains under free-stream shear around an apparent centre, for this case, of $x/l_R = 0.75$, and in a time-mean sense is practically enclosed (whilst this perspective masks the complex nature of transitional

Journal of Physics: Conference Series **525** (2014) 012018 doi:10.1088/1742-6596/525/1/012018

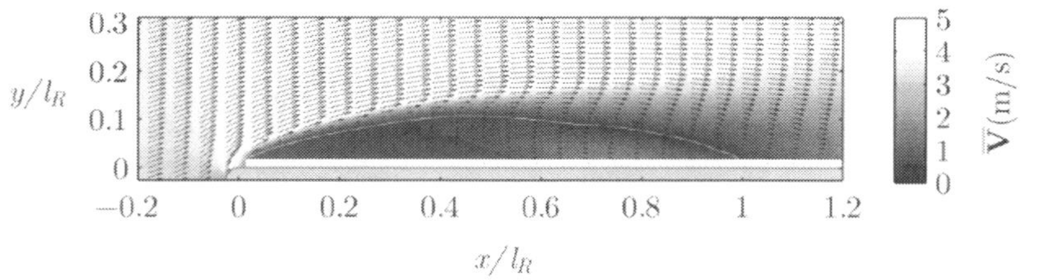

Figure 5: Time-mean PIV absolute velocity vector-contour map of the 0° flap bubble. The $\overline{u}$ zero-contour is shown by the white line and the plate surface is coloured in grey. $l_R = 39$ mm.

shedding, it does serve to illustrate the recirculation mechanism responsible for inhibiting local heat transfer). At reattachment, the mean velocity profile is highly inflectional, but the shearing effect of the wall—which effects the growth of a new 'inner' layer—gradually initiates a return to fuller form (see section 3.2).

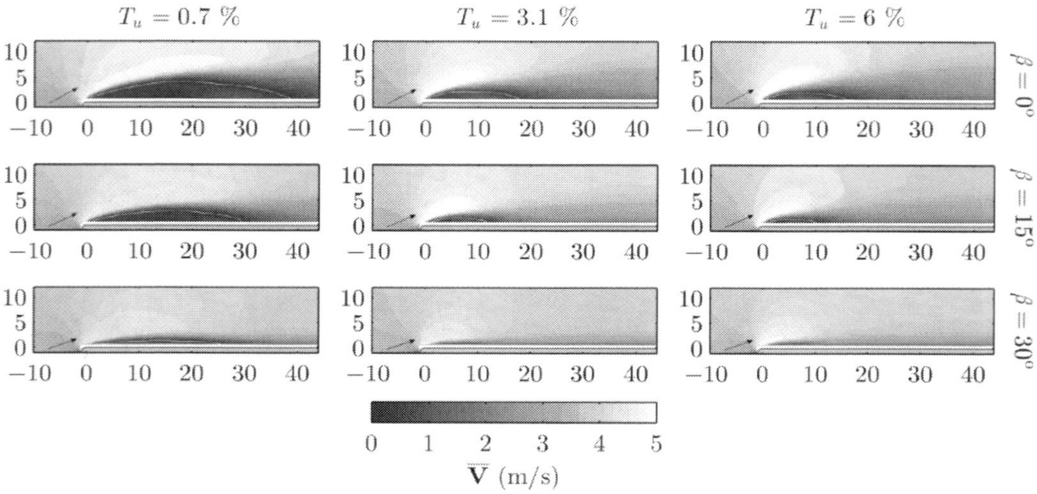

Figure 6: Time-mean absolute velocity contours of all series I bubbles (see table 1). Leading-edge flow incidence α is indicated by an arrow in each case. As per figure 5, $\overline{u}$ zero-contours are shown in white and the plate is coloured in grey. Axis units are [mm].

The bubbles for all series I cases are shown in figure 6. It is evident that increasing either the flap setting or turbulence level causes the time-mean bubble to shrink. The effect of the flap, and thus separation incidence, is an inertial one: at high flap settings, associated with shallow approach angles, 'overshoot' of the separated layer is milder and the mean reattachment length shortens. The wall also inhibits spreading of the layer [5]. Present measurements indicate that in order to remove the bubble entirely, a mildly negative approach incidence would be required—but that the bubble will, in practical terms, have diminished to immeasurably small proportions by the attached flap setting of 45°. The effect is dependent upon both leading- and trailing-edge conditions, however, and will therefore differ between facilities.

The influence of free-stream turbulence on separation bubbles has been studied by many workers (e.g. [6]) and in all cases, its presence has shortened the time-mean bubble. The operating mechanism depends on the Reynolds number, however. In the present transitional

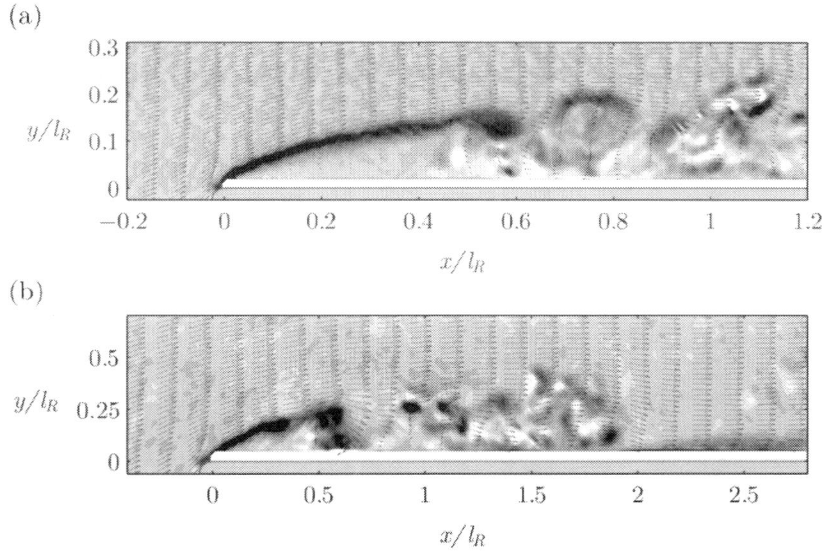

Figure 7: Snapshots of absolute velocity (vectors) and vorticity (contours) for the 0° flap bubble: (a) background and (b) high free-stream turbulence. $l_R = 39$ mm and 16.2 mm, respectively.

regime, the effect results primarily from advanced breakdown in the separated shear layer and is analogous, at least qualitatively, to the 'bypass' scenario of the attached case [7]. This is illustrated by the instantaneous flow fields of figure 7 for the 0° flap setting: under elevated free-stream turbulence, breakdown initiates at $x/l_R \approx 0.25$—as opposed to $x/l_R \approx 0.45$ in the background turbulence case—and the shear layer returns to the wall more quickly in the mean. But as shown by the contours of figure 6, the effect is one of diminishing returns and the bubble cannot be removed entirely whilst separation is forced at the leading edge.

3.2. Boundary layer development (series II)

Figure 8 shows mean boundary layer velocity profiles for all series II cases, in customary bubble-length (x/l_R) spacing. Normalised layer thickness δ_{99}/l_R (at which $\overline{u}/U_\infty = 0.99$) and Prandtl's one-seventh turbulent profile are also shown for downstream stations.

Several observations may be made on the form of the reattached layers: first, in no case do the profiles begin to settle until the initial velocity 'dip' at $y/l_R \approx 0.25$—effected by the sudden action of wall shear on the inflectional profiles at reattachment—has been smoothed out. Here, this has largely taken place by $x/l_R \approx 6$. Second, for a fixed flap setting, the addition of free-stream turbulence marginally extends the 'overshoot' of the reattachment ($x/l_R = 1$) profile. This is a result of the greater mean distortion induced by the lower aspect ratio bubble at high T_u (*cf.* figure 6). Third, by $x/l_R = 10$, both 22.5° flap layers are further from Prandtl's profile than those at the 0° flap setting. Downstream momentum thickness Reynolds numbers Re_θ ($= U_\infty\theta/\nu$, where θ is the momentum thickness) are also low—on the order of twofold lower, in fact, than those usually observed for orthodox inchoate turbulent boundary layers [8].

Figure 9 shows the shape factor H_{12} ($= \delta^*/\theta$, where δ^* is the displacement thickness and θ is as above) for each reattached wall layer. The shape factor is a measure of 'fullness' across the mean velocity profile and is well suited to the region of rapid change after reattachment (termed 'post-reattachment' herein). Under background turbulence, the 22.5° flap case is the quicker to develop initially. But as further shown by its mean velocity profiles, it is insensitive to further change after $x/l_R \approx 3$ and has settled to a high non-turbulent shape factor of 1.6 by $x/l_R = 10$.

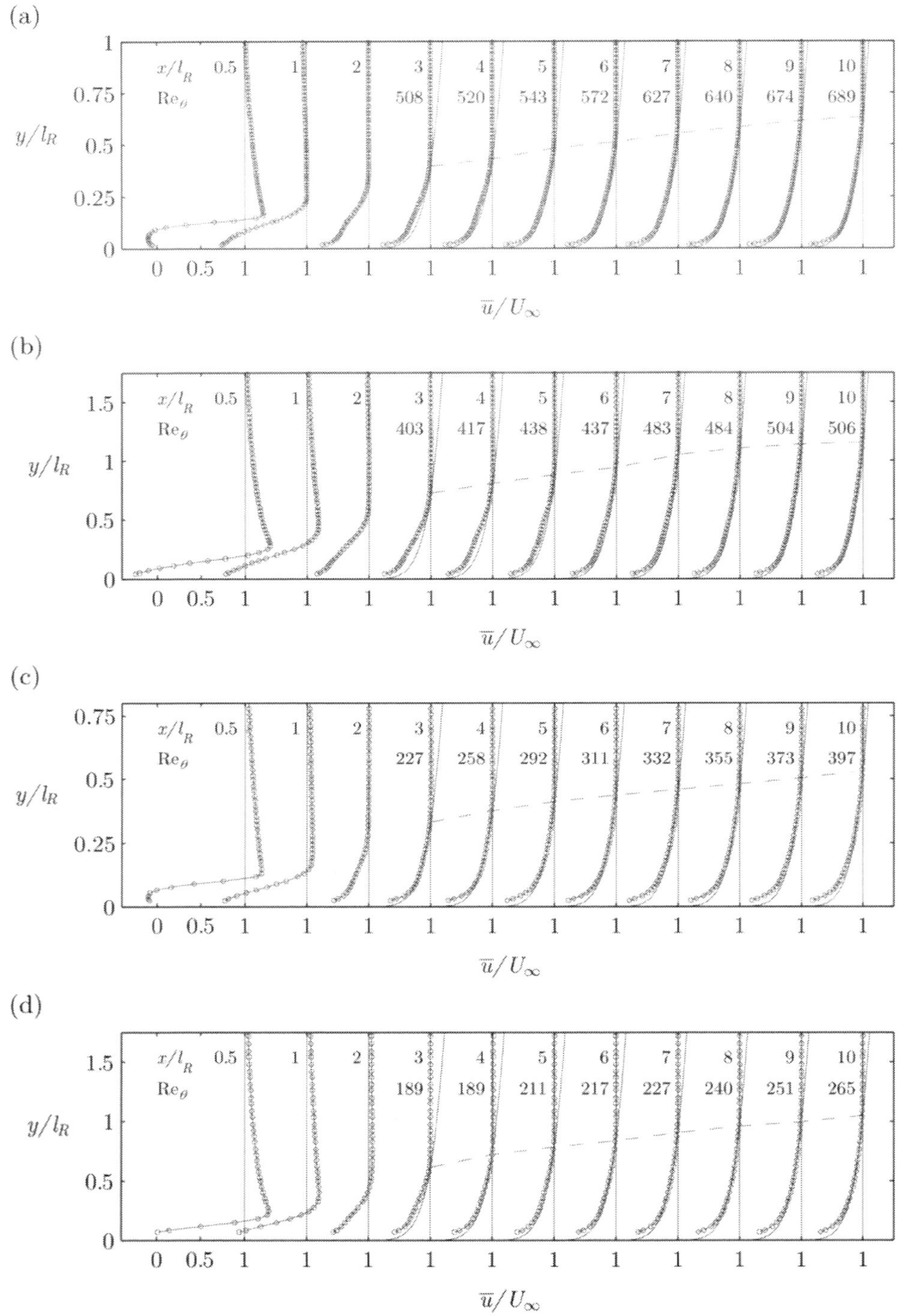

Figure 8: Normalised velocity profiles for each reattached layer. (a) $\beta = 0°$ at background T_u; (b) $\beta = 0°$ at $T_u = 5.7$ %; (c) $\beta = 22.5°$ at background T_u; (d) $\beta = 22.5°$ at $T_u = 5.7$ %. In each figure: – – – δ_{99}/l_R; —— Prandtl's turbulent profile (scaled to δ_{99}/l_R at each station). Re_θ is the boundary layer momentum thickness Reynolds number and the profile shift is $\overline{u}/U_\infty = 0.7$.

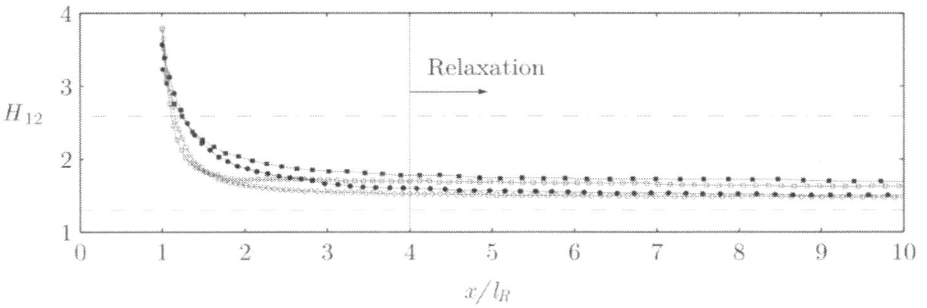

Figure 9: Decline in shape factor H_{12}. ○ $\beta = 0°$ at background T_u; ● $\beta = 0°$ at $T_u = 5.7$ %; □ $\beta = 22.5°$ at background T_u; ■ $\beta = 22.5°$ at $T_u = 5.7$ %. — · — Laminar; – – – Turbulent.

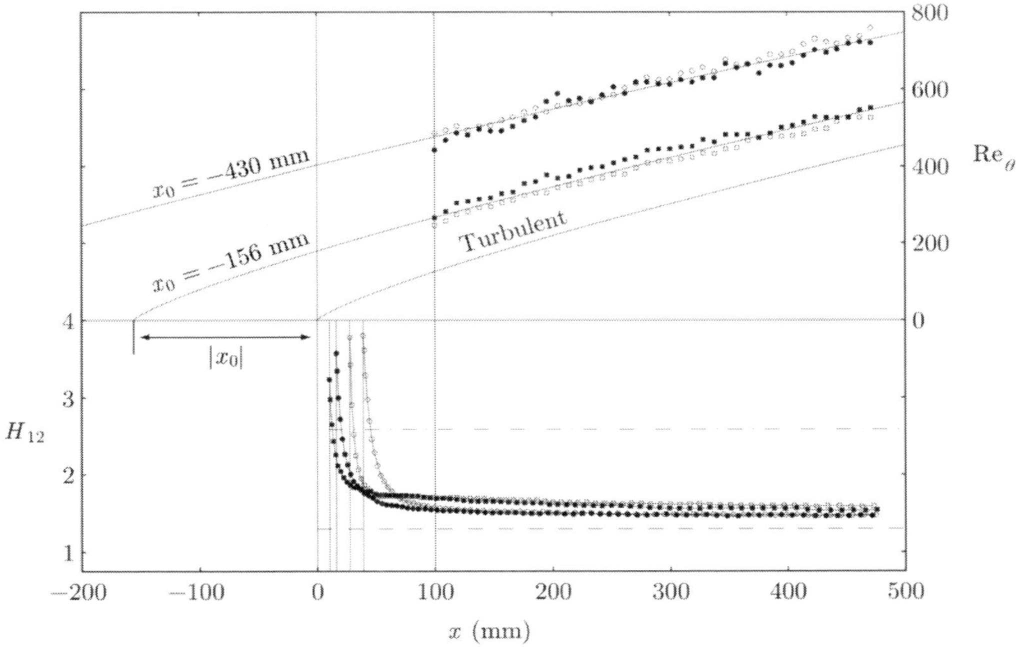

Figure 10: Evolution of momentum thickness Reynolds number Re_θ and shape factor H_{12} with physical streamwise distance x. Upper plot: —— Equation 1 (for various virtual origins x_0). In the lower plot, vertical lines across $0 < x < 50$ mm signify reattachment for each case. See figure 9 for marker symbols.

Under free-stream turbulence, both cases develop similarly at first but (a) level out more slowly and (b) eventually plateau to higher shape factors than the background turbulence cases. This result—that free-stream turbulence should mildly *delay* mean development in bubble-length terms—is somewhat counter-intuitive, but has also been observed over blunt-nosed flat plates e.g. Castro & Epik [2] found that the aforementioned velocity dip, which corresponds to a higher shape factor, was more pronounced under turbulence (*cf.* figure 8).

Downstream of post-reattachment, further change is practically undetectable in all cases and as shown in figure 10, none has reached a turbulent value ($H_{12} \approx 1.3$ for Prandtl's profile, although often nearer 1.4 in practice) by the last station.

Although bubble-length scaling (i.e. x/l_R) is frequently used in separation-reattachment flows, here it masks the fact that downstream of post-reattachment each flap setting behaves similarly in dimensional terms. This is illustrated in figure 10 with a stacked plot of momentum thickness Reynolds number Re_θ and shape factor H_{12}. The data are plotted against streamwise coordinate x from the leading edge. Also shown for each Re_θ set is the standard formula for an attached turbulent boundary layer [9] but with the inclusion of a virtual origin x_0:

$$\mathrm{Re}_\theta = 0.036\left(\frac{U_\infty}{\nu}\right)^{\frac{4}{5}}(x - x_0)^{\frac{4}{5}}. \tag{1}$$

U_∞ is the nominal free-stream velocity *above* the plate (which differs slightly from U_I due to the acceleration around the leading edge and bubble) and is approximately constant at 4 m/s for $x > 100$ mm. Three origins are shown: (i) $x_0 = 0$ mm (a regular turbulent boundary layer growing from the leading edge); (ii) $x_0 = -156$ mm; and (iii) $x_0 = -430$ mm.

At both flap angles, Re_θ develops almost independently of the turbulence level and although none of the constituent cases—especially those at the 22.5° setting—is canonically turbulent in the mean, streamwise growth can be correlated by shifted turbulent curves with origins (ii) and (iii). In combination with corresponding shape factors from the lower part of the figure, it follows that neither an accelerated transition (*cf.* figure 7) nor additional downstream forcing significantly affect growth of the layer beyond post-reattachment. The area-averaged drag coefficient C_f for $100 < x < x_{max}$ (mm) is therefore almost identical in both cases.

To a first order, it is also possible to estimate the heat transfer characteristics of the reattached layer by the *Reynolds-Colburn* analogy. Local skin-friction coefficient C_{fx} can be related to the Stanton number St_x by [10]:

$$\mathrm{St}_x\mathrm{Pr}^{\frac{2}{3}} = \frac{C_{fx}}{2}. \tag{2}$$

Pr is the Prandtl number. Local skin-friction coefficients were thus derived by the Clauser chart method [11] and a Coles-type fitting procedure [12]. This was possible only for the 0° flap cases for $x > 100$ mm, however, since the standard inner log law ($u^+ = 2.44\ln y^+ + 5.0$; [9]) did not exist for most of both 22.5° flap cases—despite apparent similarities in outer growth to a shifted turbulent boundary layer.

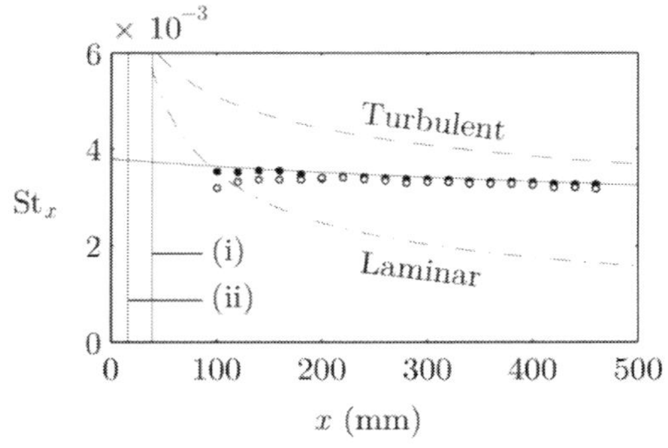

Figure 11: Development of local Stanton number St_x with distance x for both 0° flap cases. — · — Equation 3; – – – Equation 4; —— Equation 5. See figure 9 for marker symbols. Note (i) and (ii) are the bubble lengths for background and elevated turbulence cases, respectively.

Figure 11 shows the corresponding Stanton number distributions for $x > 100$ mm alongside the iso-flux laminar (L):

$$\left(\mathrm{St}_x\right)_L = 0.453\mathrm{Pr}^{-\frac{2}{3}}\left(\frac{U_\infty}{\nu}\right)^{-\frac{1}{2}}x^{-\frac{1}{2}} \tag{3}$$

and turbulent (T) relations:

$$\left(\mathrm{St}_x\right)_T = 0.0308\mathrm{Pr}^{-\frac{2}{3}}\left(\frac{U_\infty}{\nu}\right)^{-\frac{1}{5}} x^{-\frac{1}{5}} \tag{4}$$

for attached flat-plate boundary layers growing from the leading edge [10]. Also plotted for comparison is the shifted turbulent (TS) curve:

$$\left(\mathrm{St}_x\right)_{TS} = 0.0308\mathrm{Pr}^{-\frac{2}{3}}\left(\frac{U_\infty}{\nu}\right)^{-\frac{1}{5}} (x - x_0)^{-\frac{1}{5}} \tag{5}$$

where $x_0 = -430$ mm, as per figure 10.

The results for both reattached layers lie above the laminar line. From a heat transfer perspective, the question therefore arises as to whether it might be beneficial, for those flows at sufficiently high Reynolds numbers that would otherwise be attached and laminar, to *force* separation (e.g. with a flap or actuated surface) and trip the boundary layer into a 'quasi-turbulent' state. This would effect a faster transition than that induced by free-stream turbulence alone (indeed the turbulent correlation, although shown here for comparison, would not usually apply until further downstream). The corollaries, though, are: (a) a probable rise in system pressure drop, and (b) a reduction in heat transfer around the bubble and post-reattachment zones due to circulation and low mean wall gradients. And although, as shown in section 3.1, free-stream turbulence serves to shrink these bubble regions, it has minimal impact beyond post-reattachment (figures 10 and 11) and should probably be avoided if an additional energy penalty must be incurred.

4. Summary & Conclusions

High-resolution Particle Image Velocimetry (PIV) data have been presented for the separation-reattachment flow over a round-nosed flat plate. Focus is given to the effects of flow approach angle (manipulated via a trailing edge flap) and free-stream turbulence (generated by grids) on both mean aerodynamic and heat transfer characteristics. Salient conclusions are as follows:

(i) In the first series of measurements (I) it was found that an increase in either flap setting (corresponding to a decrease in the approach angle) or free-stream turbulence level caused the time-mean separation bubble to shrink. The effect was monotonic in both variables and one of diminishing returns.

(ii) Under free-stream turbulence, instantaneous PIV snapshots revealed that an *accelerated* transition mechanism was responsible for shortening the bubble.

(iii) In the second series of measurements (II) it was found that in all test cases, relaxation of the reattached boundary layer unfolded rapidly at first but reached a plateau approximately three bubble-lengths downstream of reattachment.

(iv) Local momentum thickness Reynolds and Stanton numbers for each flap setting were found to develop independently of the free-stream turbulence level for $x > 100$ mm and could be described by shifted turbulent correlations. In this region, heat transfer performance is expected to be superior to regular laminar flow.

A detailed perspective on the underlying mechanics of this separated flow will be presented in a future paper.

Acknowledgments

Funding support from the Irish Research Council (IRC) is gratefully acknowledged. Bell Labs Ireland would also like to thank the Industrial Development Agency (IDA) Ireland for their continued support.

References

[1] Ota T, Asano Y and Okawa J 1981 *B. JSME* **24** p 944
[2] Castro I P and Epik E 1998 *J. Fluid Mech.* **374** 91–116
[3] Ruderich R and Fernholz H H 1986 *J. Fluid Mech.* **163** 283–322
[4] Roach P E 1987 *Int. J. Heat Fluid Fl.* **8** 82–92
[5] Bearman P W and Morel T 1983 *Prog. Aerospace Sci.* **20** 97–123
[6] Kalter M and Fernholz H H 2001 *J. Fluid Mech.* **446** 271–308
[7] Langari M and Yang Z 2013 *Phys. Fluids* **25**
[8] Castro I P 1984 *J. Fuid Eng.—T. ASME* **106**
[9] Schlichting H 1979 *Boundary Layer Theory* (New York: McGraw-Hill)
[10] Holman J P 2010 *Heat Transfer* (New York: McGraw-Hill)
[11] Clauser F H 1954 *J. Aeronaut. Sci.* 91–108
[12] Coles D E 1962 *The Turbulent Boundary Layer in a Compressible Fluid* R–403–PR (RAND: California)
[13] Mayle R E and Bellows W J 1986 *J. Turbomach.* **108** 131–136

The use of segregated heat sink structures to achieve enhanced passive cooling for outdoor wireless devices

K O'Flaherty, J Punch
CTVR, Stokes Institute, University of Limerick, Limerick, Ireland.

Email: kevin.oflaherty@ul.ie

Abstract. Environmental standards which govern outdoor wireless equipment can stipulate stringent conditions: high solar loads (up to 1 kW/m^2), ambient temperatures as high as 55°C and negligible wind speeds (0 m/s). These challenges result in restrictions on power dissipation within a given envelope, due to the limited heat transfer rates achievable with passive cooling. This paper addresses an outdoor wireless device which features two segregated heat sink structures arranged vertically within a shielded chimney structure: a primary sink to cool temperature-sensitive components; and a secondary sink for high power devices. Enhanced convective cooling of the primary sink is achieved due to the increased mass flow within the chimney generated by the secondary sink. An unshielded heat sink was examined numerically, theoretically and experimentally, to verify the applicability of the methods employed. Nusselt numbers were compared for three cases: an unshielded heat sink; a sink located at the inlet of a shield; and a primary heat sink in a segregated structure. The heat sink, when placed at the inlet of a shield three times the length of the sink, augmented the Nusselt number by an average of 64% compared to the unshielded case. The Nusselt number of the primary was found to increase proportionally with the temperature of the secondary sink, and the optimum vertical spacing between the primary and secondary sinks was found to be close to zero, provided that conductive transfer between the sinks was suppressed.

1. Introduction

The recent increase in demand for mobile data has presented significant challenges for contemporary wireless communications systems. Active antenna arrays, remote radio heads, pico- and femto- cells are typical artefacts of today's wireless communication infrastructure. These devices are subject to substantial packaging restrictions, strict reliability requirements, severe environmental conditions, the imperative of passive cooling, strict constraints on weight (~15-20 kg) and volume, and strict Ingress Protection (IP) ratings. Furthermore, environmental standards which govern outdoor wireless equipment stipulate stringent conditions: high solar loads (up to 1 kW/m^2), ambient temperatures as high as 55°C and negligible wind speeds (0 m/s) [1]. These challenges impose limitations on the power dissipation within a given envelope, as a consequence of the limited heat transfer rates achievable with passive cooling. Most contemporary outdoor wireless artefacts feature a cast metal casing which serves as a heat sink: moreover, most wireless artefacts feature two broad classes of device, each with different thermal characteristics: relatively low power, temperature-sensitive devices such as logic ICs; and high power devices such as Power Amplifiers (PAs) which can reliably tolerate high operating temperatures. These two classes of device may be segregated on to different assemblies on account of their function, and they may also be separated in terms of their heat sinking – for example, on opposing sides of a cast 'clamshell' structure. Finally, some outdoor equipment also features shields to mitigate solar ingress, which can act to induce conditions of chimney flow over the casing.

Content from this work may be used under the terms of the Creative Commons Attribution 3.0 licence. Any further distribution of this work must maintain attribution to the author(s) and the title of the work, journal citation and DOI.
Published under licence by IOP Publishing Ltd

The configuration under investigation in this paper features two conventional heat sink structures within a shield that acts to induce chimney flow. Bar-Cohen and Rohsenow [2] investigated heat sinks as a series of vertical parallel plates under conditions of constant heat flux and constant temperature, and proposed a correlation for the Nusselt number which revealed an optimum fin spacing. Malhammer [3] built upon this correlation, incorporating the effect of the base plate on the optimum spacing. Shabany [4] developed relations for the view factor of a plate-fin heat sink. Khor *et al.* [4] experimentally investigated these relations for thermal radiation and view factor of plate-fin heat sinks. The use of a chimney structure to enhance flow is not uncommon; one of the earliest works by Moore [5] examined a natural draft heat exchanger system for large dry cooling towers. Lee and Yan [6] investigated laminar natural convection for an unheated entry and unheated exit with a heated middle section; they found that the induced volume flow rate increased with both decreasing unheated entry length and increasing total length of heated and unheated exit. Naylor and Tarasuk investigated numerically [7] and experimentally [8] a divided vertical channel with heated walls on the chimney and a heated dividing plate. The position of the divided plate, relative to the inlet, strongly influenced the heat transfer rate at lower Rayleigh numbers. The average Nusselt number for a plate at the inlet was nearly double that for a plate located at the exit. Fischer *et al.* [9] formulated an analytical solution for a natural convection draft heat sink system, based on previous experimental work. A porosity term was used to capture effects from a wider chimney and to account for the potential reduction in mean velocity at the exit and the mean channel velocity. It was found that a reduction in heat sink height can be compensated by higher local heat transfer rates induced by the chimney effect; achieved by a large temperature (and thus density) difference between the fluid inside and outside the chimney.

The preceding literature review indicates the potential to utilize a chimney structure to enhance heat transfer; however the use of multiple independently placed heat sinks is rare. This paper aims to address the potential to utilize two segregated heat sink structures arranged vertically within a shield or chimney structure: a lower, or primary, sink to cool temperature-sensitive components; and an upper, or secondary, sink for the high power components which can tolerate higher temperatures. Enhanced convective cooling of the primary sink is achieved due to the increased mass flow within the chimney generated by the secondary sink. The specific practical focus of the paper is the improvement of heat dissipation in a representative remote radio head by exploiting the chimney effect and a secondary segregated structure. Experimental details are presented in the following section.

2. Experimentation

Two identical heat sinks, representative of a typical remote radio head, were machined in aluminium 6082T6 to the dimensions listed in table 1 and illustrated in figure 1.

Table 1: Dimensions of the heat sink structure and shield.

Term	Value
Width, W	149.5 mm
Length, L	300.0 mm
Height, H	60.0 mm
Fin thickness, t	1.9 mm
Base thickness, B	20.0 mm
Fin spacing, S	11.5 mm
No. fins, N (one side)	12
Hole spacing, L_{h1}	25.0 mm
Hole spacing, L_{h2}	50.0 mm

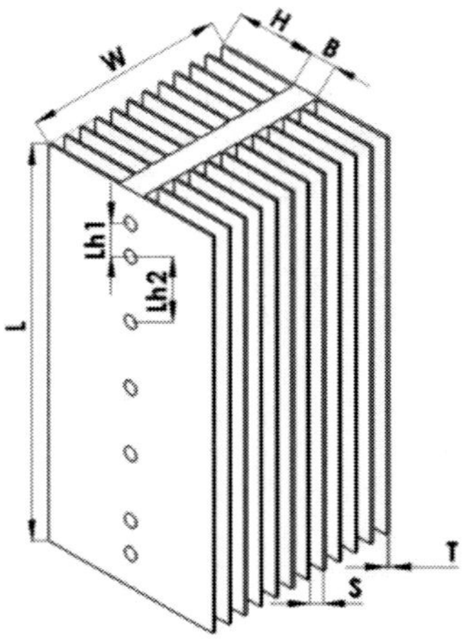

Figure 1: Dimensions of the heat sink.

Seven 10mm diameter, 150mm long cartridge heaters, each coated in thermal paste with a nominal thermal conductivity of 0.67 W/m.K, were embedded into the base plate of each heat sink. The sinks were double-sided (fins both sides of the base plate) in order to obviate errors due to losses from the back plate of a single-sided sink structure. In addition, the sinks were anodized black to ensure a known emissivity to accurately quantify radiative losses. Three different configurations were tested, as shown in figure 2; a single heat sink with no shield (unshielded); a single heat sink within a shield (chimney); and two heat sinks in a shield (segregated structure, primary and secondary). The shield, which was made from polycarbonate, featured a length of 900mm, width of 161mm, thickness of 5mm, and clearance of 5.75mm from the surfaces of the sink. The shield cover neither the top nor bottom of the heat sink, in order to allow air to flow freely through the sink. All testing was carried out in a polycarbonate enclosure, five times the length of the heat sink, to reduce uncertainties due to air currents in the laboratory. The segregated structure was further investigated by quantifying the impact of the gap distance on heat transfer from the primary heat sink, as shown in figure 3.

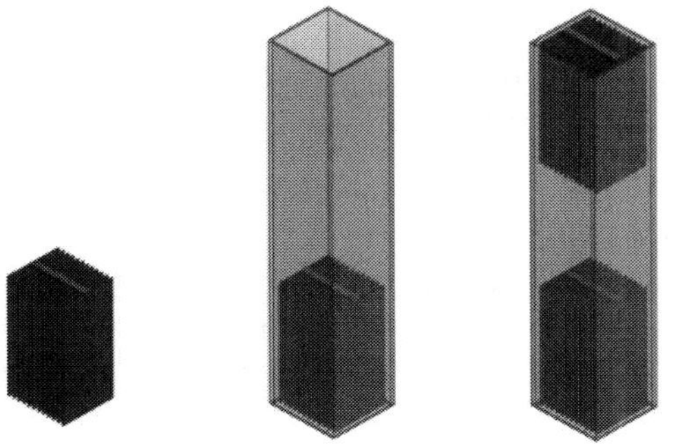

Figure 2: The three different configurations under consideration: unshielded (left), shielded (centre) and segregated (right).

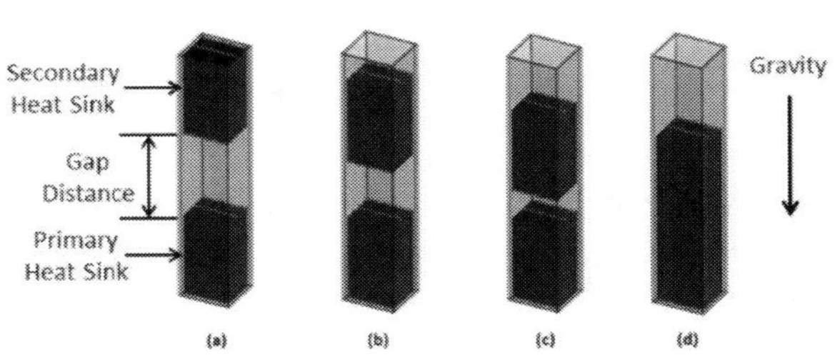

Figure 3: Segregated heat sink structures with varied heat sink gap; (a) 300mm (b) 200mm, (c) 100mm and (d) 5mm.

To enable the quantification of radiative losses, a thermocouple probe was attached to the heat sink at a range of locations and monitored at different temperatures using a FLUKE Ti32 handheld thermal imager. The emissivity of the heat sink was calculated using a process outlined in Khor *et al.* [4], and found to have an average value of 0.97. The polycarbonate had an emissivity of 0.95.

Temperature was monitored using National Instruments TBX-68T and USB-TC01 cards reading K-type thermocouples, calibrated to ±0.1°C, placed in different locations and depths within the base of the heat sink. The average temperature of the base plate of the heat sink was used to calculate values for Nusselt number, Elenbaas number and the film temperature. Three power supplies – a BSI PSM 3/2A, a Sorensen DCS 80-37 and a Delta Elektronika SM 3004-D – were used to provide controlled voltage to each cartridge heater. Equation (1) was used to calculate the power dissipated. Labview 2011 provided PID control to each individual heater, using feedback from a thermocouple to vary the duty cycle of a pulse-width modulated digital output. A National Instruments PCI-6229 card and a CB-68LP connector block generated the pulse-width modulated digital output which switched a Crydom D2D07 relay on and off.

$$P = \sum V_h \frac{V_r}{R} \tag{1}$$

The voltage drop across the heaters (V_h) and the average voltage drop in each resistor (V_r) were measured using a calibrated FLUKE 8842a multimeter, connected as close as possible to the cartridge heater, and a National Instruments PCI-6229 card with a CB-68LP connector block to record the average voltage drop in each resistor (V_r), of known resistance (R), connected in series with each heater. A circuit diagram for three heaters and resistors is shown in Figure 4. V_r was averaged over a 10 minute period after the system reached thermal equilibrium.

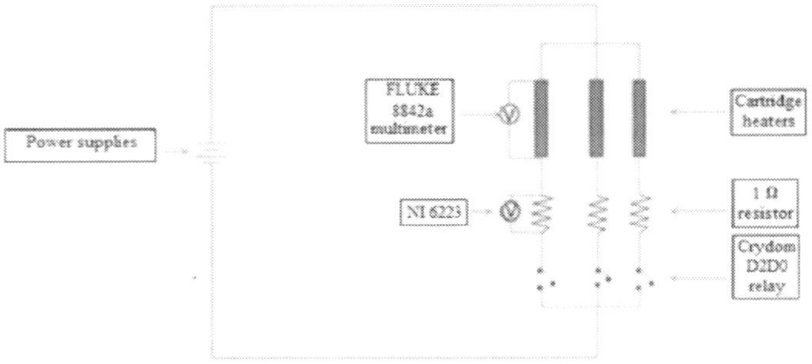

Figure 4: Schematic of the electrical setup.

The Nusselt number was calculated by removing radiation heat dissipation, calculated using the relations presented in section 3.2, to isolate convective dissipation in order to obtain an average heat transfer coefficient. The primary sink was maintained at a constant temperature of 25°C above ambient during the experimentation, while the temperature of secondary heat sink was increased sequentially above the temperature of the primary heat sink. The method of Kline and McClintock [10] was used to conduct an uncertainty analysis on the measured data. The maximum uncertainty in Nusselt number was found to be 8.2%, with most cases below this value.

3. Theory

Theoretical models, using established literature, were used to quantify the radiative and convective heat transfer from an unshielded heat sink with the characteristic dimensions of the remote radio head defined in section 2. The heat sink was optimized at an operating temperature of 80°C in an ambient temperature of 55°C, as stipulated by a commonly-used industry standard [1].

3.1. Convective heat transfer

The convective heat transfer from a heat sink approximated as a series of parallel plates is governed by the Rayleigh number (Ra_L) based on the length of the heat sink (L):

$$Ra_{\mathrm{L}} = \frac{g\beta(T_s - T_{amb})L^3}{\gamma^2} Pr \tag{2}$$

Where g is acceleration due to gravity (~9.81m/s^2) and the properties of air are as follows: the expansivity, β (K^{-1}); the kinematic viscosity, ν (m^2/s); and Prandtl number, Pr (-). The plates are assumed to be at a uniform temperature T_s in an ambient of temperature T_{amb}. For heat sink structures, the Rayleigh number can be modulated by fin spacing, S, to yield the Elenbaas number:

$$El_S = Ra_{\mathrm{L}}(S^4/L^4) \tag{3}$$

Equation (4) [11] is used to calculate the optimum spacing for the parallel plates of a heat sink, with equation (5) [3] applied to incorporate the influence of the base plate:

$$S_{\mathrm{opt}} = 2.71\left(\frac{Ra_L}{L^4}\right)^{-\frac{1}{4}} \tag{4}$$

$$S_{opt} = \frac{2\mathrm{HS}}{2\mathrm{H}+\mathrm{S}} \tag{5}$$

The optimum spacing (S_{opt}) for parallel plates is modified in equation (5) to a revised spacing (S) for a finned heat sink with a fin height (H). The convective heat transfer for an unshielded heat sink can be determined using equation (6) [2]:

$$Q_{conv} = NQ_{fin} + \overline{h_{base}}A_{\mathrm{base}}(T_s - T_{amb}) \tag{6}$$

The total heat removed by convection (Q_{conv}) is a sum of the number of fins (N) times the heat dissipation in a single fin (Q_{fin}), calculated using equation (8), and a base plate dissipating heat via convection. The heat dissipated from the base plate is calculated using the heat transfer coefficient ($\overline{h_{base}}$), evaluated using equation (7), for the unfinned base area (A_{base}):

$$\overline{h_{base}} = 0.59\text{Ra}_{\text{L}}^{0.25}\frac{k_f}{L} \tag{7}$$

The heat transfer coefficient is calculated using the thermal conductivity (k_f) of the air at the film temperature and the Rayleigh number. The heat dissipation from a single fin is as follows:

$$Q_{fin} = \overline{h_{fin}}A_{fin}(T_s - T_{amb}) \tag{8}$$

The average Nusselt number ($\overline{Nu_S}$) over the finned surface of the heat sink is calculated using the Rayleigh number referenced to fin spacing (Ra_S), and the fin efficiency (η_{fin}), calculated using equation (10). This Nusselt number yields the heat transfer coefficient over the surfaces of the fin:

$$\overline{Nu_S} = \left[\frac{576}{\left(\frac{\eta_{fin}Ra_S S}{L}\right)^2} + \frac{2.87}{\left(\frac{\eta_{fin}Ra_S S}{L}\right)^{0.5}}\right]^{-0.5} = \frac{\overline{h_{fin}}S}{k_f} \tag{9}$$

Finally, the fin efficiency is calculated from Incropera *et al.* [12]:

$$\eta_{fin} = \frac{\tanh(aL_c)}{aL_c} \tag{10}$$

Where the fin parameter, $m = \left(2\overline{h_{fin}}/k_{fin}\,t\right)^{1/2}$ and the corrected fin length, $L_c = L + (t/2)$.

3.2 Radiative heat transfer

An established model expressed in Shabany [13] was used to quantify the effects of radiation in order to isolate the convective heat dissipation in the experimental data:

$$Q_{rad} = \eta_{fin}\left(Q_{top} + Q_{bot} + Q_{side} + (\text{N} - 1)Q_{channel}\right) \tag{11}$$

The total heat removed by radiation (Q_{rad}) is calculated as the contributions from the top (Q_{top}), bottom (Q_{bot}), side (Q_{side}) and channel ($Q_{channel}$) surfaces. N is the number of fins and η_{fin} accounts for the temperature variation in the fins:

$$Q_{top} = Q_{bot} = \sigma\varepsilon(HtN + BW)(T_s^4 - T_{amb}^4) \tag{12}$$

$$Q_{side} = \sigma\varepsilon(4H + 2B + tN)L(T_s^4 - T_{amb}^4) \tag{13}$$

The heat dissipated by radiation from the top, bottom and side are evaluated using the Stefan-Boltzmann constant (σ), the base (T_s) and ambient (T_{amb}) temperatures, the emissivity (ε) of the heat sink, the fin height (H), fin thickness (t), base thickness (B), heat sink width (W), length (L) and number of fins (N).

$$Q_{channel} = \sigma\hat{F}(S + 2H)L(T_s^4 - T_{amb}^4) \tag{14}$$

The heat removed by radiation from each channel is a function of the channel gray body shape factor ($\hat{F}$) which is a function of F_{s-surr}, the total view factor between the walls and base of the channel and their surroundings:

$$\hat{\text{F}} = \frac{1}{((1-\varepsilon)/\varepsilon) + (1/F_{s-surr})} \tag{15}$$

$$F_{s-surr} \approx 1 - \frac{2(H/S)\left[(1 + (L/S)^2)^{1/2} - 1\right]}{2(H/S)(L/S) + [(1 + (L/S)^2)]^{1/2} - 1} \tag{16}$$

It should be noted that the fin efficiency values used in the convective and radiative calculations in this paper were found to be close to unity, as would be expected for a passively cooled heat sink.

4. Numerical modelling

In order to complement the experimental data, numerical modelling was performed using Star-CCM+ v6.02.007 from CD-adapco, running on a 64-bit Windows 7 PC. The modelling was used to examine the influence of two parameters associated with the segregated structure: the vertical spacing between the primary and secondary heat sinks, and the fin count of the secondary heat sink. In all cases, the heat sink and shield combinations, outlined in section 2, were modelled as a constant density, three-dimensional structure with segregated solid energy and surface-to-surface grey thermal radiation. The air was modelled as a three dimensional, laminar flow using constant density and invoking the Boussinesq approximation. Steady-state conditions were assumed throughout.

The heat sink, shown in figure 1, was represented with an emissivity of 0.968, thermal conductivity of 180W/m.K and density of 2700 kg/m^3. The shielding material was made from polycarbonate with an emissivity of 0.953, conductivity of 0.19W/m.K and density of 1200 kg/m^3. The numerical analysis used values obtained experimentally for the ambient temperature and the temperature difference. The properties of air were taken at the film temperature using values interpolated from table 3.

In the analysis presented in this paper, a fixed temperature difference above ambient is imposed on the heater locations, shown as L_{h1} and L_{h2} in figure 1. The simulation monitored heat dissipation of the heat sink and the maximum exit velocity from the primary heat sink. The solution criteria required the maximum velocity to have a standard deviation of less than 0.003m/s over 50 samples and the heat dissipation to be less than 0.05W in 100 samples. The meshing was checked against experimental values obtained by Oosthuizen [14] to ensure the simulation results were mesh independent.

Firstly, a single heat sink was numerically simulated in isolation and compared to experimental data and established correlations. The heat sink was then modelled at the base of a shield, 900mm in length, and simulated using the ambient temperatures and temperature difference listed in table 2. An unheated secondary heat sink was then added to the shielded structure, at the top of the shielded structure as shown in figure 2.

Table 2: Ambient temperature and temperature difference for an unshielded, shielded and secondary heat sink with no heat.

Unshielded		**Shielded**		**Secondary No Heat**	
Ambient (°C)	$T_p - T_\infty$ **(°C)**	**Ambient (°C)**	$T_p - T_\infty$ **(°C)**	**Ambient (°C)**	$T_p - T_\infty$ **(°C)**
19.2	5.1	17.0	5.1	17.9	5.1
19.1	10.0	17.1	10.1	18.7	10.1
19.1	15.0	17.0	15.0	18.8	15.0
19.2	20.0	16.3	20.0	19.2	20.0
19.3	25.0	16.8	25.0	19.2	24.9
19.4	34.9	17.2	34.9	16.4	34.8
19.4	44.8	17.6	44.8	19.2	44.8
20.3	54.8	18.1	54.7	17.6	54.8
20.5	64.7	18.4	64.5	18.7	65.4

A heated secondary heat sink, temperature given in table 3, was numerically calculated at a primary-secondary spacing of 0mm up to 300mm in steps of 50mm; in addition, a spacing of 5mm was also evaluated. At a gap distance of 0mm, both heat sinks were in physical contact, however, they were not permitted to conduct heat to each other. The primary heat sink was maintained at a temperature difference of 24.9°C above ambient throughout.

Finally, based on the results from the location of the heated secondary heat sink, the number of fins on the secondary heat sink was varied. The fin thickness was kept constant while the number of fins was varied from 2, in steps of 2, up to 14 fins; additional variants with 7, 13 and 15 fins were tested. The thermal properties were evaluated at an ambient temperature of 19.5°C, a primary heat sink temperature of 25.0°C above ambient and a secondary heat sink temperature of 64.4°C.

Table 3: Thermal properties of a heated secondary heat sink.

Secondary Heat Sink		
Gap Distance (mm)	**Ambient (°C)**	$T_{sec} - T_{\infty}$ (°C)
0	19.5	64.4
5	14.8	64.3
50	19.5	64.4
100	20.8	64.2
150	19.5	64.4
200	15.4	64.3
250	19.5	64.4
300	19.5	64.4

5. Results and discussion

Section 5 is split into three sub-sections which address, respectively: 5.1, the unshielded heat sink; 5.2, the shielded heat sink (within the chimney); and, 5.3, the segregated structure. For the segregated structure, a Nusselt ratio – a comparison against the baseline case of the unshielded heat sink – is also presented with a heated secondary heat sink present. The influences of the spacing between the primary and secondary heat sinks, and the fin count of the secondary sink, are also investigated.

5.1. Unshielded

Correlated, numerical and experimental values for the Nusselt number of the primary heat sink are plotted as a function of the Elenbaas number in figure 5. Error bars represent the uncertainty in the experimental values, which exhibited maxima of 9% and 7% respectively in the Nusselt and Elenbaas numbers. The experimental and numerical values are within 10% and 6% of the correlation values respectively, an indication of the adequacy of the investigative methods used in this paper.

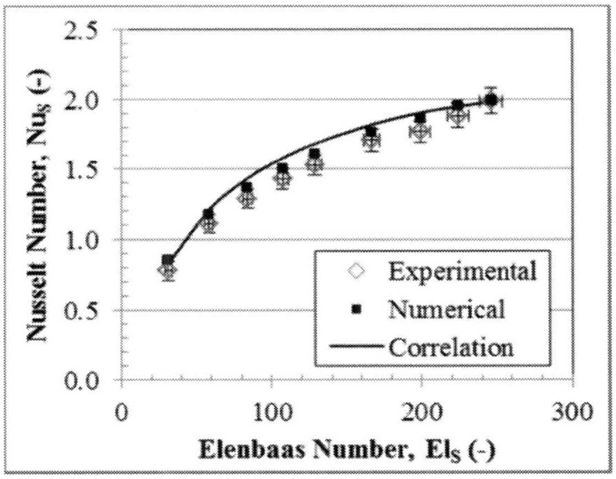

Figure 5: Nusselt Number as a function of Elenbaas Number for a single unshielded heat sink.

5.2. Chimney structure

Figure 6 presents the Nusselt number as a function of Elenbaas number for a shielded and an unshielded heat sink. The unshielded values are the numerical values presented in figure 5. The experimental and numerical values for a shielded heat sink agree within 8% of one another. The use of a shield augments the convective heat transfer by an average of 64% in comparison with an unshielded sink. At lower Elenbaas numbers, the shielded structure offers the most improvement, greater than 70%, however as the Elenbaas number increases, the enhancement diminishes to below 58%.

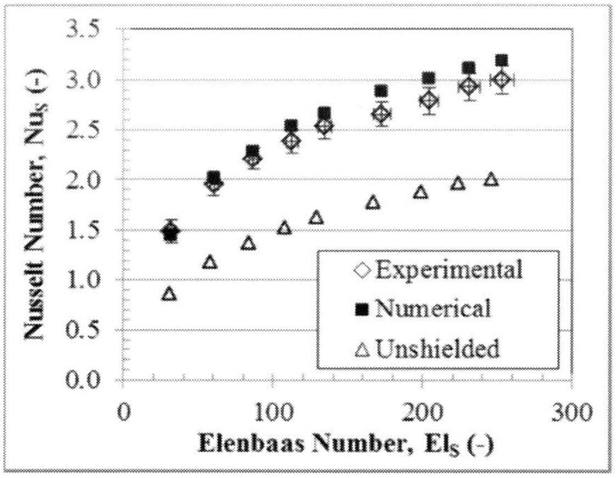

Figure 6: Nusselt number as a function of Elenbaas number for a shielded heat sink.

5.3. Secondary segregated structure

Figure 7 presents the numerical (squares) and experimental (diamond) Nusselt-Elenbaas data for a segregated structure with an unheated secondary heat sink. For comparison, numerical results for an unshielded (triangles) and a shielded (crosses) heat sink, replicated from figure 6, are also presented. A reduction of greater than 30% in Nusselt number in comparison with a shielded case is evident, however the addition of an unheated secondary heat sink in a shielded structure still outperforms an unshielded sink. The additional flow resistance offered by the secondary sink within the shield evidently diminishes the Nusselt number of the primary heat sink.

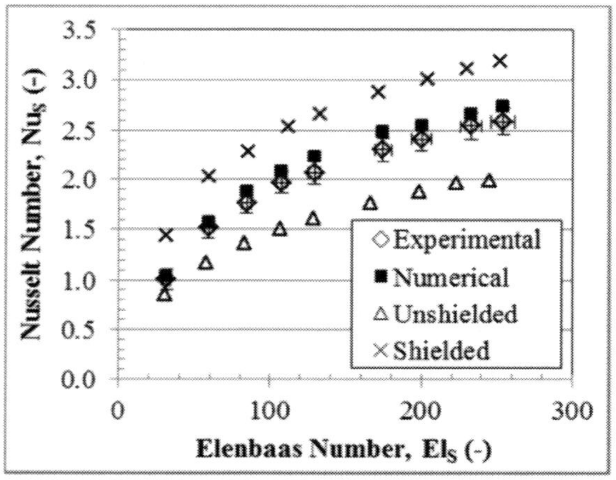

Figure 7: Nusselt number as a function of Elenbaas number for a secondary segregated structure with an unheated secondary heat sink.

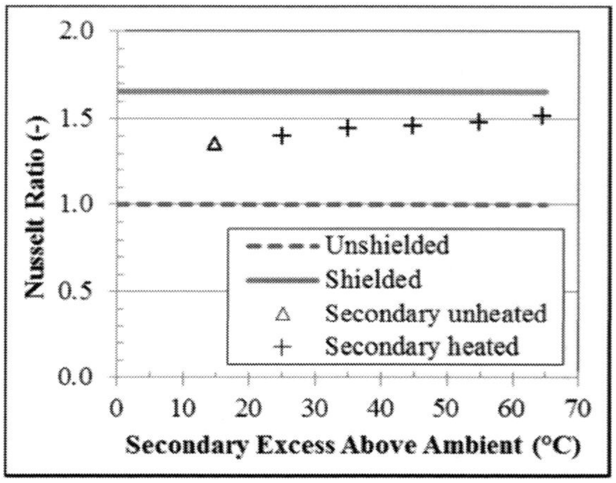

Figure 8: Experimental Nusselt ratio for a primary heat sink plotted against the secondary heat sink temperature excess over ambient.

In figure 8, the Nusselt ratio – a comparison against the baseline case of the unshielded heat sink – is plotted as a function of the base plate temperature of the secondary heat sink for four cases: unshielded, shielded, unheated secondary; and heated secondary. The Nusselt ratio exhibits an apparently linear increase in proportion to the temperature of the secondary heat sink, due to the additional mass flow induced by the secondary sink. At a secondary sink temperature of 65°C above ambient, the limit of the variable under consideration in this paper, an enhancement in the Nusselt number of ~50% was observed in comparison with the baseline unshielded sink. This finding has practical utility: there is an evident advantage in segregating heat sinks within a representative outdoor wireless device to carry temperature sensitive and power devices on primary and secondary heat sinks, within a shielded structure that promotes chimney flow.

Figure 9 presents the Nusselt ratio, based on numerical data, as a function of the spacing between the primary and secondary heat sinks. The secondary sink is fixed at a 65°C over ambient throughout, but an unheated secondary configuration (at 300mm spacing) is also included for reference. Baseline cases of unshielded (blue dashed line) and shielded (red solid line) are also presented. The idealised case of the secondary heat sink touching, but not conducting heat to, the primary heat sink provides an increase of over 75% in the Nusselt ratio. A spacing of 50mm provides the largest improvement for a practical scenario, surpassing the Nusselt ratio of a single heat sink within a shield.

Nusselt ratio, evaluated numerically, is presented in Figure 10 as a function of the number of fins of the secondary heat sink. Numerical data for an unshielded sink, a shielded sink and a segregated structure with an unheated secondary heat sink are also shown, each with 12 fins on each side (24 in total). The optimum fin count for the secondary heat sink was found to be 8, which would appear to be a counterintuitive finding (the form of equation 4 would suggest that the higher the temperature of a sink, the smaller the optimum spacing). However, the analysis presented here considers the behaviour of the primary heat sink in a segregated ensemble: moreover, the local air temperature at the inlet to the secondary sink is considerably higher than ambient, due to the influence of the primary sink.

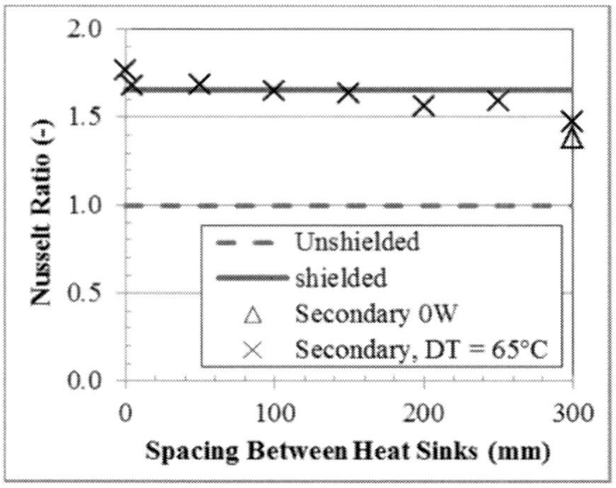

Figure 9: Nusselt ratio as a function of the spacing between the primary and secondary heat sinks.

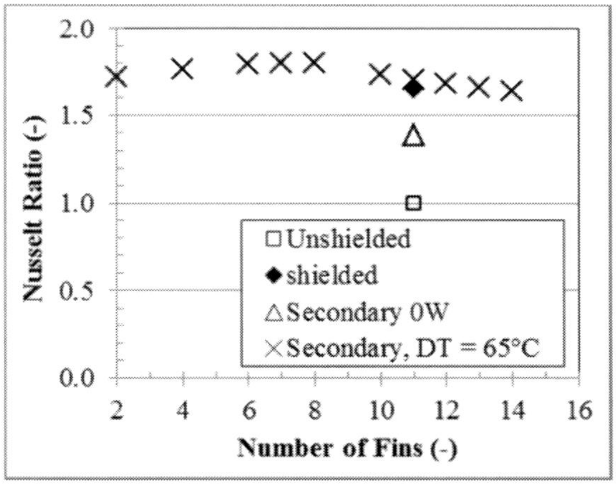

Figure 10: Nusselt ratio as a function of the number of fins on the secondary heat sink with a fixed primary-secondary spacing of 50mm.

6. Conclusions

In this paper, heat sink geometries representative of a remote radio head were examined in terms of the enhancement in the Nusselt-Elenbaas behaviour of the primary sink. The following improvements were found when compared to the Nusselt number of an unshielded heat sink:

- A single heat sink, when placed at the inlet of a shield three times its length, enhanced the Nusselt number by an average of 64% compared to the unshielded case.

- A segregated structure, with an unheated secondary heat sink, reduces the Nusselt number of the primary sink by over 30% compared to a shielded heat sink. The Nusselt number of the primary sink increases in proportion to the temperature of the secondary heat sink.
- The vertical spacing between primary and secondary heat sinks within a segregated structure could be adjusted to achieve further enhancements in the Nusselt number of the primary heat sink. The optimum location was found to occur at zero spacing, but this required the suppression of conduction between the sinks. Consequently, a spacing of ~50mm represented a practical choice.
- Variations of the number of fins on the secondary heat sink at a fixed temperature excess of 65°C above ambient, in a segregated structure with 50mm primary-secondary spacing, offered an enhancement of 80% compared to an unshielded sink. The maximum enhancement occurred at 8 fins, resulting in a fin spacing of 19.2 mm.

In conclusion, the outcome of this paper is an improved understanding of a novel passive thermal management solution which is of practical relevance for the design of passively-cooled outdoor wireless communication equipment.

Acknowledgements
The authors acknowledge the financial support of Science Foundation Ireland under Grant No. 10/CE/I1853.

References

[1] Telcordia Technologies, "Generic Requirements for Electronic Equipment Cabinets," *GR-487-CORE,* no. 3, 2009.

[2] A. Bar-Cohen and W. M. Rohsenow, "Thermally optimum spacing of vertical, natural convection cooled, parallel plates," *Journal of Heat Transfer,* no. 106, pp. 116-123, 1984.

[3] A. Malhammer, "Optimum Sized Air Channels for Natural Convection Cooling," in *Telecommunications Energy Conference*, Stockholm, 1987.

[4] Y. K. Khor, Y. M. Hung and B. K. Lim, "On the role of radiation view factor in thermal performance of straight-fin heat sinks," *International Communications in Heat and Mass Tranfer,* vol. 37, pp. 1087-1095, 2010.

[5] F. K. Moore, "Minimum size of large dry cooling towers with combined mechanical and natural draft," *J. Heat Transfer,* vol. 95, no. 3, 1973.

[6] K. T. Lee and W. M. Yan, "Laminar natural convection between partially heated vertical parallel plates," *Wärme-und Stofföbertragung,* no. 29, 1994.

[7] D. Naylor and J. D. Tarasuk, "Natural convective heat transfer in a divided vertical channel. I: Numerical study.," *Journal of heat transfer,* vol. 2, no. 115, 1993.

[8] D. Naylor and J. D. Tarasuk, "Natural convective heat transfer in a divided vertical channel. II: Experimental study.," *Journal of heat Transfer,* vol. 2, no. 115, 1993.

[9] T. S. Fisher, K. E. Torrance and K. K. Sikka, "Analysis and optimization of a natural draft heat sink system.," *IEEE Transactions on Components, Packaging, and Manufacturing Technology Part A* , vol. 2, no. 20, 1997.

[10] S. J. Kline and F. A. and McClintock, "Describing uncertainties in single-sample experiments," *Mechanical engineering,* vol. 1, no. 75, 1953.

[11] A. Bar-Cohen and W. M. Rohsenow, "Thermally optimum spacing of vertical, natural convection cooled, parallel plates," in *Heat transfer in electronic equipment*, Washington, 1981.

[12] F. Incropera, D. P. DeWitt, T. L. Bergman and A. S. Lavine, Fundamentals of heat and mass transfer, Hoboken, NJ: John Wiley & Sons, Inc., 2006.

[13] Y. Shabany, Heat Transfer; Thermal Management of Electronics, Boca Raton Florida: Taylor & Francis Group, 2010.

[14] P. H. Oosthuizen, "A numerical study of laminar free convective flow through a vertical open partially heated plane duct," *ASME HTD,* no. 32, 1984.

Journal of Physics: Conference Series **525** (2014) 012020 doi:10.1088/1742-6596/525/1/012020

High Density Die Casting (HDDC): new frontiers in the manufacturing of heat sinks

Andrea Sce, Lorenzo Caporale

Aavid Thermalloy S.r.l. - Via del Fonditore 4, 40138 Bologna, Italy (www.aavid.com)

sce.a@aavid.com, caporale.l@aavid.com

Abstract. Finding a good solution for thermal management problems is every day more complex, due to the power density and the required performances. When a solution suitable for high volumes is needed, die-casting and extrusion are the most convenient technologies. However designers have to face the well-known limitations for those processes. High Density Die Casting (HDDC) is a process under advanced development, in order to overcome the extrusion and traditional die casting limits by working with alloys having much better thermal performances than the traditional die-casting process, while keeping the advantages of a flexible 3D design and a low cost for high volumes. HDDC offers the opportunity to design combining different materials (aluminium and copper, aluminium and stainless steel) obtaining a structure with zero porosity and overcoming some of die-casting limits, as shown in this paper. A dedicated process involving embedded heat pipes is currently under development in order to offer the possibility to dramatically improve the heat spreading.

1. Introduction

Thermal management problems are becoming more and more challenging: power density is constantly growing in order to overcome the limits and develop a powerful application in a reduced space (or giving a more compact final product). Furthermore, power sources often require improved performances in order to ensure them a longer lifetime.

Thermal engineers try to find suitable solutions by using the most common technology, by respecting all the customer's requirements (including the target cost). There are many technologies available for manufacturing; however, when high volumes are involved for mass production, extrusion and die-casting are the most common alternatives; however, during the design phase it is necessary to remember their well-known limitations:

- Extrusion is only a 2D process;
- It can use only 6xxx alloys;
- It has limitations in fin density and geometry (as shown in figure 1);
- It cannot have any different material inserted.

Die-casting can be done only with some kind of aluminum alloys, characterized by a very poor thermal conductivity (normally between 90 and 140 W/m-K): this further constraint could be a bottleneck where thermal performances are difficult to be reached and in all the applications where the heat transfer by conduction has a relevant role in the overall heat transfer phenomena ([1], [3] and [5]). The most common constraints of die-casting (including geometry limits) are summarized in figure 2.

Content from this work may be used under the terms of the Creative Commons Attribution 3.0 licence. Any further distribution of this work must maintain attribution to the author(s) and the title of the work, journal citation and DOI.
Published under licence by IOP Publishing Ltd

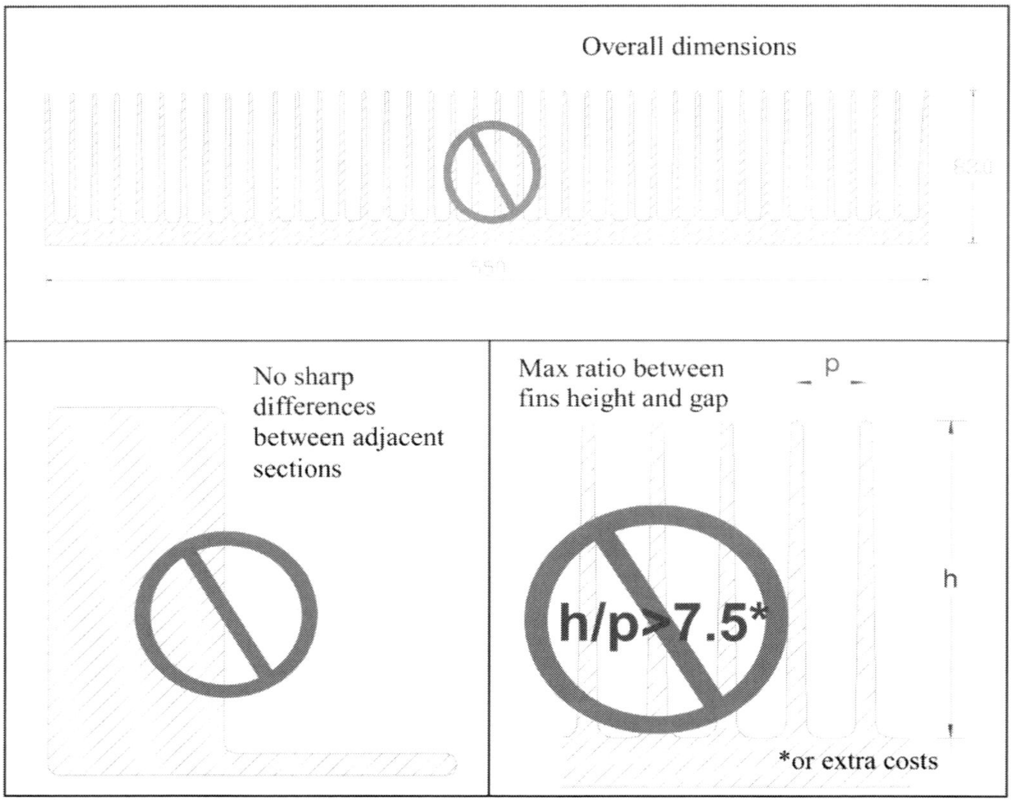

Figure 1. Extrusion limits

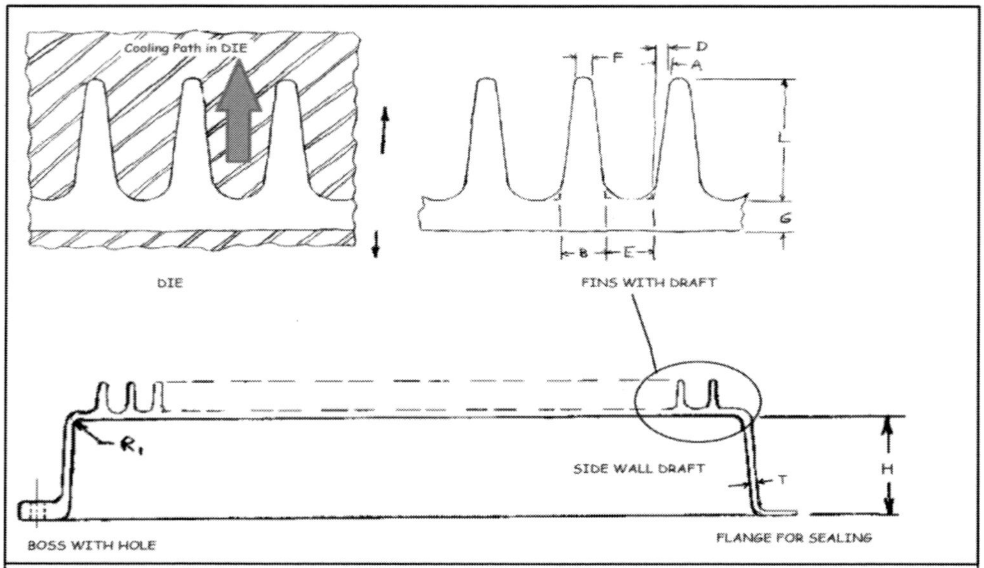

Easier to cast alloys have higher copper content ⇨ limited corrosion resistance for outdoor applications
Die cast parts have porosity ⇨ not well suited for electrochemical finishes (anodizing)
The typical high silicon content lowers the melting point ⇨ metal joining processes (brazing) cannot be used

Figure 2. Die-casting limits

2. High Density Die Casting (HDDC)

2.1. General description
High Density Die Casting (HDDC) is a process under advanced development, able to overcome extrusion limits and some of die-casting limits, in addition to the opportunity to work with high thermal conductive aluminum alloys. The process keeps the advantages of a flexible 3D design and a low cost for high volumes due to the manufacturing process. HDDC is a hybrid process that combines elements of casting and hot forging, offers the opportunity for cost effective heat sink manufacturing in near net shapes. This process enables optimized heat sink designs that incorporate cavities, bosses, and thin fins for low weight. In addition, the process is flexible to allow heat sink fins to be forged in place as integral parts of an electronic housing or enclosure.

The HDDC process used to produce aluminum heat sinks from molten aluminum is shown in figure 3. The process involves firstly transferring a pre-fixed amount of molten aluminum from a melt furnace to a mold cavity using a ladling robot. The top section of the die, the punch, is then lowered into the cavity and pressed downwards with high pressure to form the molten metal into the shape of the heat sink. When the molten metal has solidified, the punch is raised, the heat sink is ejected from the cavity and a pick-up robot grabs the heat sink and places it onto an unloading area. The whole cycle is then repeated for continuous production.

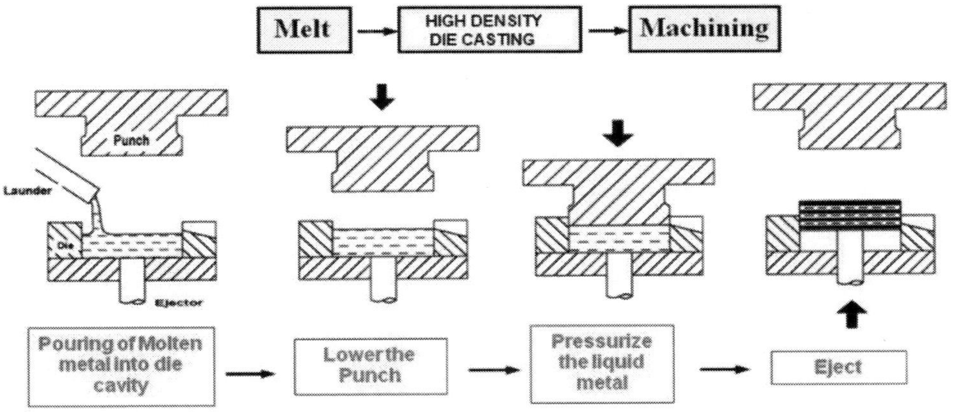

Figure 3. Steps in the HDDC process

2.2. Differences against die-casting
With its capability of utilizing high thermal conductivity wrought aluminum material like Al 6063 or Al1070, HDDC heat sinks can achieve much better heat dissipation efficiencies than aluminum die cast heat sinks. With the application of direct high pressure during the solidification process, porosity free, leak tight products with good mechanical strength can be produced.

Different fin shapes such as flat, oblique, elliptical, round pin arrays and radial plate designs can be made to take full advantage of the three-dimensional geometry capability of the process and to direct the airflow as required. This can greatly reduce back pressure by creating turbulence only where needed. The re-design of the fin arrangement facilitated by the HDDC process can enable reduced flow resistance to incoming air or liquid streams and enhance turbulence between the fins for improved thermal performance.

Advantages and features of HDDC products can be summarized as follows:

- High thermal conductivity Al alloys (200+ W/m-K) as well as normal die-cast Al alloys as needed
- Fine grain structure, zero porosity or shrinkage defects due to robust process control during solidification.
- Thin and high aspect ratio fin features
- Defect free surface with lower roughness values
- Feasible to do all kind of electrochemical surface treatment process
- Embedding of copper, graphite as well as aluminum material within product

- High feasibility of brazing products (application like liquid cold plates and housings)

Some of these aspects are summarized in table 1.

Table 1. Preliminary comparison for heat sink with 40-60mm tall fins between HDDC and die-casting

Description of features	Die-casting	HDDC
Fin Geometry (Plate, pin etc.)	All Types	All Types
Minimum Enclosure wall thickness (mm)	2-4	1.5-3.5
Minimum fin to fin spacing (mm)	5	3-4
Minimum fin tip thickness (mm)	2.5-3	1.0-1.5 plate fin 2.0 pin fin
Minimum fin angel (degree)	1-1.5	0.5-1.0
Porosity (%)	3-7	0
3D near net shape rating (5 is best)	5	4
Surface quality rating (5 is best)	2	4
Leak tight rating (5 is best)	2	5
Aluminium Alloys	ADC 10-12 or 3xxx	1xxx, 6xxx, 3xxx

Some of the characteristic dimensions need a range as the limit is fixed in combination with others: for example, the minimum wall thickness can be variable according to the overall shape and geometry of the part (fins height is very important in order to define the other parameters).

The latest three characteristics have been evaluated through a qualitative evaluation rate, where values are coming from a comparison between the two products (made with two different alloys).

HDDC offers the opportunity to work with different materials (aluminium and copper, aluminium and stainless steel) where the added material is embedded in a high density aluminium "body" made by casting process.

A dedicated process involving embedded heat pipes is currently under evaluation and development in order to offer the possibility to have solutions with heat spreading.

2.3. Development process

The development of this technology has started from some thermal simulations (made with different CFD software's such as Icepak and QFin), (as shown in figure 4) for some applications. They have been performed by considering a fully ducted air flow and the same air speed (2 m/s), ambient temperature overall dimensions and power density applied on the base.

The results show an improvement of performances (Rth is about 40% less) due to the increased thermal conductivity of the material (from 120 to 200 W/m K) and the exchange surface (from 14 to 20 fins).

After a first and detailed research phase ([2] and [4]), a pilot plant has been built for the manufacturing of the first prototypes: the aim of these pieces was to conduct a detailed metallographic study in order to understand if through this process it was possible to get the expected good results.

The results (shown in figure 5) have been good: there is no porosity in the final product. The figure shows the difference between the two technologies: while the traditional casting process is characterized by some porosity distributed in the volume (with a relevant impact on the leak prevention of the final product), HDDC products have pore free microstructure. This means that HDDC can be used to make leak tight enclosures, including leak tight liquid cold plates.

Journal of Physics: Conference Series **525** (2014) 012020 doi:10.1088/1742-6596/525/1/012020

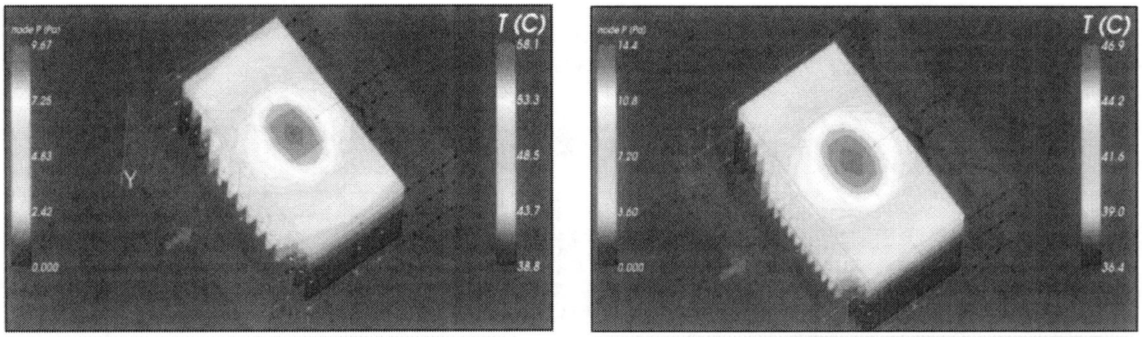

Tmax=58.1 °C Rth=0.281 °C/W ΔP=9.67 Pa **Tmax=46.9 °C Rth=0.169 °C/W ΔP=14.4 Pa**

Figure 4. CFD comparison between traditional DC (left) and HDDC (right); HDDC allow more fins

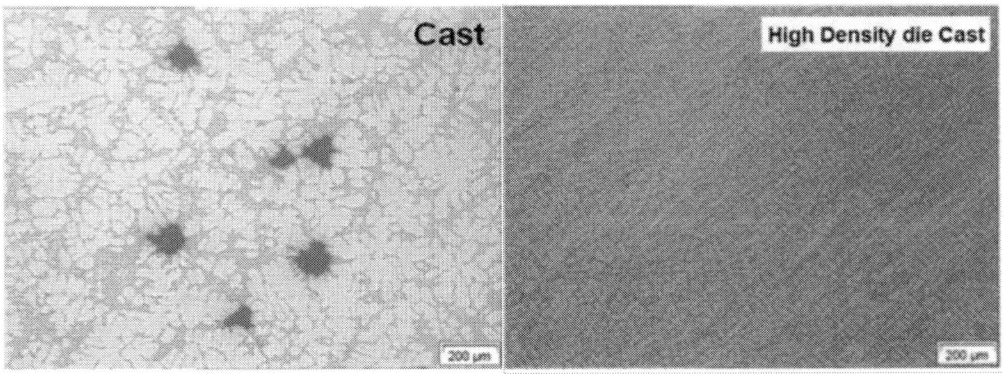

Figure 5. Metallographic analysis of HDDC structure vs die-casting structure

The pore free microstructure is able to increase also mechanical strength: it has been properly validated as shown in figure 6 (through the classic tensile test). HDDC products are similar or better than extruded ones made with the same alloy and the

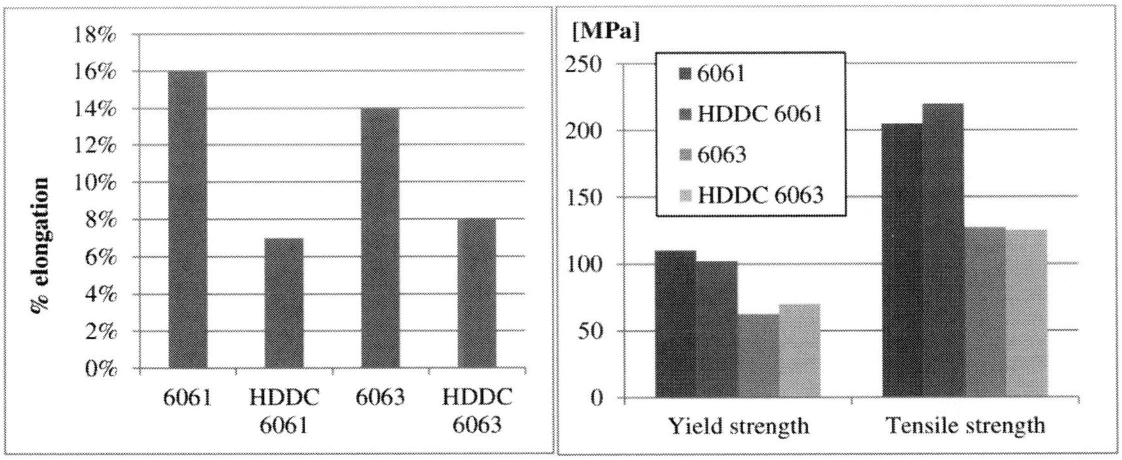

Figure 6. Mechanical Properties for Al 6xxx-T4 Series: test results comparison between HDDC and extruded material

Based on the above results and after different manufacturing tries (in order to verify the theoretical limits), scale-up has gone on and a plant for mass production has been built.

Capabilities are still under an improvement process: it is started with simple products and it is oriented to more complex parts, with embedded different materials and heat pipes (as indicated in figure 7).

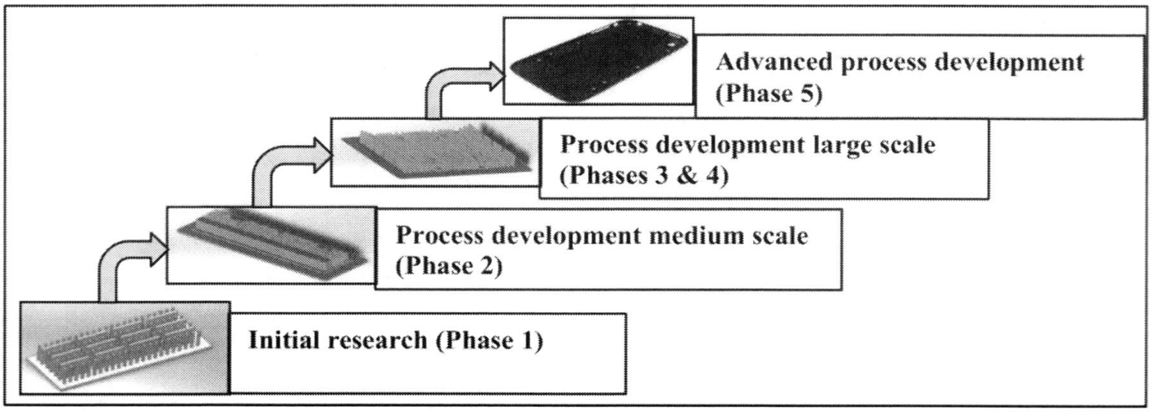

Figure 7. Capabilities development path

After a first step where small parts (up to 80x200 mm for the base) have been studied and manufactured (with a parallel development of the tool manufacturing), efforts have been concentrated on the dimensions of the parts, from medium scale to large scale (up to 500x500 mm for the base).
These steps have given a more realistic idea of the dimensional limits and the size of the tools to be developed for future manufacturing.
More complex products (with different materials and heat pipes) still have to be further developed, in order to use in the best way this technology.

3. Results and applications

3.1. Products feasible with HDDC

Based on the above tests and analysis, HDDC can be considered as a real improvement of traditional die-casting: it is possible to work with better aluminum alloys (from thermal point of view) and to overcome some of the die-casting limits.
Particularly:

- Successfully formed a full size heat sink 450x450x60 mm with embed continuous extrusion fins in manual mode
- Bonding between fin and base is quite strong, pull strength is 80 N/mm^2
- Thermal performance is as good as solid extruded heat sink
- Overall mass reduced significantly by using thinner fins/walls with an improvement of cooling performances

HDDC has been already used to made different kind of products (enclosures, pin fin structures, etc.) with good results; some example are in figure 8. Dimensions are variable from 80x200 mm (the first pieces made after the research stage – figure 8a) to 450x450 mm (figure 8d), passing through 150x450 mm (intermediate stage).

Concerning the tolerances level, dimension tolerances are usually positive and are approximately 0.5 % of the dimension. Die closure tolerances are in the direction of opening and closing, and range from 0.1mm to 1.5mm in the vertical direction depending on the precise volume control of the melt from the ladling.

High density die cast wrought Al heat sinks can achieve a surface roughness Ra better than 0.8 microns and they can be treated for aesthetic appearance and surface protection using wet electrochemical processes such as anodizing, nickel plating, e-coating and chromate treatment. In addition, the parts can be finished using finishing processes like electrostatic powder coating and polyurethane coating that are conventionally used for aluminum die castings.

Due to the possibility to use different alloys than traditional die-casting process, HDDC wrought Al heat sinks can be incorporated into more complex or larger assemblies by fusion processes like soldering, CAB and vacuum brazing, and fusion welding as well by non-fusion processes like friction stir welding.

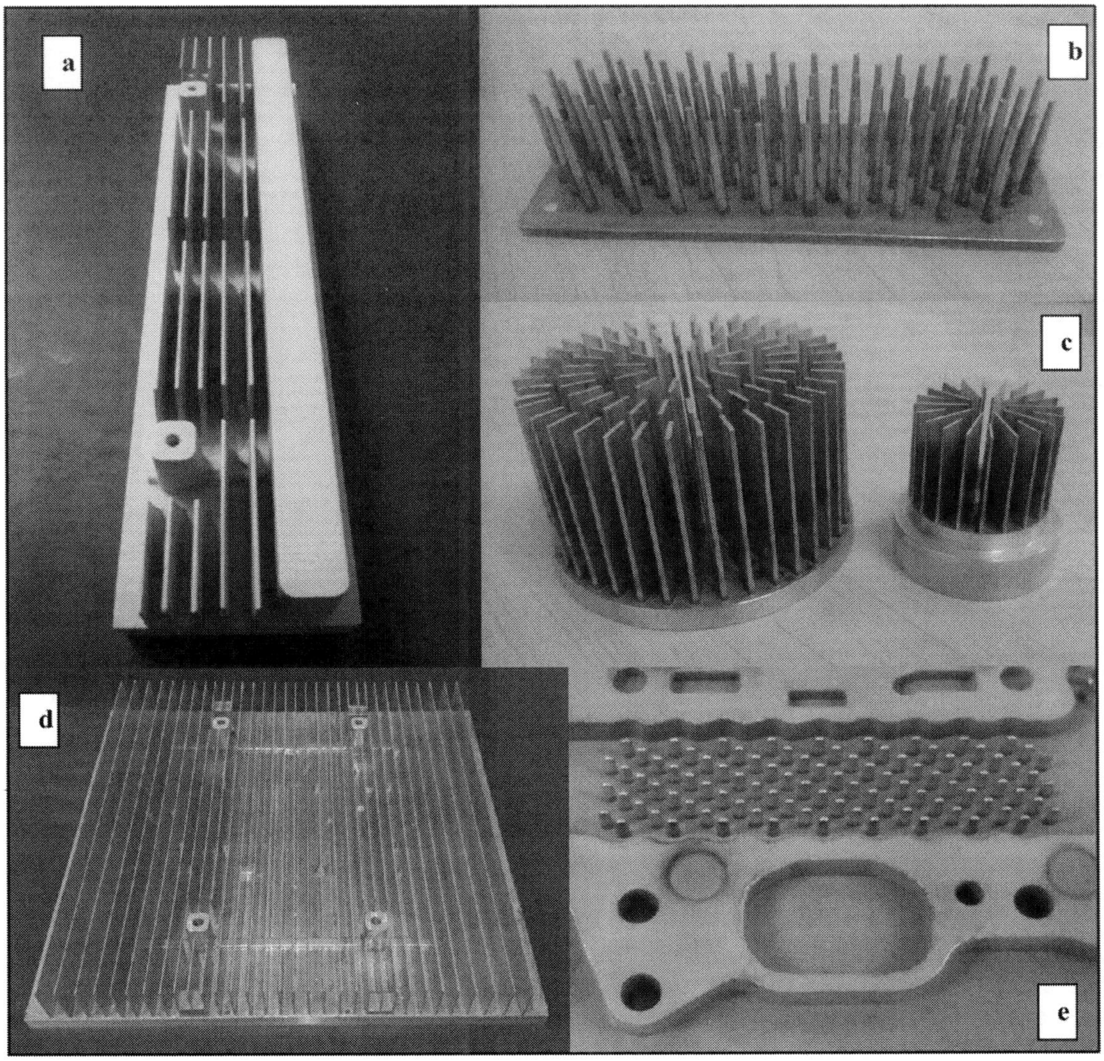

Figure 8. Examples of HDDC products

3.2. Working with different materials

Test and manufacturing processes have been launched by using different materials, such as aluminum and copper

Cuts, microscopic analysis and ultrasonic scans been made and the results are shown in figures 9 and 10.

Figure 9. Examples of HDDC products with Al and Cu (cut section – 60x17x40 mm)

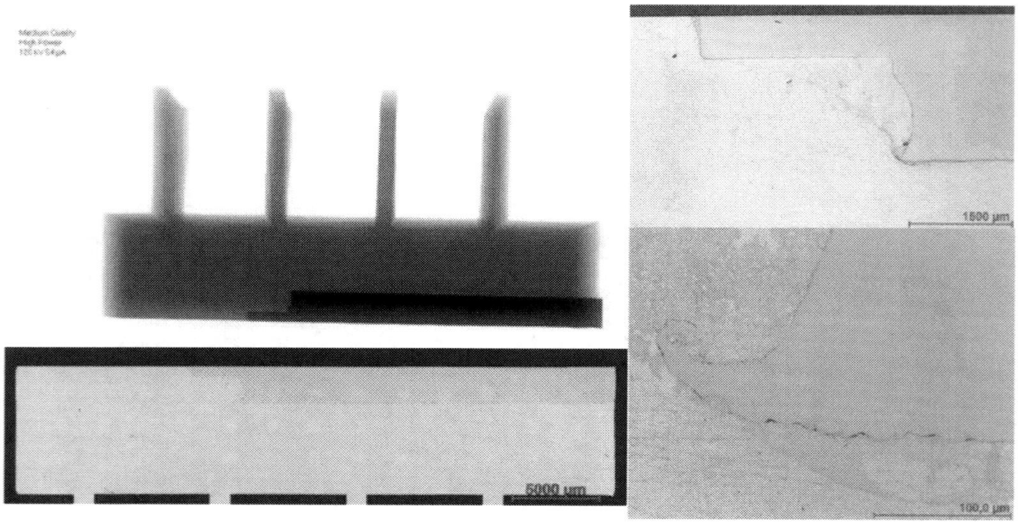

Figure 10. Microscopic and ultrasonic scan on HDDC made with Al and Cu

This step has been developed by study and manufacture a small part with aluminum and copper: the theoretical study has been used to fix the geometry and the detail of the shape in the contact area between the two metals. The expected results were to confirm the absence of porosity or micro gap int the mentioned contact area.

Then, the part has been made through HDDC process and a section where the two metals were in contact deeply analyzed in order to verify the results in terms of porosity (in the interface area), strength and behavior of the two materials during the HDDC process itself.

The above results have been used to fix the process and to make bigger parts, like the heat sink shown in figure 11.

Dimensions are much bigger than the part used for the first tests (400x500x60 mm vs 60x40x17 mm) and the results show that the process is moving in the right direction.

Figure 11. A big part made by HDDC with Al and Cu

4. Conclusions

High density die casting gives the opportunity to make heat sinks with several important advantages over their die cast and extruded counterparts. The increased thermal performance coupled with the ability to expand the fin configurations and shapes of the heat sinks are unique advantages for example in the design of pin fins, oblique or elliptical fins etc. and radial fin arrays. Some of these features cannot be achieved via extrusion or die casting techniques. Lastly, HDDC allows the harnessing of wrought aluminum alloy in a liquid state to form heat dissipation fins on housings and enclosures with intricate features.

The benefits related to the new limits on shapes and geometry are mainly visible according to the ventilation conditions: a HDDC part optimized for natural convection will be totally different than another one designed for forced convection. Even parts made for forced convection applications will be different according to the boundary conditions (available flow rate, maximum pressure drop, cabinet sizes and shape, etc.).

The application of high direct pressure helps eliminate hot tearing and creates products with superior mechanical properties and pore-free structure with improved thermal performance. The elimination of porosity allows the parts to be finished using wet electrochemical processes like anodizing and the use of wrought alloys enables the parts to be brazed and welded to create more complex assemblies.

5. Next steps

Now it is planned to go some steps ahead during 2014; the plan is:

- Embedded heat pipes
- Thin product development with multi-cavity tooling

The first development is a very interesting application for HDDC as heat pipes are common where a better heat spreading is needed or it is necessary to move the heat from a region to another one where there is space for the finned area.

It is not easy, as heat pipes have to be carefully handled: they have a thin copper wall and the contact surface between them and aluminum base is crucial in order to have a good heat transfer process. At 680 °C (melting temperature of aluminum alloys), the pressure of the vapor inside the tube can reach almost 100 bar, causing the bumping of the tube; in order to prevent this phenomena, a special process has been studied and the theoretical study is finished. First experimental test are currently planned.

The other development is important in order to manage very high volumes: multi-cavity tooling would allow the manufacturing of more pieces in a single shot. Of course, the number of cavities is strictly related to the dimensions of the part we are going to design: efforts will be first concentrated on small parts and according to the results of this step, bigger products could be analyzed too.

References

[1] NADCA Product Specification Standards for Die Castings, NADCA Publication #402, 7th Edition, 2009
[2] This M. C. Flemings, R. G.Riek, and K. P.Young, "Rheocasting," Materials Science & Engineering", Vol. 25, pp. 103-117, 1976
[3] "Another Design of an Electronic Enclosure for Die casting Production," Die Casting Development Council, Publication 9302/310, 1999 (accessed as "Application Brief No T010" at: http://www.cwmdiecast.com/resrce_briefs.cfm)
[4] J. L. Jorstad, "Future Technology in Die Casting," Die Casting Engineer, pp. 18-25, September 2006.
[5] D. S. Steinberg, Cooling Techniques for Electronic Equipment, Wiley Interscience, 1980

Journal of Physics: Conference Series **525** (2014) 012021 doi:10.1088/1742-6596/525/1/012021

Benchmarking of a novel contactless characterisation method for micro thermoelectric modules (µTEMs)

S. Hickey[1], **J. Punch**[1], **N. Jeffers**[2]

[1] CTVR, Stokes Institute, University of Limerick, Limerick, Ireland
[2] Thermal Management Research Group, Bell Labs, Alcatel-Lucent Ireland, Blanchardstown, Dublin 15, Ireland

E-mail: seamus.hickey@ul.ie

Abstract. Significant challenges exist in the thermal control of Photonics Integrated Circuits (PICs) for use in optical communications. Increasing component density coupled with greater functionality is leading to higher device-level heat fluxes, stretching the capabilities of conventional cooling methods using thermoelectric modules (TEMs). A tailored thermal control solution incorporating micro thermoelectric modules (µTEMs) to individually address hotspots within PICs could provide an energy efficient alternative to existing control methods. Performance characterisation is required to establish the suitability of commercially-available µTEMs for the operating conditions in current and next generation PICs. The objective of this paper is to outline a novel method for the characterisation of thermoelectric modules (TEMs), which utilises infra-red (IR) heat transfer and temperature measurement to obviate the need for mechanical stress on the upper surface of low compression tolerance (~0.5N) µTEMs. The method is benchmarked using a commercially-available macro scale TEM, comparing experimental data to the manufacturer's performance data sheet.

1. Introduction

Photonics Integrated Circuits (PICs) increasingly feature in today's optical communication systems, in order to realise devices with greater spectral efficiency and reduced power losses. PICs can represent a stringent packaging challenge, however, particularly in terms of their requirements for thermal control. Devices such as laser arrays can feature tight temperature limits (±0.1K), low operating temperatures (as low as 45°C), moderate heat loads (~1W) but very high heat fluxes (over 10^2 W/cm^2). Addressing the areas of high heat flux directly using micro Thermoelectric Modules (µTEMs) reduces the need for heat spreading and reduces the energy required to maintain temperature control in PICs. Characterising commercially-available µTEMs is essential to determine their suitability for PIC applications. Literature has documented the characterisation of macro scale TEMs extensively [1-8] but has not yet reported the precise characterisation of micro scale devices. TEM characterisation consists of two distinct approaches in published works. The first is the characterisation of thermoelectric parameters [1-5] such as Seebeck coefficient (α), module conductivity (K_M) and module electrical resistance (R), leading to the calculation of the thermoelectric figure of merit (Z). This approach is concerned with the fundamental thermoelectric material properties and their efficiency in TEM operation. The second approach is considered a more top level view of TEM application and is

Content from this work may be used under the terms of the Creative Commons Attribution 3.0 licence. Any further distribution of this work must maintain attribution to the author(s) and the title of the work, journal citation and DOI.
Published under licence by IOP Publishing Ltd

concerned with measuring the performance characteristics [6-9] for TEMs which can be used determine the suitability of a device for particular operating conditions. The relevant performance characteristics are the temperature difference across the device, (T_h - T_c = ΔT), and heat pumped from the cold side of the module, (Q_c), both of which are a function of the current (I) through the TEM. It is the latter approach which is pursued in this study to determine the suitability of μTEMs for use in PIC applications. Problems exist however, preventing the use of conventional TEM characterisation methods, chief among them being the low compressive load bearing capacity of μTEMs (~0.5N) compared to conventional TEMs (~100N). Most characterisation methods [2-9] rely on compressing the TEM between a heat source and sink in order to measure the characteristics of the device. Methods which obviate the need for contact with the upper surface of the μTEMs avoid potential damage to the devices while providing relevant performance data. To this end, an experimental setup is developed in this paper which utilizes an infra-red (IR) heating element to deliver heat to the μTEMs and a calorimeter to quantify the heat pumped through them. It is proposed that this novel method of characterisation be proven on macro scale TEMs and compared to existing standards as part of an overall body of work investigating the suitability of μTEMs for use in photonics packaging.

The objective of this paper is to accurately characterise the thermal performance of a TEM using contactless heat application in the form of an IR source. The method is subjected to an uncertainty analysis and is benchmarked using quoted performance data from the manufacturer. The outcome is the demonstration of a novel characterisation method for TEMs which does not require thermal contact with the device.

2. Experimentation

An experimental characterisation setup, illustrated in figure 1, was devised to deliver heat to the upper surface of a test TEM in a contactless fashion via an IR heat source, while measuring the heat pumped through the device (Q_c) using a calorimeter and temperature difference across it (ΔT) with an IR temperature sensor and thermistors. The setup was placed in a vacuum to eliminate condensation on the upper TEM surface at low temperatures and to minimise losses due to convection. This section details the components of the experimental apparatus assembly and test procedure. An analysis of experimental measurement and uncertainty is discussed in section 3, data reduction.

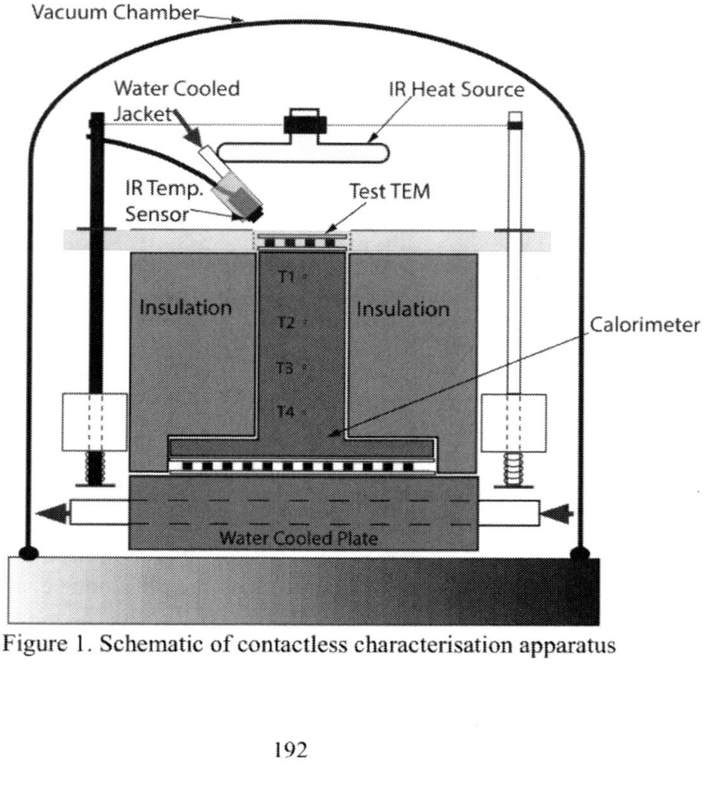

Figure 1. Schematic of contactless characterisation apparatus

2.1 Apparatus

A calorimeter was machined to the dimensions of 15 x 15 x 40mm with a base of 30 x 30 x 5mm using oxygen free copper (395 W/m K). Four 0.82mm holes were drilled 3mm deep at 5, 15, 25 and 35mm from the surface in which were embedded four 0.8mm diameter, 1.2mm long, 5kΩ Epcos B57540G502F thermistors in order to measure the temperature gradient along the length of the calorimeter. Temperature measurements from the embedded thermistors were recorded using Labview software. From this temperature gradient, the heat transfer from (Q_h), and temperature of (T_h), the lower surface of the TEM, referred to as the hot side, was determined. The upper face of the copper calorimeter was polished to a mirror finish ensuring that planarity was maintained. Polyurethane foam insulation (0.035 W/m K) was placed around the calorimeter to minimise thermal losses. The copper calorimeter was placed on a secondary TEM, European Thermodynamics model number ET-241-14-15, in order to hold T_h at 25°C (298K) via PID control using Labview software. This was then mounted on a water cooled aluminium block with water circulated at 10°C by a Lauda E100 water bath. All interfacing surfaces were coated with a thin layer of Electrolube HTSP heat transfer compound to reduce thermal resistance at the interfaces.

Benchmarking tests were carried out on a Multicomp MCPF-031-10-25 single stage TEM of dimensions 15 x 15 x 4.8mm. The upper surface of the TEM, referred to as the cold side, was coated with a 20μm layer of matte black paint to increase the IR absorptivity of the ceramic surface. The test TEM was fixed in place by a spring clamping mechanism. Polycarbonate sheet, 5mm thick, was machined to hold the TEM at each of the four corners, leaving a minimal area of the TEM surface in contact with the clamp and almost the entire face of the TEM exposed. The clamp was fixed to two spring loaded stainless steel rods which applied the downward force. Two steel springs, each 2.5kN/m stiffness, applied a minimum of 50N downwards force. This imposed a pressure of 222kPa over the area of the test TEM to minimise interfacial thermal resistance.

The contactless heat source for the apparatus was a 125W Ceramicx IR quarter flat ceramic heater which was fixed 45mm above the test TEM using steel guide rods and heat resistant rubber rings in order to adjust its height. The heater was connected to a variable power source in order to adjust the heat supplied to the test TEM. Low emissivity (< 0.1) aluminium foil was used where appropriate to reduce stray radiation heat transfer from the heater to the calorimeter and electrical wires within the bell jar. Temperature on the cold side of the module (T_c) was measured by a Raytek MI3 miniature IR sensor with close focus sensing head targeting a spot size of 1mm^2 on the centre of the TEM surface. The temperature reading was processed and recorded by Raytek's Data Temp Multidrop software. The sensor was positioned using a specially designed mounting bracket machined in polycarbonate, which was angled to the precise angle of view and distance from the test surface to produce a 1mm^2 spot size for temperature measurement. Due to the proximity of the miniature IR sensor to the heater, steps were taken to prevent the sensing head from heating up excessively. To this end, a water-cooled aluminium jacket was machined and placed around the sensing head, with water circulated through it by the water bath at a temperature of 10°C.

Constant current was conveyed to the TEM under test using a TTi QL355TP power supply, with measurements of voltage and current taken using certified calibrated Fluke 45 and Fluke 37 multimeters respectively. Voltage measurements were taken across the test device as close as possible to the positive and negative terminals to minimise uncertainty attributed to resistance in the electrical wires. The experimental setup was placed on an aluminium baseplate 380mm in diameter and covered by a glass bell jar 320mm in diameter and 360mm in height. This was evacuated to an absolute vacuum of approximately 400Pa, the lowest achievable vacuum for this setup using a PVR PHV-5 vacuum pump. At this vacuum, moisture is removed from the air preventing condensation build up on the upper TEM surface at low temperatures. The purpose of this was to prevent moisture in the air condensing on the cold side surface of the TEM at low temperatures and to reduce convective losses. Vacuum rated water fittings were used to supply the water cooled plate and IR sensor cooling jacket with water from the water bath. Electrical power supply wires and sensing wires were connected across the aluminium base plate by vacuum rated electrical fittings with connectors on both sides.

Vacuum losses were reduced by using a rubber sealing ring around the bottom of the bell jar and Dow Corning high vacuum silicone grease.

Having described the experimental apparatus in detail, the experimental procedure is outlined in the following subsection.

2.2 Test procedure

The objective of the test procedure was to benchmark the apparatus using a commercially available TEM. According to the data sheet provided by the manufacturer, the test conditions for the quoted data were for T_h maintained at 25°C over a current range of *0.4A – 2.0A* in steps of *0.4A*, using the AC four terminal method [1]. In order to measure the thermal performance characteristics of the TEM using the contactless apparatus detailed above, the following test procedure was developed to be highly repeatable while minimising areas of variance and possible uncertainty:

- The water bath was switched on one hour in advance of each test to allow the water to cool from room temperature to the test condition of 10°C and to enable the water cooled plate and IR sensor jacket to reach steady state.
- The test TEM was fitted in place on the upper surface of the calorimeter with a thin layer of heat transfer compound on the contact surface. The spring loaded clamping mechanism was then applied to hold it in place.
- The miniature IR sensor was positioned using its custom mounting bracket. Positional repeatability was ensured using physical stops to place the bracket.
- The IR heater was then put in place above the test TEM surface. Initially the heater remained off for the case of minimum heat flow, 0W.
- The bell jar was then placed over the apparatus and sealed to the vacuum plate. The vacuum release valve was then closed and the vacuum pump was switched on.
- The test TEM was powered with a constant current set to 0.4A. The system was allowed to reach steady state with T_h maintained at 25°C; on average this took between 15 to 20 minutes.
- Electrical measurements of TEM voltage and current were then taken.
- Cold side temperature and the temperature gradient along the calorimeter were recorded and averaged over a five minute interval.
- Test TEM current was then raised in steps of 0.4A as far as 2A, allowing each step to reach steady state. Electrical and temperature measurements were again recorded for each current step.
- Upon collecting a full set of data for the current range, the heater power was then increased in steps of 20W up to a maximum of 120W. Increasing the heater power also increased the time required to reach steady state thermally, when compared to adjusting TEM current, requiring approximately 45 minutes.
- The procedure was repeated for each current setting until a complete set of data was attained for all heater power settings.

The experimental measurements were ultimately used to compare the TEM performance characteristics provided by the manufacturer and those obtained using the contactless apparatus. In order to use the data collected, a formal data reduction and uncertainty analysis was required, and this is detailed below in section 3.

3. Data reduction

Presented in this section are all the calculated variables derived from experimental data with each of their related primary variables. Their relevance to the overall TEM characterisation is discussed to give a better understanding of the process.

Journal of Physics: Conference Series **525** (2014) 012021 doi:10.1088/1742-6596/525/1/012021

To compare the thermal performance of TEMs, the established method [7-10], is to plot measures of temperature difference generated across the module (*ΔT*) versus heat pumped through the module (Q_c) for a given value of current (*I*) through the device. Q_c is defined in equation 1, where Q_h is the total heat flow through the calorimeter from the hot side of the TEM and *P* is the electrical power supplied to the module:

$$Q_c = Q_h - P \quad (1)$$

Heat flow through the calorimeter, presented in equation 2, is determined using a best approximation least squares regression fit based on thermistor temperature and spatial positioning, (dT_t/dx) and multiplying by the thermal conductivity (K) and surface area (A) of the copper calorimeter:

$$Q_h = KA\frac{dT_t}{dx} \quad (2)$$

Electrical power delivered to the module is measured as the product of TEM voltage (*V*) times TEM current (*I*) and is seen in equation 3:

$$P = VI \quad (3)$$

Temperature difference across the module, seen in equation 4, is the calculation of difference between the constant temperature maintained at the hot side of the module and the temperature measured by the IR sensor on the cold side of the module:

$$\Delta T = T_h - T_c \quad (4)$$

The hot side temperature of the TEM is again determined using a best approximation least squares regression fit similar to Q_h, which is in turn used to calculate the temperature intercept at the surface of the calorimeter [12]. This is computed in real time during the test using Labview software. The cold side temperature is a direct measurement taken by the IR sensor over a spot size of 1mm^2 at the centre of the upper TEM surface.

An uncertainty analysis was also employed in order to quantify the uncertainties associated with each primary measurement and its effect on the calculated properties in the TEM performance characteristics. Each measured quantity and its uncertainty are listed in table 1.

Table 1. Uncertainty in measured quantities

Measured Quantities	**Uncertainty**
Thermistor temperature (T_t)	±0.005°C
Thermistor location (x)	±50μm
Calorimeter area (A)	±1.2825x10^{-6}m^2
Calorimeter thermal conductivity (K)	±5% (W/m K)
IR sensor temperature (T_c)	±0.5°C
TEM Current (I)	±0.01A for 0.3A-3A
TEM Voltage (V)	±1x10^{-4}V for 1V-3V

Many of the measured quantities listed in table 1 were subjected to calibration, and various other steps in order to minimise uncertainty.

- The thermistors used were calibrated, while connected through the vacuum fittings, in the Lauda E100 water bath using a Fluke 1504 reference thermistor, for a temperature range of 15 — 40°C.
- The location of each thermistor along the length of the calorimeter was found using a digital microscope for an optical measurement, which determined the thermistor position to a resolution of 50μm. The area of the calorimeter surface was found in a similar manner.
- Goodfellow supplied the copper used in the calorimeter with an uncertainty of ±5% in the material's thermal conductivity. This value is the largest contributor to uncertainty in heat flow measurements. The author intends to independently characterise the material for further work.
- The cold side temperature measurement is a non-contact IR measurement which was rigorously calibrated in-situ under replicated test conditions. For each increase in heater power setting, the increase in reflected radiation is accounted for in the temperature measurement of the sensor. This was achieved using a copper meter bar with its surface painted with a 20μm layer of black paint, consistent with test TEM, with two 1.2mm diameter, 2mm long, Epcos B57550G502F thermistors calibrated for a temperature range of -30˚C to 30˚C to a certified accuracy of 0.05˚C, embedded 2mm from the surface. The temperature read by the IR sensor was compared to the actual reading of the thermistors in the meter bar, for a range of emissivity values. The temperature range was established on the surface of the meter bar using the secondary TEM to raise and lower temperature. Limiting the accuracy of the IR sensor is the ±0.5˚C repeatability quoted in the device data sheet.
- Current through the test TEM and voltage across it, were measured using certified calibrated meters.

Knowing the accuracy of the measured values, the uncertainties for all calculated values were resolved using the Kline and McClintock method [11] for determining uncertainty in single sample experiments. An example of the maximum uncertainty for each calculated value is displayed in table 2.

Table 2. Maximum uncertainties for calculated quantities

Calculated Quantities	**Maximum Uncertainty**
Power (P)	±1.25%
Heat flow from hot side (Q_h)	±5%
Heat flow from cold side (Q_c)	±5.15%
Hot side temperature (T_h)	±0.25°C
Temperature difference (ΔT)	±0.56°C

This section has discussed the calculated and measured quantities associated with the thermal performance characterisation of TEMs. The uncertainty of measured and calculated quantities were also discussed with uncertainty values given for the relevant measurements in the experimental testing. The data collected is presented and discussed in section 4.

4. Results and discussion

The following results were obtained using the Multicomp MCPF-031-10-25 TEM in a working vacuum of 400Pa. Results are presented and discussed in terms of heat removed from the cold side of the TEM, the temperature difference across it and electrical current through the module. The data is discussed in three subsections for purposes of clarity, beginning with the overall TEM performance characteristics. The special cases of maximum TEM performance, heat pumping from the cold side and temperature difference, are then discussed.

4.1 TEM performance characteristics

Determining the performance characteristics of the test TEM is the primary objective of this study. Figure 2 plots the experimental results of performance in terms of temperature difference as a function of cold side heat pumping for the TEM current range of 0.4A – 2.0A. Included on the plot are the manufacturer's performance curves for the same TEM current to provide a direct comparison. Error bars for cold side heat pumping and temperature difference illustrate the relative uncertainty associated with both measurements.

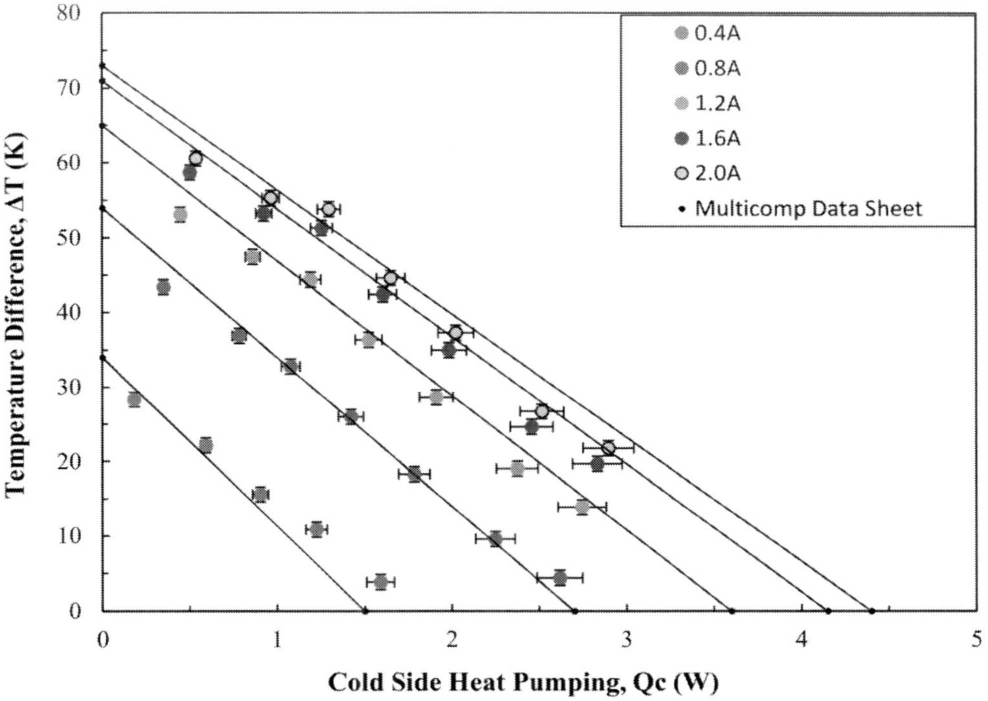

Figure 2. Temperature difference as a function of cold side heat pumping and TEM current, measured data and manufacturer's characteristic curves

Initial impressions from figure 2 confirm the experimental data displayed takes the characteristic linear form expected for thermoelectric performance characteristics seen in [7-10]. There is good agreement between experimental and quoted characteristic behaviour, with slight discrepancies at the highest and lowest current values. Higher current values, 1.6A and 2.0A, show good agreement at higher temperature difference but the data trends towards an under-prediction of cold side heat pumping at lower temperature differences, when compared to the manufacturer's characteristic curves. The differences in trends, especially at lower and higher currents, can be attributed to a fundamental difference in the characterisation method used by the manufacturer. The AC four terminal method, used by Multicomp, is a material level characterisation using electrical measurements to characterise the bulk thermoelectric material which does not fully account for the thermal contribution of the ceramic substrates sandwiching the thermoelectric material and the electrical connections between the thermoelectric pairs in the module. The overall data fit, however, is considered within the expected range.

Areas of particular interest in figure 2 are the cases of maximum TEM performance. These occur at the intersection of the data with the x and y axes, with maximum temperature difference at the y intercept where heat pumping is zero and maximum cold side heat pumping at the x intercept where temperature difference is zero. For the latter case, only the data sets of lower current values 0.4A and 0.8A cross the x-axis. This is due to the IR heater power limit of 125W, with insufficient heat being supplied to increase the temperature of the upper TEM surface at higher currents. Higher heater power would enable values for temperature difference to approach zero, however this is beyond the range required for the overall objective of the work to characterise μTEMs which have lower heat pumping capabilities and lower temperature differences. It is deemed appropriate for validation purposes to extrapolate a best fit trend line from the data for the x-axis intercept. For the case of maximum temperature difference, the data set, again, does not cross the y-axis. This is due to contributions of radiation heat transfer from surroundings within the evacuated bell jar to the surface of the TEM providing heat for the cold side to pump. It is difficult in this configuration to completely eliminate heat flow from the cold side so an extrapolation of a best fit trend line is necessary to establish the y-axis intercept and maximum temperature difference. These cases of maximum performance are presented in the following sub-sections.

4.2 Maximum heat pumping from TEM cold side

The case of maximum heat pumping from the cold side occurs when the temperature difference across the TEM is zero. As explained in subsection 4.1, the data presented in figure 3 is derived from extrapolated least squares fits of each current setting to the point of the x-axis intercept. Figure 3 plots cold side heat pumping of the test TEM at zero temperature difference across the module, for the current range of 0.4A – 2.0A. Again, experimental and manufacturer's data are compared.

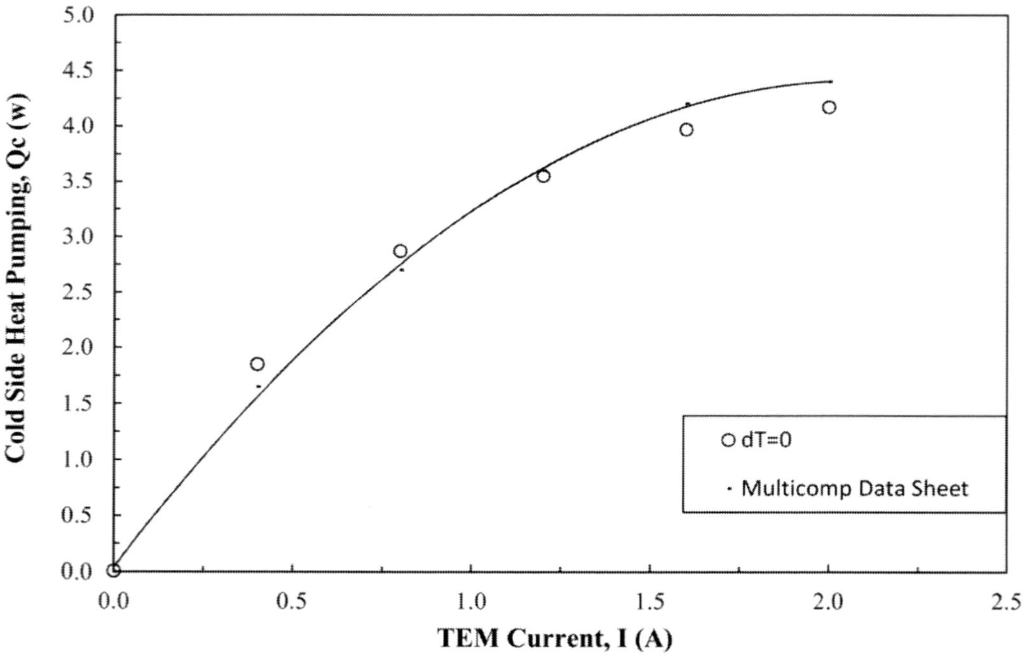

Figure 3. Heat pumped from cold side of TEM as a function of current for the zero temperature difference across module case

Direct comparison of experimental and manufacturer's data for maximum cold side heat pumping for TEM current shows good overall alignment. Slight differences are evident, however, at lowest and highest current values, with the result of marginal difference in trends among measured and quoted data. This is most likely accounted for by the difference in characterisation methods mentioned previously, which is undesirable in a benchmarking process. Despite this, there is good agreement between experimental and quoted data for the current range of 0.4A – 1.2A, with a slightly higher calculation of heat pumping seen for 0.4A of 0.2W. Lower values of heat pumping at the higher values of current 1.6A and 2.0A, both amount to a difference of 0.23W from manufacturer's quoted data. This represents a 5.2% discrepancy at the maximum heat pumping value at 2.0A, falling marginally outside maximum uncertainty for cold side heat pumping established in section 3. This reflects the expected thermal losses associated with interfacing materials in the TEM, which are approximated in the method used by the manufacturer.

4.3 Maximum temperature difference across TEM

The case of maximum temperature difference across the TEM occurs when heat pumping from the cold side is zero. As explained in subsection 4.1, the data presented in figure 4 is derived from extrapolated least squares fits of each current setting to the point of the y-axis intercept. Figure 4 plots temperature difference across the test TEM versus the current through the module for the condition of zero heat flow through the module. This case represents the maximum temperature difference achievable for each given current value, which is of importance in optical telecommunications applications where the control temperature required can be significantly below the allowable sink temperature. Manufacturer's quoted data is compared to the experimental data.

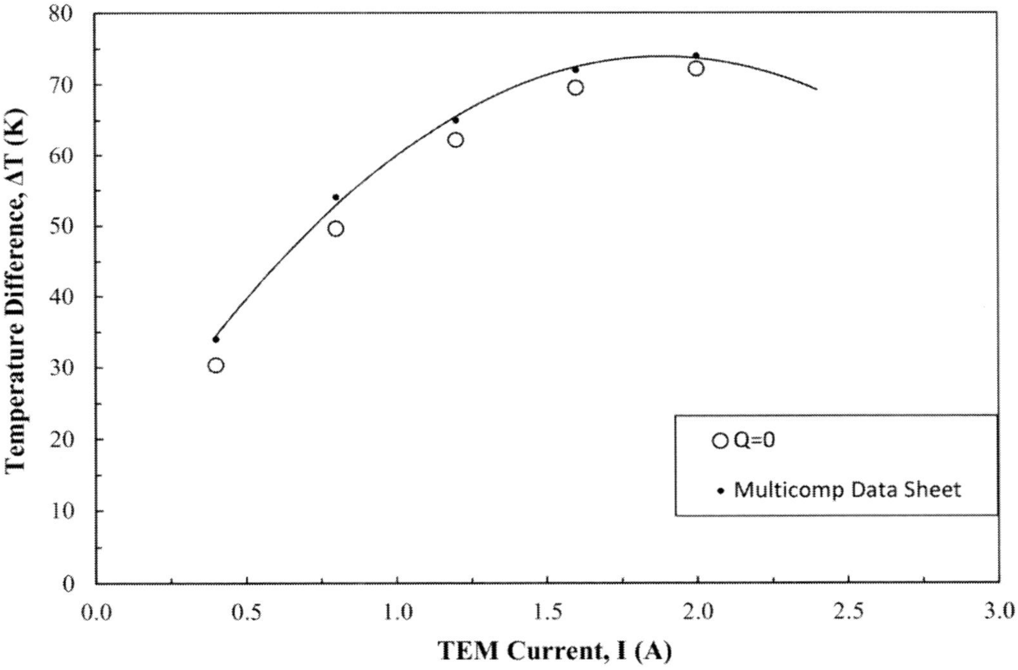

Figure 4. Temperature difference across TEM as a function of current for the zero cold side heat pumping case

Experimental data for temperature difference compare very well with quoted values by the manufacturer. All experimental data points fall less than 3°C from the manufacturer's data series, with that value decreasing for increasing current values. Maximum temperature difference measured at 2.0A by the contactless apparatus falls within 0.9°C of that claimed by the manufacturer. Considering that the Multicomp characterisation method measure temperature difference across the bulk material this is an encouraging outcome that, again, gives confidence in the experimental apparatus and method. Having detailed and discussed the experimental results, the conclusions drawn and further work will be presented.

5. Conclusions

In this paper, a novel contactless method for the characterisation of thermoelectric modules (TEMs) is presented. The method was benchmarked using a commercially available TEM of dimensions 15 x 15 x 4.8mm. Measured data from the experimental apparatus is compared to manufacturer's data sheets, with a view to proving the merit of the novel concept in order to implement the method in a characterisation study of commercially available μTEMs. The following conclusions apply:

- Overall data comparison with manufacturer performance curves yields good agreement, with slight discrepancies evident due to difference in characterisation method.
- The novel contactless characterisation method yields maximum heat pumping from the cold side of the TEM within 5.2% of the manufacturer's quoted data and maximum temperature difference across the TEM within 0.9°C.
- Maximum uncertainty associated with Q_c and ΔT, was calculated at ±5.15% and ±0.27°C respectively.

A contactless method for the characterisation of thermoelectrics was benchmarked successfully allowing further work on μTEMs to proceed with confidence. Future work will include minor adaptations of the apparatus for the measurement of μTEM array performance as well as efforts to further minimise uncertainty associated with primary and calculated measurements, including the characterisation of oxygen free copper to reduce uncertainty in its thermal conductivity.

Acknowledgements

The Author would like to thank Bell Labs, Irish Research Council (IRC) and Science Foundation Ireland (SFI). Bell Labs would like to thank the Industrial Development Agency (IDA) Ireland for their continued support.

References

[1] J. de Boor and V. Schmidt, "Complete Characterization of Thermoelectric Materials by a Combined van der Pauw Approach", *Advanced Materials* **38**, 4303-4307 (2010).
[2] D. Mitrani et al., "Methodology for extracting thermoelectric module parameters", *IEEE Tranactions on Insrumentation and Measurement* 54, 1548, (2005).
[3] T. C. Harman, "Special techniques for measurement of thermoelectric properties", *Journal of Appied Physics* **29**, 1373 (1958).
[4] E. Sandoz-Rosado and R. J. Stevens, "Experimental characterization of thermoelectric modules and comparison with theoretical models for power generation", *Journal of Electronic Materials* 38, 1239 (2009).
[5] S. Dalola et al., "Characterization of thermoelectric modules for powering autonomous sensors", *IEEE Transactions on Instrumentation and Measurement* 58, 99 (2009).
[6] J. D'Angelo and T. Hogan, "Long term thermoelectric module testing system", *Review of Scientific Instruments* **80**, 105102 (2009).
[7] Z. S. Wang et al, "A practical method for measuring thermal conductance and cooling power of thermoelectric modules", in Proc. 20th Int. Conf. on Thermoelectrics, ICT2001, Beijing, China, pp. 515-519, (2001).
[8] R. Ahıska and K. Ahıska, "New method for investigation of parameters of real thermoelectric modules", *Energy Conversion Management* **51**, 338 (2010).
[9] R. J. Buist, "Methodology for testing thermoelectric materials and devices," in CRC Handbook of Thermoelectrics, ed. D. M. Rowe, Boca Raton, FL: CRC Press, ch. 18, (1995).
[10] A. D. Kraus and A. B. Bar-Cohen, Thermal Analysis and Control of Electronic Equipment, McGraw-Hill, ch. 18, (1983)
[11] S. J. Kline & F.A. McClintock, "Describing Uncertainties in Single Sample Experiments" *Mechanical Engineering* **75**, pp. 3-8, (1953)
[12] R. Kempers et al, "A high-precision apparatus for the characterization of thermal interface materials", in Review of Sceintific Instruments **80,** 095111, pp. 1-11, (2009)

Heat Transfer Characteristics of Liquid-Gas Taylor Flows incorporating Microencapsulated Phase Change Materials

J A Howard[1,3,4] and P A Walsh[2]

[1] Bell Labs, Thermal Management Research Group, Alcatel-Lucent, Dublin, Ireland
[2] Stokes Institute, Department of Civil Engineering, University of Limerick, Ireland

Email: james.howard@Alcatel-Lucent.com

Abstract. This paper presents an investigation on the heat transfer characteristics associated with liquid-gas Taylor flows in mini channels incorporating microencapsulated phase change materials (MPCM). Taylor flows have been shown to result in heat transfer enhancements due to the fluid recirculation experienced within liquid slugs which is attributable to the alternating liquid slug and gas bubble flow structure. Microencapsulated phase change materials (MPCM) also offer significant potential with increased thermal capacity due to the latent heat required to cause phase change. The primary aim of this work was to examine the overall heat transfer potential associated with combining these two novel liquid cooling technologies. By investigating the local heat transfer characteristics, the augmentation/degradation over single phase liquid cooling was quantified while examining the effects of dimensionless variables, including Reynolds number, liquid slug length and gas void fraction. An experimental test facility was developed which had a heated test section and allowed MPCM-air Taylor flows to be subjected to a constant heat flux boundary condition. Infrared thermography was used to record high resolution experimental wall temperature measurements and determine local heat transfer coefficients from the thermal entrance point. *30.2%* mass particle concentration of the MPCM suspension fluid was examined as it provided the maximum latent heat for absorption. Results demonstrate a significant reduction in experimental wall temperatures associated with MPCM-air Taylor flows when compared with the Graetz solution for conventional single phase coolants. Total enhancement in the thermally developed region is observed to be a combination of the individual contributions due to recirculation within the liquid slugs and also absorption of latent heat. Overall, the study highlights the potential heat transfer enhancements that are attainable within heat exchange devices employing MPCM Taylor flows.

1. Introduction

Two phase flows consist of a dispersed immiscible phase within a base fluid and they can take many different geometric configurations which are driven by balances between interfacial tension with viscous, inertial and gravitational forces. Liquid-gas flows are the primary focus of this work, specifically investigating non-boiling flows. In mini and micro channels, where the hydraulic diameter is less than the Capillary length of the fluid, slug or Taylor flow regime is typically observed. This is a well ordered segmented flow comprised of alternating liquid slugs and gas bubbles. Closely controlled flow rates can be used to develop trains with varying slug lengths where the gas bubble is separated

[3] To whom correspondence should be addressed.
[4] Work was conducted while part of Stokes Institute, University of Limerick.

Content from this work may be used under the terms of the Creative Commons Attribution 3.0 licence. Any further distribution of this work must maintain attribution to the author(s) and the title of the work, journal citation and DOI.
Published under licence by IOP Publishing Ltd

from the channel wall by a liquid film [1]. Non-boiling flows are simply generated by pumping both immiscible phases into the channel which allows for greater control. A key component of such flows is the void fraction which is the fraction of channel occupied by the gas phase. This can typically be controlled by the inlet volumetric quality, or simply the ratio of gas flow rate to total flow rate within the system.

The primary benefit associated with Taylor flows is significantly enhanced transport phenomena. The separation of the two phases into discrete fluid packets significantly affects fluid dynamics and as a result, the heat transfer rates attainable. One of the most notable effects is the enhancement of radial transport observed in laminar flows due to flow confinement within alternate slugs and bubbles. If viewed from a moving wall reference frame, this radial convection is observed as an internal circulation [2]. This fluid motion is presented in figure 1 and results in modified velocity profiles that have been shown to increase heat transfer rates. Many authors have investigated specifically the enhancement associated with liquid-gas Taylor flows, both experimentally [3, 4] and numerically [5, 6]. A study [7] which examined liquid-gas Taylor flows, confirmed liquid slug length to be one of the most important parameters related to heat transfer enhancement. Increased performance is observed with a reduction in slug length but data needs to be reduced to account for the gaseous voids within the channel. A comprehensive model was also put forward to predict for local heat transfer rates. A later study [8] expanded the applicability of this model to a greater range of liquid phases.

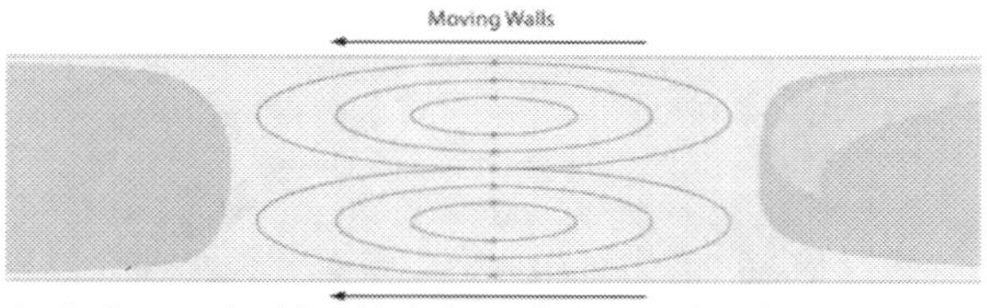

Figure 1. Schematic highlighting fluid recirculation within liquid slugs when observed from a moving wall reference frame.

Alternative investigations into heat transfer enhancements have looked into the incorporation of specifically engineered suspension fluids. These are typically either micro or nano scale in size and suspended within a base fluid. However, recent advances in microencapsulation technologies have enabled the development of microencapsulated phase change material (MPCM) suspensions. These are suspensions where a solid core is encapsulated within a polymer shell and suspended within a base fluid as demonstrated in figure 2. MPCM suspensions offer significantly enhanced convective area for heat transfer to the core while the resultant flows have only a modest increase in viscosity. Phase change particles absorb latent heat of fusion as they melt and have been reported to demonstrate enhancement in thermal management applications [9]. Latent heat absorption results in an apparent increase in specific heat capacity at the phase transition temperature. This increase in specific heat capacity can be many times greater than the sensible heat of the suspension which increases its apparent thermal capacity. This technology enables devices which are subjected to isoflux conditions to theoretically maintain a near constant temperature within the fluid during the phase change process.

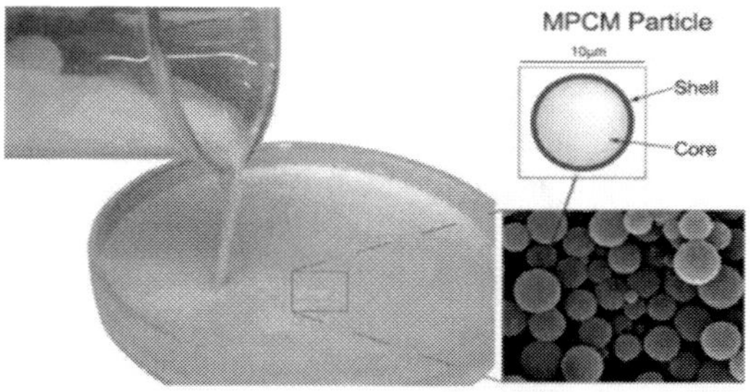

Figure 2. Microencapsulated phase change material (MPCM) suspension.

Journal of Physics: Conference Series **525** (2014) 012022 doi:10.1088/1742-6596/525/1/012022

Many investigations have examined the heat transfer characteristics of microencapsulated phase change material suspensions, particularly under laminar flow conditions. Studies looking at local heat transfer rates are typically numerical [10, 11]. Experimental investigations are commonly focused on global performance and *2 – 4* fold increases in Nusselt number have been reported [12]. Other investigations have demonstrated significant deviation in local Nusselt number values from the conventional Graetz solution [13]. Many authors have identified the concentration of particles within the base fluid as being the one of the most dominant parameters [14]. Microencapsulated phase change materials are conventionally a separate field of research from Taylor flows with no literature available on segmented type MPCM or particulate liquid-gas flows. A large gap exists to segment MPCM flows with a gas phase as it would theoretically combine the benefits associated with both flow types. These include the increased heat transfer rates observed for Taylor flows with the increased thermal capacity attributable to the phase change process. The goal of this investigation is to combine these flow technologies and study the overall thermal performance.

The investigation addresses a significant deficit in the literature. Previous publications have investigated local heat transfer rates associated with Taylor flows and MPCM particulate flows and this work combines both those technologies. An experimental study was conducted to investigate liquid-gas Taylor flows incorporating MPCM materials and analyse the effects associated with the primary Taylor flow parameters. These include liquid slug length and gas void fraction. Overall, this investigation aims to further enhance the fundamental understanding associated with two phase flows and provide steps in developing future thermal solutions.

2. Phase Change Characteristics of MPCM Suspensions

Microencapsulated phase change materials are solid micron sized particles that are dispersed within a liquid phase to form a suspension. The presence of particulates result in modified bulk thermo-physical properties associated with the base fluid. It is necessary to be able to predict such properties when using analytical models or investigating MPCM flows. These suspensions are dominated by the phase change process and this is principally defined by the onset melt temperature and latent heat of fusion. The melt temperature is simply the point at which phase change occurs within the particles while latent heat represents the quantity of heat which can be absorbed throughout the transition. These properties are determined with the use of differential scanning calorimetry (DSC). Figure 3 demonstrates a DSC plot of the MPCM particles used in this investigation, where a heating/cooling rate of $5°C/min$ was used. It clearly indicates the onset melt temperature, T_l, and the latent heat of fusion, h_{fs}, which are equal to $36.1°C$ and $160.7 J/g$, respectively.

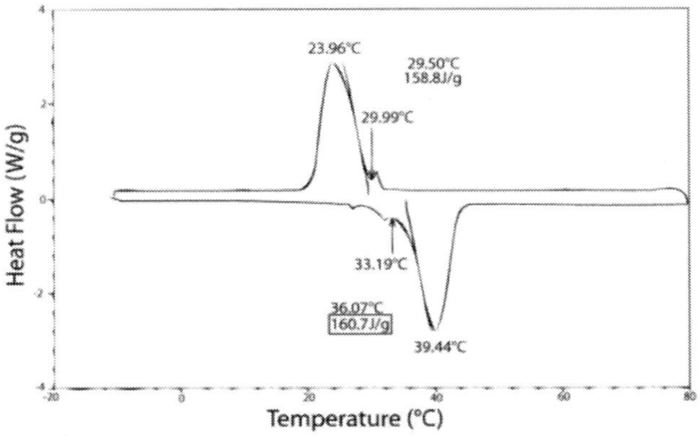

Figure 3. Differential scanning calorimetry (DSC) plot of MPCM particles, highlighting the onset melt temperature and latent heat of fusion.

Developing a specific heat capacity model is required when analysing MPCM flows as it embodies both sensible and latent heat capacity of the suspension. It is typically modelled as a piece wise equation where latent heat is absorbed as an effective enhancement in specific heat capacity over the melt temperature range [15]. Sensible heat capacity is employed for temperatures below and above T_1 and T_2, respectively, which is described using a mass weighted average expression (1).

$$c_{p,mpcm} = c_{p,p}\phi_m + (1-\phi_m)c_{p,bf} \tag{1}$$

Typically, specific heat capacity of the particles can be lower than the base fluid, hence, the primary benefit associated with MPCM flows is latent heat attributed to the solid-liquid phase transition. The latent heat of the MPCM particles is measured using DSC. This is obtained by integrating the change in specific heat capacity with respect to temperature defining the phase change. Hence, the effective increase in specific heat capacity over the melt range, $c_{p,pc}$, as given by (2).

$$c_{p,pc} = \frac{h_{fs}}{T_2 - T_1} \tag{2}$$

A piece wise expression is used to describe the specific heat capacity, with a step change observed over the phase change region and this technique of accounting for the latent heat of MPCM suspensions is widely used by many authors [16, 17]. The use of different profiles within the phase change region, such as a sine wave was also suggested [16].

3. Theoretical Models

The primary benefit associated with Taylor flows is the potential they offer to enhance transport phenomena or more specifically heat transfer. Local dimensionless heat transfer rates for such flows are characterised by local Nusselt number (3)

$$Nu_x = \frac{h_x D}{k} = \frac{q''D}{k\left(T_{x,w} - T_{x,bm}\right)} \tag{3}$$

This is a function of dimensionless longitudinal position or inverse Graetz number (4).

$$x^* = \frac{x}{D\,\mathrm{Re}\,\mathrm{Pr}} = \frac{x}{dPe} \tag{4}$$

The Graetz solution describes local heat transfer values occurring in hydrodynamically developed flows [18, 19] but this was simplified to a piecewise solution [20]. The solution is derived from an energy balance and fully defines the temperature field which results in an expression governing variation of local Nusselt number with dimensionless position and spans both the thermally developing and fully developed regions. A further simplification to the piecewise expression was achieved with asymptotic limits [21] where an expression is defined for the thermal entrance region while fully developed values are constant. These limits were combined using the addition of asymptotic limits approach [22] to develop (5).

$$Nu_{x,Pois} = \left[\left(\frac{1.302}{x^{*\,1/3}}\right)^5 + (4.36)^5\right]^{1/5} \tag{5}$$

A key feature of Taylor flows is the void fraction within the channel and this is related to the volumetric gas quality. Volumetric quality, β, is defined as the ratio of gas flow rate to total volume flow rate of the phases. $\beta = 0$ represents single phase liquid flow while $\beta = 1$ represents single phase gas flow, with intermediate values indicating specific flow rate fractions of the dispersed phase. Void fraction is the fraction of the channel occupied by the gas phase. $\varepsilon_{L(\beta)}$ indicates the proportion of the channels axial length occupied by the gas phase which is approximately equal to the volumetric gas quality (6). It is a very important parameter in two phase flows as it is used when evaluating overall flow performance or reducing experimental data.

$$\varepsilon_{L(\beta)} \approx \beta = \frac{Q_g}{Q_l + Q_g} \tag{6}$$

When considering heat exchange systems incorporating MPCM slurries, phase change within the particles leads to deviations of theoretical predictions from conventional single phase flows. The primary difference is observed in the bulk mean temperature as illustrated in figure 4. This is due to absorption of latent heat as the phase change particles melt. For conventional liquid cooling flows subjected to a constant heat flux, bulk mean temperature increases linearly which can be calculated using a simple energy balance. However, with MPCM flows this is not the case due to the change in specific heat capacity over the melt region. Phase change occurs over a temperature range, T_1 to T_2.

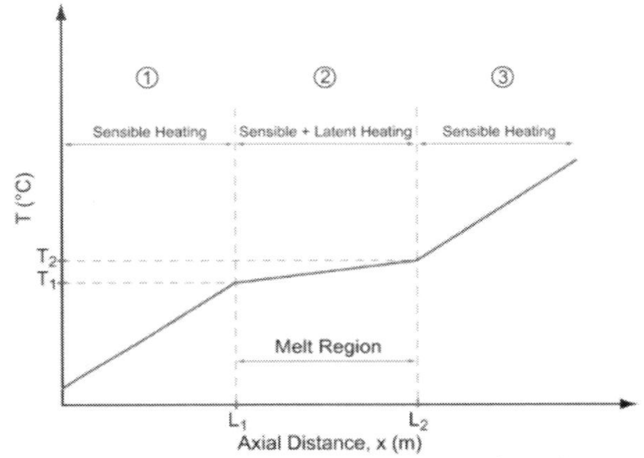

Figure 4. Theoretical bulk mean temperature profile of MPCM suspension flows.

The bulk mean temperature increase with respect to axial distance can be divided into three distinct regions: pre-melt region where $T < T_1$, melt region where $T_1 < T < T_2$ and post-melt region where $T_2 < T$. Within the pre- and post-melt regions, temperature rise is dictated by sensible heating based on the specific heat capacity of the suspension but the effective phase change heat capacity, $c_{p,pc}$, within the melt region results in a lower temperature increase along the channel. An energy balance can be used to develop analytical correlations describing these distinct regions (7).

$$T_{x,bm} = \begin{cases} T_{in} + \dfrac{q''\pi D x}{c_{p,mpcm}\dot{m}} & x < L_1 \\ T_1 + \dfrac{q''\pi D(x - L_1)}{\left(c_{p,mpcm} + \phi_m c_{p,pc}\right)\dot{m}} & L_1 < x < L_2 \\ T_2 + \dfrac{q''\pi D(x - L_2)}{c_{p,mpcm}\dot{m}} & L_2 < x \end{cases} \tag{7}$$

The energy balance is based on the applied heat flux, q'', Diameter, D, specific heat capacity, c_p, and the mass flow rate, $\dot{m}$. L_1 and L_2 are the axial distances downstream from the thermal entrance point which indicates onset and end of the phase change process, (8, 9), respectively. They are defined using a balance based on temperatures T_1 and T_2.

$$L_1 = \frac{(T_1 - T_{in})\dot{m}c_{p,mpcm}}{q''\pi D} \tag{8}$$

$$L_2 = \frac{(T_2 - T_1)\dot{m}(c_{p,mpcm} + \phi_m c_{p,pc})}{q'' \pi D} \tag{9}$$

By employing the Graetz solution (5), to predict wall temperatures in conjunction with (7), a reduction in local wall temperature rise is predicted. This highlights a common issue with MPCM suspension flows where particles nearest the wall reach the melt temperature, T_l, before the remainder of the channel, hence, this results in a moving melt boundary, known as the 'Stefan Problem'. It leads to propagation of a phase change boundary moving down stream with transition also dependent on the radial position [23]. This means that the phase melt range occurs at the wall before the bulk mean of the fluid reaches the phase change temperature, hence, the Graetz solution may not accurately predict experimental wall temperatures within the phase change region. Previous authors [13, 24] have shown deviation in wall temperatures from the theoretical Graetz solution but none have offered a correlation to allow its prediction.

4. Experimentation

The primary aim of this work was to investigate the thermal characteristics of liquid-gas Taylor flows incorporating microencapsulated phase change material (MPCM) suspension as the liquid phase. In order to accomplish this, an experimental setup was designed and constructed which is presented in figure 5. The rig was required to enable the development of well ordered Taylor flows and subject the flow to an isoflux boundary condition while recording experimental wall temperatures. There are three distinct regions: 1) slug generation where the well ordered segmented flow was developed, 2) thermal region where the flow is subjected to the constant heat flux boundary condition, and 3) slug visualisation where both liquid slug and gas bubble lengths can be accurately measured downstream of the test section. The MPCM particulate solution was custom made by Thies Technology. Particles contained a methyl stearate core encapsulated by a polyurethane shell and were dispersed in water. *30.2%* mass concentration of particles within the base fluid was used throughout the experimental investigation and the thermo-physical properties of this suspension are presented in Table 1.

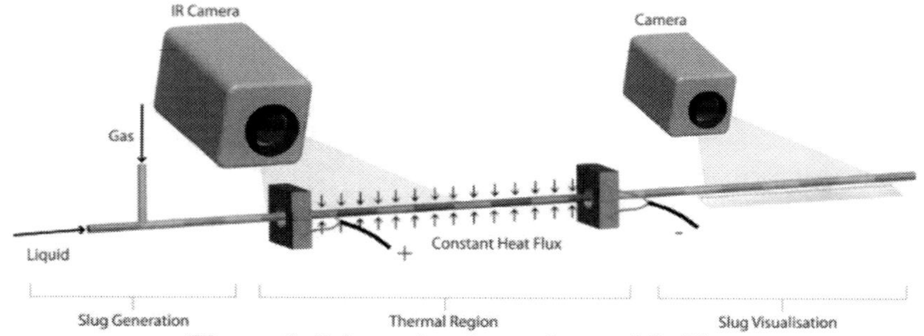

Figure 5. Schematic of experimental facility.

Steady flows were generated such that flow rates of the respective phases could be accurately controlled and this was achieved by employing two syringe pumps, pumping both a gaseous and liquid phase respectively. High precision Harvard PHD 2000 programmable syringe pumps were used and capable of accurately and independently setting liquid and gas volumetric flow rates. These were delivered from *100ml* capacity Hamilton gas tight syringes. Flow from each pump was combined in a T-junction before entering the experimental test section. A range of liquid slug and gas bubble lengths were achieved by using T-junctions with varying internal diameters. Variations in the flow structure were also achieved by changing the relative flow rate magnitudes which resulted in varying volumetric quality, β. A settling period was also allowed when generating the Taylor flows due to compressibility of the gas phase but resulted in steady flows with constant slug and bubble lengths.

Table 1. Thermo-physical properties of MPCM suspensions

	μ *mPa s*	ρ *kg/m³*	γ *mN/m*	c_p *J/kg K*	k *W/m K*	h_{fs} *kJ/kg*	T_1 *°C*	T_2 *°C*
Particles	*n/a*	*892*	*n/a*	*2400*	*n/a*	*160*	*36.1*	*38.1*
30.2%	*7.75*	*965.2*	*55.06*	*3642.4*	*0.486*	*48.3*		

The experimental test section allowed the flow to be subjected to a constant heat flux boundary condition while recording wall temperature measurements. The heated test section consisted of a *1.5mm* internal bore stainless steel tube with a wall thickness of only *250μm*. The thermal boundary condition was created by Joule heating using a D.C power supply. Electrical connections to the tube were achieved using custom made sharp edged copper ring contacts to ensure that the thermal entrance point could be clearly identified. External surface temperature measurements of the heated section were obtained using a high resolution infrared (IR) thermography system. This consisted of a FLIR Systems Thermacam Merlin series IR camera. Infrared measurements also required the external surface of the test section to be sprayed matt black to increase its surface emissivity. The value of the emissivity was determined by comparing experimental images with four thermocouple readings which were placed directly on the tube surface while single flow was pumped through the system. This was pumped at sufficiently high flow rates as to minimise the difference between thermocouples and the emissivity was found to be *0.95* which is typical for a matt black finish. The heated test section was also placed within an enclosure in order to shield it from its environment. For each test, thermal images were recorded for up to *6* minutes at a frequency of *1Hz*.

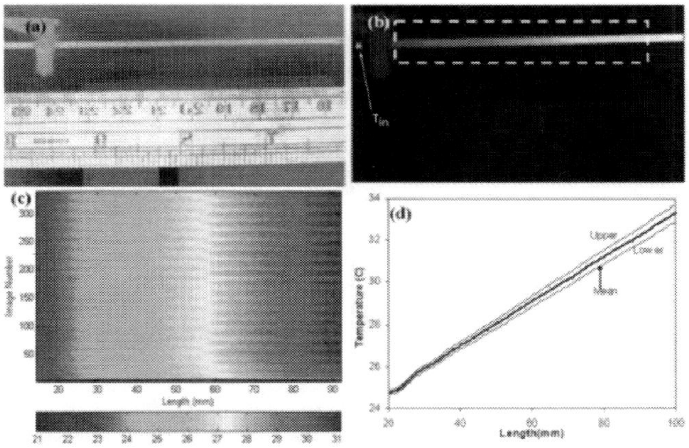

Figure 6. Steps highlighting data analysis of infrared thermography data.

An analysis code in Matlab was used for extracting the transient temperature profiles of the external tube surface from thermal images and it was a four step process as illustrated in figure 6. The first image highlights scaling which was obtained from images recorded prior to heating. Next, the region of interest was identified and an edge definition algorithm was used to locate the tube walls and hence, identify its centreline. Experimental temperature measurements were recorded at each pixel location along the tube centreline which resulted in a data point approximately every *300μm*. Temperature data was extracted from each experimental image as highlighted in figure 6 (c). The recorded image number is plotted versus the tube axial distance where image numbers are representative of a time domain through the recording frequency. The temperature in *°C* is represented by colour contours as indicated in the legend. The desired range of image numbers are selected where the flow has reached steady state and mean time averaged temperature profile was outputted as shown in image (d). Upper and lower lines represent three standard deviations of the periodically fluctuating temperature along the length of the test section. This fluctuation is indicative of the alternating flow structure associated with Taylor flows.

The entire system was calibrated using single phase flows as water was pumped through the system for a range of different flow rates with a uniform heat flux being applied. This allowed the electrical resistance of the stainless steel test section to be calculated and hence, heat flux values by assuming the water flow exactly matched the Graetz solution. Experimental wall temperatures are observed to closely follow the analytical predictions, although, slight deviation is noted immediately downstream of the thermal entrance point due to heating effects associated with the copper electrodes.

5. Results and Discussion

Dimensionless heat transfer rates were investigated for MPCM flows which were segmented by a gas phase. Only slurries with a concentration of *30.2%* were analysed since they provide the maximum enhancement over Graetz flow predictions. This concentration was held constant, as was the applied heat flux at $6.23kW/m^2$, in order to quantify the effect of primary Taylor flow parameters, i.e. dimensionless slug length, L_s^* and length void fraction, $\varepsilon_{L(\beta)}$. The thermo-physical characteristics presented in Table 1 were used throughout the analysis.

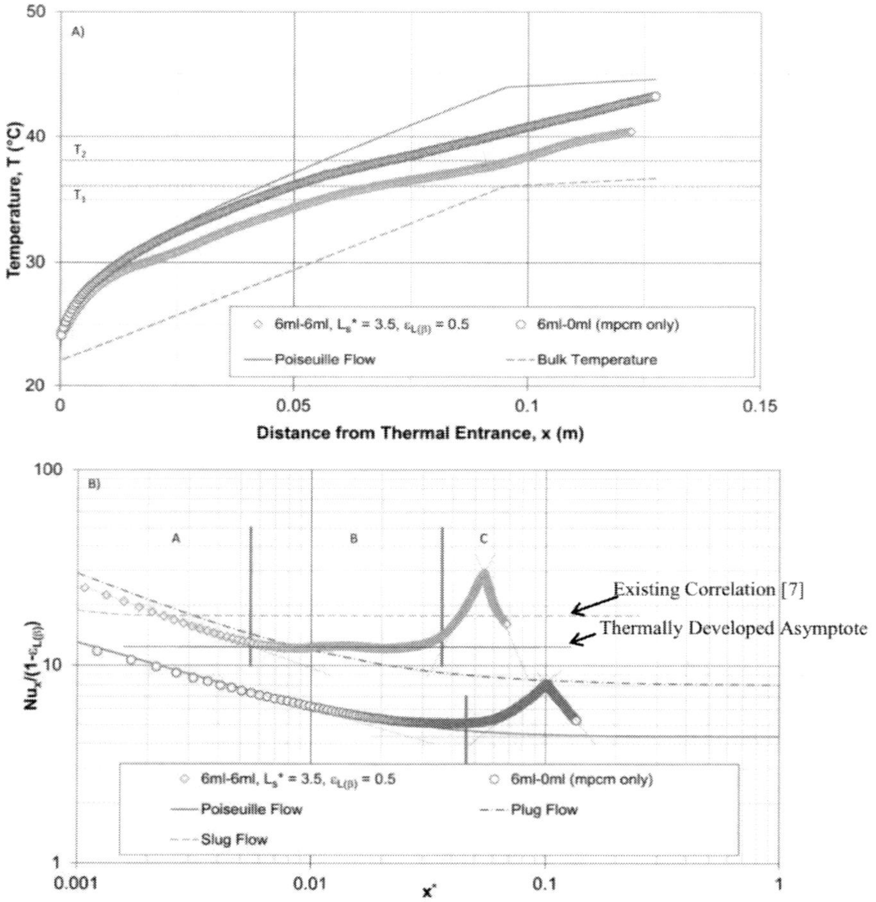

Figure 7. Local wall temperatures (a) and Local *Nu* values (b) comparing MPCM, with and without segmentation.

Experimental wall temperatures were recorded and MPCM characteristics were compared with the Graetz solution for Hagen Poiseuille flow. Figure 7 (a) compares experimental wall temperatures of single phase and two phase flows for the same mass flow rate of MPCM suspension, $Q_{mpcm} = 6ml/min$. Volumetric gas quality, β, was increased from *0* for single phase to *0.5* for the MPCM-air flow. The inlet T-junction resulted in a dimensionless liquid slug length, $L_s^* = 3.5$. Single phase MPCM flow results in reduced wall temperatures when compared with the theoretical Graetz solution for Poiseuille flow (5) and this agrees with previously published work [13]. It was also observed that the Taylor flow regime provides a further reduction in wall temperatures even when data is not reduced to take account of the gaseous voids in the flow. These are similar to observation made for short slug lengths by [7, 8] which investigated heat transfer characteristics associated with liquid-gas flows without phase change. Figure 7 (b) is developed by converting local wall temperatures into dimensionless Nusselt numbers and plotted against inverse Graetz number. The data is also reduced to account for the proportion of the channel occupied by the liquid phase, i.e. $(1 - \varepsilon_{L(\beta)})$. Two phase flow utilising MPCM suspension demonstrates elevated *Nu* values in both the thermal entrance A) and thermally developed B) regions. These are similar to trends seen for liquid-gas flows [7] but significant deviation is observed when compared with the correlation put forward by the same authors, this is indicated by a dashed line in figure 7 (b). A significant increase in dimensionless heat transfer is seen in the thermal entrance region while the thermally developed asymptote appears to be significantly below predictions. It is also observed that two phase MPCM flows exhibit the *Nu* enhancement peak C) which was previously noted for conventional MPCM flows [13]. However, in this case a reduction peak was not observed in this measurement as it occurred beyond the experimental field-of-view of the IR thermography system. An increased deviation in local *Nu* values over the phase change region is apparent when comparing the different flow regimes. A shift in dimensionless position of the enhancement peak is also observed due to the higher axial velocities, hence, lower Reynolds numbers associated with Taylor flows for similar mass flow rates of MPCM suspension.

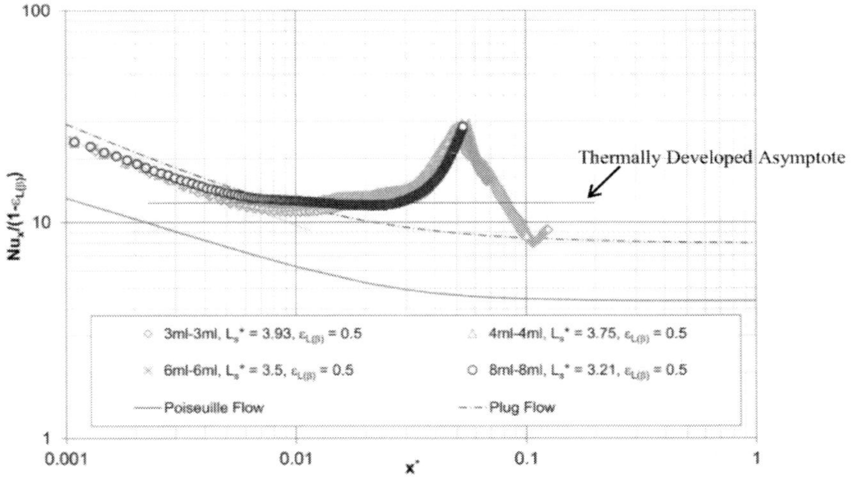

Figure 8. Experimental *Nu* vs x^* data for increasing flow rates of MPCM-gas flows.

Initially, the influence of MPCM *Re* number on dimensionless heat transfer was examined. While maintaining a constant volumetric gas quality, the volume flow rate of the MPCM suspension was increased from *3ml/min* to *8ml/min* for the same inlet T-junction. The resultant slug lengths varied from *3.2 – 3.9* diameters due to slight variations in the flow structure. Those variations are attributable to the slug-bubble curvature becoming elongated with increased two phase velocity, hence, reducing the volume of fluid within the slug core. Experimental local Nusselt numbers for this range of flow rates are plotted in figure 8. All plots demonstrate very good agreement, suggesting the effect of changing Reynolds number is adequately captured by the conventional inverse Graetz number scaling. Dimensionless profiles collapse very well in the thermal entrance region. All flow rates examined result in the same thermally developed asymptote but differences in oscillation magnitude are observed. Oscillations experienced along the *Nu* profile are due to the segmented nature of two phase flows. The dependency on dimensionless velocity suggests a transient effect which is related to heat conduction through the liquid film. This is most likely a Fourier effect and similar to damping observed in previous work [8] for flows with thick liquid films. Dimensionless profiles in figure 8 also partially capture the reduction peak typically observed in MPCM flows.

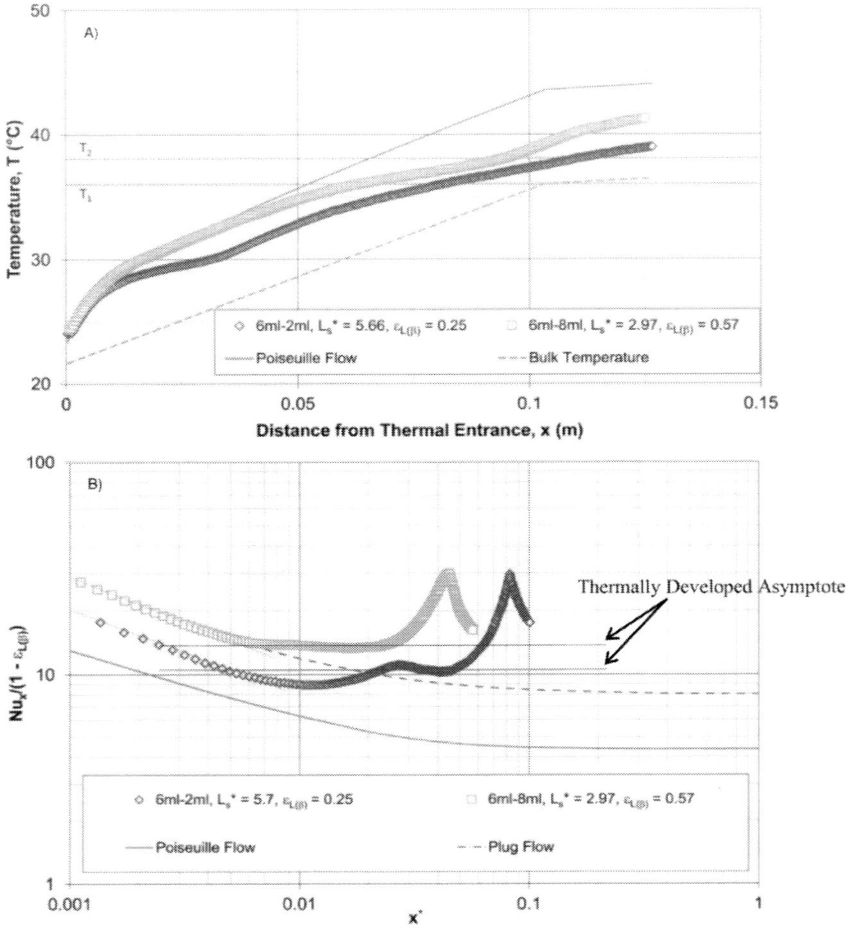

Figure 9. Local wall temperature (a) and local *Nu* vs x^* (b) data of MPCM-gas flows with increasing volumetric gas quality.

Investigating the influence of remaining Taylor flow parameters, the flow rate of MPCM was held constant at *6ml/min*. Three different size inlet T-junctions were used to attain a range of dimensionless slug lengths, L_s^*, while the volume flow rate of gas was varied from *2 – 10ml/min*. Varying volumetric gas quality directly resulted in varied length void fraction. Figure 9 (a) and (b) present experimental results for min and max gas flow rates which were developed using the smallest inlet T-junction. Increasing the volume flow rate of air within the system results in a larger length void fraction but reduced liquid slug length. Figure 9 (a) shows the experimental wall temperatures for MPCM-air flows with L_s^* equal to *2.97* and *5.66* while $\varepsilon_{L(\beta)}$ values are *0.57* and *0.25*, respectively. Wall temperatures shown in figure 9 (a) are not reduced to account for void fraction and both plots demonstrate similar profiles to the two phase flow characteristics already shown in figure 7 (a). Although reduced slug lengths result in enhanced heat transfer, increased void is undesirable within a system to maximise the overall heat transfer. The shorter slug length results in higher experimental wall temperatures as it has larger voids within the channel. Figure 9 (b) captures the same wall temperatures converted to dimensionless form, i.e. *Nu* vs x^*. The graph illustrates the dependence on dimensionless liquid slug length within the thermal entrance region and thermally developed values, before phase change occurs. A reduction in L_s^* results in enhanced Nu values in both the entrance and fully developed regions. Thermally developed asymptotic limits can be interpreted from oscillations in the plot. Significant differences between the thermally developed asymptote and the peak Nu values can also be observed, suggesting that it is a function of the primary Taylor flow parameters.

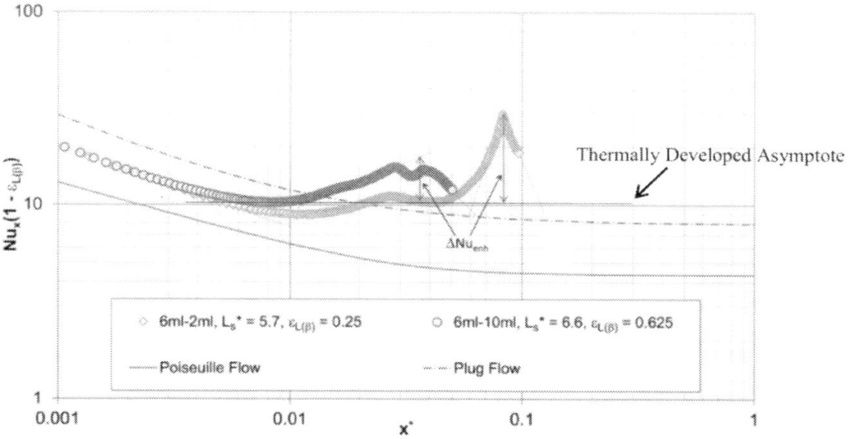

Figure 10. Local Number plot, highlighting the effects of void fraction, $\varepsilon_{L(\beta)}$, on enhancement due to phase change, ΔNu_{enh}.

The total enhancement in dimensionless heat transfer for MPCM-air flows is a summation of enhancements observed in MPCM Poiseuille flow plus an enhancement due to flow recirculation within the slugs. Results in figure 9 (b) highlight differences in Nu enhancement, ΔNu_{enh}, due to phase change, i.e. peak height above the thermally developed asymptotic limit and this is clearly identified in figure 10. This plot shows two profiles with similar dimensionless slug lengths, L_s^* equal to *5.7* and *6.6*, respectively which were developed using two different inlet T-junctions. Good agreement is observed between both profiles in the thermal entrance region. Similar slug lengths result in similar thermally developed Nu asymptotes but they have significantly different volume gas qualities, *0.25* and *0.625*, respectively which is observed to affect peak enhancement. Increased gas quality reduces the x^* distance to each peak, as previously identified but it also reduces the magnitude of peak enhancement. Enhancement due to phase change, ΔNu_{enh}, becomes a function of length void fraction. Phase change particles act as surfactants within the Taylor flow and preferentially gather at the interfaces in the flow so much of the MPCM capsules are at the leading and trailing slug interface. It is believed that these particle concentrations at the interface could lead to differences in ΔNu_{enh} observed in figure 10. It is also believed that phase change particle concentrations within the liquid film result in the elevated heat transfer rates observed in the thermal region, above the correlation of [7]. Increased amounts of surfactants or particles result in a hardening of the interface which leads to changes in the velocity profile. This modification is believed to result in some appreciable volumetric rate within the film that contributes to the heat transfer. Further work is required to fully understand the fluid dynamics within this film region and effect of phase change particles.

6. Conclusions

This investigation presented and discussed the heat transfer characteristics of MPCM particulate Taylor flows subjected to an isoflux boundary condition. Data shows segmented MPCM flows provide further enhancement than that observed for MPCM Poiseuille flows. Dimensionless liquid slug length and void fraction have been shown to have a significant effect on dimensionless heat transfer values. Total heat transfer enhancement observed with these flows is a summation of enhancement due to recirculation within the slugs and enhancement due to phase change within the particles. ΔNu_{enh} has been shown to be dependent on $\varepsilon_{L(\beta)}$ which is believed to be due to particles concentrations at the interface. Nusselt number enhancement above previous predictions within the thermal entrance region was observed for MPCM-air flows, which is believed to be attributable to modified velocity profiles and increased flow within the film. Overall, this investigation on MPCM Taylor flows has shown significant enhancements in local heat transfer rates over conventional Poiseuille flows and even higher local *Nu* values than predictions for other liquid-gas flows within the phase change region.

Acknowledgements
The authors would like to acknowledge the Industrial Development Agency (IDA) Ireland and IRCSET 'Embark Initiative' for funding this research.

References

[1] Bretherton F P 1961 *Journal of Fluid Mechanics* **10**(2) 166 – 188
[2] Taylor G I 1961 *Journal of Fluid Mechanics* **10**(2) 161 – 165
[3] Horvath C, Soloman B A and Engasser J M 1973 *Industrial & Engineering Chemistry Fundamentals* **12** 431 – 439
[4] Vrentas J, Duda L J, Lehmkuhl G D 1978 *Industrial & Engineering Chemistry Fundamentals* **17**(1) 39 – 45
[5] Lakehal D, Larrignon G and Narayanan C 2008 *Microfluidics and Nanofluidics* **4**(4) 261 – 271
[6] Narayanan C and Lakehal D 2008 *Journal of Heat Transfer* **130**(7) 074502
[7] Walsh P A, Walsh E J and Muzychka Y S 2010 *International Journal of Heat and Mass Transfer* **53** 3193 – 3201
[8] Howard J A, Walsh P A and Walsh E J 2011 *International Journal of Heat and Mass Transfer* 54 4752 – 4761
[9] Medrano M, Yilmaz M O, Nogués M, Martorell I, Roca J and Cabeza L F 2009 *Applied Energy* **86**(10) 2047 – 2055
[10] Goel M, Roy S K and Sengupta S 1994 *Inernational Journal of Heat and Mass Transfer* **37**(4) 593 – 604
[11] Alisetti E L and Roy S K 2000 *Journal of Thermophysics and Heat Transfer* **14**(1) 115 – 118
[12] Charunyakorn P, Sengupta S and Roy S K 1991 *International Journal of Heat and Mass Transfer* **37**(4) 819 – 833
[13] Howard J A and Walsh P A 2013 *Heat Transfer Engineering* **34**(2–3) 223 – 234
[14] Yamagishi Y, Takeuchi H, Pyatenko A T and Kayukawa N 1999 *AIChE Journal* **45**(4) 696 – 707
[15] Kasza K E and Chen M M 1985 *Journal of Solar Energy Engineering* **107**(3) 229 – 236
[16] Hu X and Zhang Y 2002 *International Journal of Heat and Mass Transfer* **45**(15) 3163 – 3172
[17] Roy S K and Avanic B L 2001 *International Communications in Heat and Mass Transfer* **28**(7) 895 – 904
[18] Graetz L 1883 *Annual Review of Physical Chemistry* **18** 79 – 94
[19] Graetz L 1883 *Annual Review of Physical Chemistry* **25** 337 – 357
[20] Shah R K and London A L 1978 *Laminar Flow Forced Convection in Ducts: Sussplement 1 to Advances in Heat Transfer* (New York: Academic Press)
[21] Muzychka Y S and Yovanovich M M 2004 *ASME Journal of Heat Transfer* **126** 54 – 61
[22] Churchill S W and Usagi R 1972 *AIChE Journal* **18**(6) 1121 – 1128
[23] Choi E, Cho Y I and Lorsch H G 1994 *International Journal of Heat and Mass Transfer* **37**(2) 207 – 215
[24] Zeng R, Wang X, Chen B, Zhang Y, Niu J, Wang X and Di H 2009 *Applied Energy* **86**(12) 2661 – 2670

Development of an enhanced liquid for electronic cooling

S Mossaz[1], J A Gruss[1]*, O Poncelet[2] and P Berne[2]
[1] Univ. Grenoble Alpes, F-38000 Grenoble, France
CEA, LITEN/DTBH, 17 rue des Martyrs, F-38054 Grenoble Cedex 9
[2] Univ. Grenoble Alpes, F-38000 Grenoble, France
CEA, LITEN/DTNM, 17 rue des Martyrs, F-38054 Grenoble Cedex 9

E-mail: gruss@cea.fr

Abstract: Development of NEPCM (Nanoparticle Encapsulated Phase Change Material) nanofluids using low melting temperature metals is a very pertinent area of research in the context of effective thermal management solutions. Compared to a pure fluid, these nanofluids have a higher heat capacity during the phase change and it is possible to improve the heat transfer, as a result of this phase change. To appreciate the merits, in terms of energy, an energy Performance Evaluation Criteria (*PEC*) has been defined as the ratio of heat transfer rate at fixed pumping power. The numerical results obtained show an improvement around 20% of this energetic criterion at low ΔT compared to the base fluid. A status of the development of these fluids in our laboratory is given in this paper. From an experimental point of view, a specific test loop has been developed in order to test thermo hydraulic performance in a controlled manner in laminar and turbulent conditions at imposed heat flux with a small volume of fluid (350 ml). The test loop was validated with pure water and the choice of materials used has been defined but not yet tested.

Key words: nanofluid, phase change material, physical properties, performance evaluation criteria

Nomenclature

$\dot{Q}$	heat flow rate, W	$\dot{W}$	pumping power, W	Cp	specific heat capacity, J.kg^{-1}.K^{-1}
di	inner tube diameter, m	de	outer tube diameter, m	l	tube length, m
M	mass, kg	q_m	mass flow rate, kg.s^{-1}	k	thermal conductivity, W.m^{-1}.K^{-1}
Nu	Nusselt number	Pr	Prandtl number	Pe	Peclet number
Re	Reynolds number	S	heat exchange area, m²	T	temperature, K, °C
V	velocity, m.s^{-1}	L	latent heat of fusion, J.kg^{-1}	h	heat transfer coefficient, W.m^{-2}.K^{-1}
K	consistency, Pa.s^{n}	Δp	pressure drop, Pa	Λ	Darcy coefficient
μ	dynamic viscosity, Pa.s	ρ	density, kg.m^{3}	v	Sedimentation rate, m.s^{-1}

[1]* To whom any correspondence should be addressed.
We thank ANR through the **Carnot Institute Energy Future** for funding this research

Content from this work may be used under the terms of the Creative Commons Attribution 3.0 licence. Any further distribution of this work must maintain attribution to the author(s) and the title of the work, journal citation and DOI.
Published under licence by IOP Publishing Ltd

1. Introduction

The need for convective heat transfer enhancement has produced a considerable amount of research [1, 2]. Convective heat transfer can be improved either by changing flow geometry and boundary conditions, or by modifying the thermophysical properties of the heat transfer fluid. Enhancing specific heat and thermal conductivity allows heat transfer to be increased. There has been a lot of work to develop nanofluids using solid nanoparticles dispersed in a base fluid to increase the heat transfer properties of the fluid; however from an energetic point of view, the increase of thermal conductivity is always accompanied by an increase of viscosity leading to an energetic criterion lower than 1.

NEPCM (Nanoparticle Encapsulated Phase Change Material) slurry is dispersion where the PCM is dispersed in fluid. Compared to the base fluid, these nanofluids have a higher heat capacity during the phase change and a possible enhancement, as a result of this phase change, in the heat transfer phenomenon. We can greatly increase heat capacity during the phase change and then obtain an energetic criterion greater than 1. To develop a nanofluid for heat transfer purposes, it is necessary to have a global approach. We must look at an energetic criterion and not only the apparent specific heat enhancement or improvement of heat transfer. The energetic criterion chosen is the ration of heat transfer of the nanofluid to the base fluid at constant pumping power.

The study is divided into two parts in order to examine the effects of different materials on the energetic criterion and validate the experimental installation.

- The first part focuses on the theoretical research of good materials improving the energetic criterion.
- The second part validates the experimental set-up in the turbulent and laminar regime, using demineralized water.The system operates with a heat flow imposed and between 20 and 95 °C.

2. Theoretical study

2.1. Performance Evaluation Criteria (PEC)

It is known that the addition of nanoparticles in a fluid may increase heat exchange but usually this addition will increase the viscosity of the fluid. To see if the use of a nanofluid is beneficial from an energy point of view compared to the base fluid, we have to define Performance Evaluation Criteria (*PEC*). In the literature [3, 4], several studies have established a performance ratio:

$$\eta = \frac{\dot{Q}_{nf} / \dot{W}_{nf}}{\dot{Q}_0 / \dot{W}_0} \tag{1}$$

Where $\dot{Q}$ is the heat flow rate and $\dot{W} = q_m \Delta p / \rho$ the pumping power ($_0$ denotes the base fluid and $_{nf}$ the nanofluid). If this ratio is less than 1, using a nanofluid is not recommended from an energy point of view.

We assume that we can apply to NEPCM classical laws of thermo hydraulic taking the mixing properties of the nanofluid, this has been demonstrated in previous studies.

As the intended applications are in turbulent regime, we can express the Nusselt numbers (*Nu*) determined with Colburn's correlation, Reynolds number (*Re*), Prandtl number (*Pr*), exchange coefficient *h*, friction coefficient *f*:

$$Nu = 0.023.\mathrm{Re}^{0.8}.\mathrm{Pr}^{0.33} \qquad \mathrm{Re} = \frac{\rho.V.d_i}{\mu} \qquad f = 0.316.\mathrm{Re}^{-0.25} \qquad \mathrm{Pr} = \frac{Cp.\mu}{k} \qquad h = \frac{Nu.k}{d_i}$$

and therefore $h = 0.023.\mathrm{Re}^{0.8}.\mathrm{Pr}^{0.33}.\frac{k}{d_i} = 0.023.\rho^{0.8}.V^{0.8}.d_i^{-0.2}.\mu^{-0.47}.Cp^{0.33}.k^{0.67}$ (2)

As defined in [4], the performance factor becomes:

$$\eta = h_{nf} / h_0 = h' = PEC \tag{3}$$

The performance factors thus allow us to evaluate the gain or loss of a nanofluid relative to a base fluid for hydraulic or thermal criterion. We can establish an energetic *PEC* based solely on the physical properties of nanofluids.
We will use here a *PEC* based on the heat exchange coefficient ratio where the geometry and the pumping power are fixed. The *PEC* used is:

$$PEC = \rho'^{0.582}.\mu'^{-0.543}.Cp'^{0.33}.k'^{0.67} \quad (4)$$

With ρ' being the ratio of nanofluid to base fluid densities, μ' the ratio for the viscosities, Cp' the ratio for apparent heat capacity (including phase change).

2.2. Model

To predict this performance factor we need to know the fluid properties.

2.2.1. Density

Typically, the mass conservation law is used to evaluate the density of a nanofluid [5] :

$$\rho = \rho_0(1-\phi) + \rho_{np}\phi = \rho_0\left[1+\phi(\rho_{np}/\rho_0 - 1)\right] \quad (5)$$

With ϕ as the volume fraction of nanoparticles, ρ as the density ($_{np}$ denotes the nanoparticles).

2.2.2. Specific heat

To calculate the heat capacity of a nanofluid, it is common to take the formula adapted for liquid / solid dispersions:

$$C_p = C_{p0}(1-\phi_{np}) + C_{pnp}\phi_{np} = \frac{\phi\rho_{np}C_{pnp} + (1-\phi)\rho_0 C_{p0}}{\phi\rho_{np} + (1-\phi)\rho_0} \quad (6)$$

With ϕ_{np} as the mass concentration of the nanoparticle.
However, when the phase change is involved in estimating the heat transfer performance, the effect of latent heat must be taken in account so an effective specific heat is used [6] , which can be written as:

$$C_{pef} = C_p + (\phi_{np}.L/\Delta T) \quad (7)$$

With C_{pef} as the effective heat capacity, L as latent heat of fusion, and ΔT as the temperature range of the change of state. Note that the effective specific heat is dependant of the temperature range.

2.2.3. Thermal conductivity

The suspension may be treated as a homogeneous material, we can use the Hamilton and Crosser formulation [7] to predict the thermal conductivity:

$$\frac{k}{k_0} = \frac{k_{np} + (n-1)k_0 - (n-1)\phi(k_0 - k_{np})}{k_{np} + (n-1)k_0 + \phi(k_0 - k_{np})} \quad (8)$$

Where $n=3/\Psi$ is the scale factor and $\Psi = 1$ for spherical particles. This model is known to be conservative model that gives lower results than experimental measurements.

2.2.4. Dynamic viscosity

Einstein's model [8] determines the viscosity of a fluid containing particles in monodisperse suspension without interactions between them:

$$\mu = \mu_0.(1+2.5.\phi_{vh}) \text{ valid for } \phi < 5\ \% \quad (9)$$

Where μ is the dynamic viscosity of the nanofluid, μ_0 the dynamic viscosity of the base fluid, ϕ_{vh} the hydrodynamic volume fraction of the particles: it takes into account the solvation shell around the particles in contrast to ϕ.

2.3. Results

This study gives an approach of the energetic criteria for several base fluid/PCM combinations in the range 20-100°C using organics, salts and metallic compounds with melting temperatures of less than 100°C. Heat transfer simulations have been performed with various water/materials. A transient model for melting of the PCM particle has been developed to verify the complete melting of the particle passing in the heat source.

Figure 1 show the results of this study with C_{pef} calculated with ΔT=20°C. The thermophysical properties of the materials are given in Table 1.

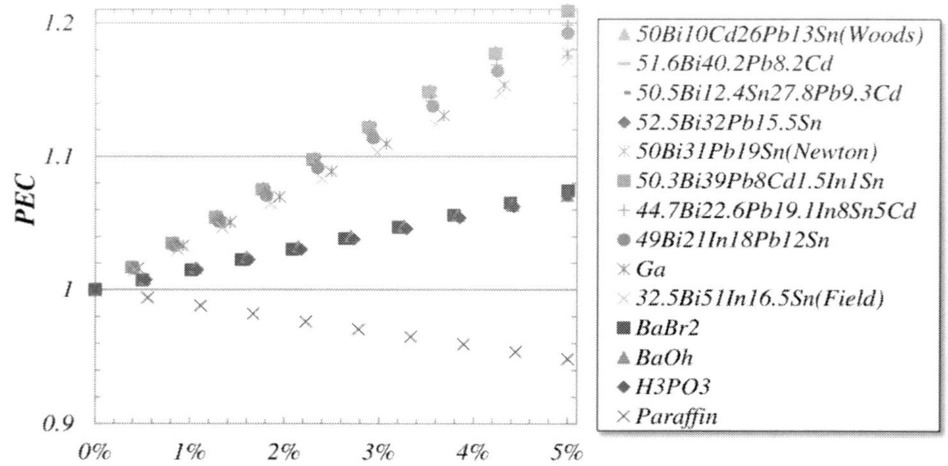

Figure 1. Influence of volume concentration of nanoparticles on the *PEC* for various materials.

In this figure we can see that paraffin, material commonly used for heat storage, is very bad for the flowing fluids. This is due to the fact that the viscosity increases with the volume concentration while other physical properties increase with mass concentration. As can be seen in Figure 2, where the volume concentration was fixed at 5%, the more dense material will give better results. The heat of fusion is an important criterion because it increases the apparent *Cp*. Metallic compounds having a low melting point, are the best candidates for this. Paraffins are eliminated, their *PEC* being less than 1.

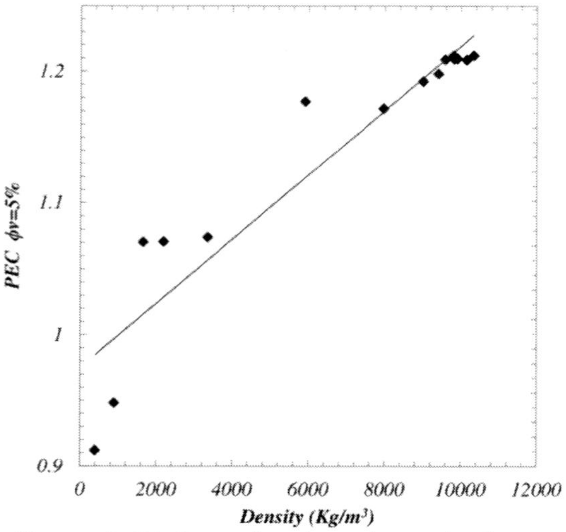

Figure 2. *PEC* as a function of Density for ϕ_v=5%

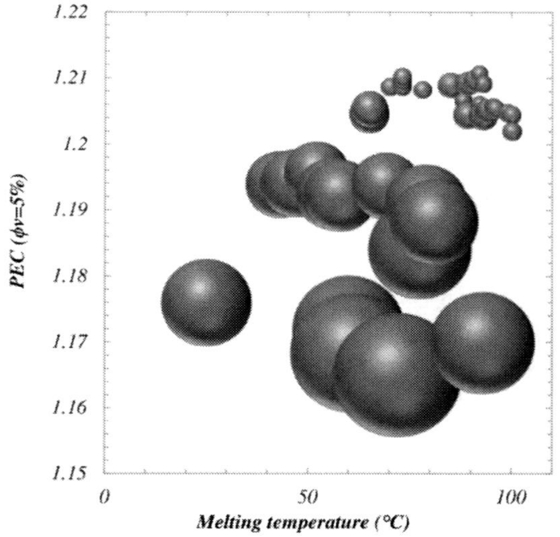

Figure 3. *PEC* as a function of Melting temperature. The bubble surface is the price.

Figure 1 shows the evolution of the *PEC* for different materials but does not take into account the cost and volume of the capsule. Table 1 lists the various PCM properties. Their *PEC* is given for a volume concentration of 5%. Note that the price reflects the cost of bulk material [9], calculated with the mass of constituent materials. The price of manufacturing nanoparticles should be considered in further studies. These prices therefore enable us to give a first classification of different materials but not to estimate the real cost of nanofluid. Figure 3 attempts to synthesize this table. These shows the *PEC* as a function of melting temperature, the surface of the bubbles is the price of the alloy in *€/Kg*. It is found that the smaller the melting temperature, the more expensive the alloy is.

Table 1. Property existing alloys for lower melting temperature to 100°C.

Alloy Description	*Liquidus*	*Solidus*	*Cp*	*Lpm*	*K*	*Density*	*PEC*	*Price*
	[°C/°F]	*[°C/°F]*	*[J/Kg/k]*	*[J/Kg]*	*[W/m/K]*	*[Kg/m3]*	*Φv=5%*	*[€/Kg]*
62.5 Ga, 21.5 In, 16 Sn	10.7/51	10.7/51	318	65851	56	6360	1.172	405
75.5 Ga, 24.5 In	15.7/60	15.7/60	336	67685	53	6200	1.171	471
95 Ga, 5 In	25/77	15.7/60	363	77805	43	5962	1.176	335
100 Ga	29.78/85.6	--/--	370	80400	41	5904	1.18	300
42.9Bi, 21.7Pb, 18.3In, 7.97Sn, 5.1Cd, 4Hg	43/110	38/100	164	41031	34	9285	1.194	194
44.7 Bi, 22.6 Pb, 19.1 In, 8.3 Sn, 5.3 Cd	47/117	47/117	165	42245	36	9165	1.194	203
44.7 Bi, 22.6 Pb, 16.1 In, 11.3 Sn, 5.3 Cd	52/126	47/117	165	43167	35	9165	1.196	173
49.14 Bi, 20.89In, 17.92 Pb, 11.55 Sn,0.5 Cd	56/133	54/129	166	43225	33	9015	1.192	221
49 Bi, 21 In, 18 Pb, 12 Sn	58/136	58/136	166	43182	33	9005	1.192	221
51 In, 32.5 Bi, 16.5 Sn	60/140	60/140	200	41530	52	7965	1.172	518
61.72 In, 30.78 Bi, 7.5 Cd	61.5/143	61.5/143	203	38179	57	8035	1.168	624
47.5 Bi, 25.4 Pb, 12.6 Sn, 9.5 Cd, 5 In	65/149	57/134	159	45292	33	9445	1.205	63
48.0 Bi, 25.63 Pb, 12.77 Sn, 9.6 Cd, 4 In	65/149	61/142	159	45483	33	9472	1.205	53
49 Bi, 18 Pb, 18 In, 15 Sn	69/156	58/136	166	44104	33	9005	1.194	192
50 Bi, 26.7 Pb, 13.3 Sn, 10 Cd	**70/158**	**70/158**	**200**	**36757**	**53**	**8000**	**1.209**	**14**
66.3 In, 33.7 Bi	72/162	72/162	200	36757	52	8149	1.165	670
50.5 Bi, 27.8 Pb, 12.4 Sn, 9.3 Cd	**73/165**	**70/158**	**154**	**45774**	**31**	**9642**	**1.209**	**14**
50 Bi, 25 Pb, 12.5 Sn, 12.5 Cd	**73/165**	**70/158**	**157**	**46751**	**32**	**9565**	**1.210**	**14**
50 Bi, 24.95 Pb, 12.5 Sn, 12.5 Cd, 0.5 Ag	**73/165**	**70/158**	**158**	**47260**	**35**	**9525**	**1.210**	**14**
H3PO3	74	74	200	156000	20	1650	1.071	70
BaBr2	75	75	256	61143	9.3	3340	1.074	20000
48.5 Bi, 41.5In, 10 Cd	77.5/171	77.5/171	185	43233	44	8492	1.184	426
50 Bi, 34.5 Pb, 9.3 Sn, 6.2 Cd	**78/173**	**70/158**	**148**	**43441**	**28**	**9875**	**1.208**	**13**
BaOh	78	78	**585**	280000	1.2	2180	1.071	4000
Parraffin	**78**	**78**	**2500**	**265000**	**0.25**	**900**	**0.948**	**5**
57 Bi, 26 In, 17 Sn	79/174	79/174	176	47686	34	8545	1.191	274
54.02 Bi, 29.68 In, 16.3 Sn	81/178	81/178	179	46741	36	8470	1.188	310
50.31 Bi, 39.2 Pb, 8 Cd, 1.49 In, 1 Sn	**85/185**	**81/178**	**143**	**41218**	**28**	**10155**	**1.209**	**27**
42.5 Bi, 37.7 Pb, 11.3 Sn, 8.5 Cd	**88/190**	**71/160**	**152**	**42694**	**32**	**9820**	**1.207**	**12**
50.31 Bi, 39.2 Pb, 7.99 Cd, 1.5 Sn, 1 In	**89/192**	**80/176**	**143**	**41369**	**28**	**10155**	**1.209**	**22**
50.9 Bi, 31.1 Pb, 15 Sn, 2 In, 1 Cd	89/192	80/176	150	44121	27	9635	1.205	33
51.08Bi, 39.8 Pb, 8.12 Cd, 1 In	**91/196**	**87/188**	**142**	**41100**	**27**	**10216**	**1.210**	**22**
51.6 Bi, 40.2 Pb, 8.2 Cd	**92/197**	**92/197**	**140**	**41227**	**27**	**10260**	**1.211**	**12**
52Bi, 31.67Pb, 15.33Sn, 1 Cd	**92/198**	**83/181**	**148**	**44460**	**26**	**9695**	**1.206**	**13**
50 Bi, 39 Pb, 7 Cd, 4 Sn	**93/200**	**73/165**	**143**	**41789**	**28**	**10110**	**1.209**	**12**
51.45 Bi, 31.35 Pb, 15.2 Sn, 2 In	93/200	87/190	149	44018	27	9645	1.204	33
44 In, 42 Sn, 14 Cd	93/200	93/200	231	45389	78	7470	1.170	448
52 Bi, 31.7 Pb, 15.3 Sn, 1 In	**94/201**	**90/195**	**148**	**44164**	**26**	**9680**	**1.205**	**23**
52.5 Bi, 32 Pb, 15.5 Sn	**95/203**	**95/203**	**148**	**44330**	**26**	**9710**	**1.205**	**13**
52 Bi, 32 Pb, 16 Sn	**95.5/204**	**95/203**	**148**	**44362**	**26**	**9690**	**1.206**	**13**
52 Bi, 30 Pb, 18 Sn	**96/204**	**96/204**	**150**	**45088**	**26**	**9600**	**1.205**	**14**
50 Bi, 31 Pb, 19 Sn	**99/210**	**93/200**	**151**	**44849**	**27**	**9580**	**1.205**	**13**
50 Bi, 28 Pb, 22 Sn	**100/212**	**100/212**	**154**	**45939**	**28**	**9450**	**1.205**	**14**
46 Bi, 34 Sn, 20 Pb	**100/212**	**100/212**	**165**	**49092**	**32**	**8995**	**1.202**	**15**

The materials giving good results are very high density metal alloys materials are very high density metal alloys. Sedimentation problems may appear if the particle diameter is not small enough.
Figure 4 shows the sedimentation rate v according to temperature, calculated using the Stokes's Law:

$$v = {d^2 . g . \Delta\rho}/{9\mu} \tag{10}$$

Where g is the gravity and $\Delta\rho$ is the density difference between the particle and the fluid.
The effect of the particle diameter is clearly visible in this figure .

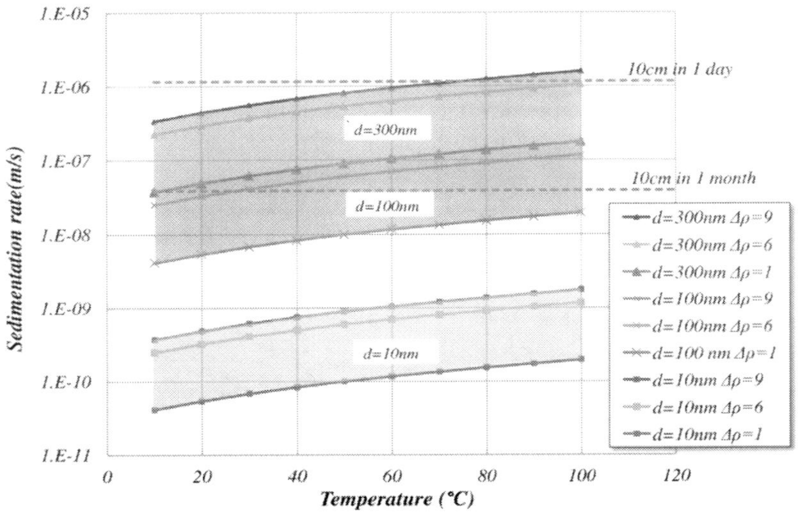

Figure 4. Sedimentation rate as a function of temperature

2.4. Conclusion

Different fluid/material *PEC* simulations have led to a choice of a water/metal nanofluids that give a *PEC*=1.2, representing a 20% heat transfer increase at same pressure loss. A summary table is given. This allows us to select the right alloy having an adequate melting temperature. This melting temperature should be between the minimum and maximum temperature of the system to benefit from the increase in the apparent *Cp*. The diameter of nanoparticle should be as small as possible to limit sedimentation, but greater than 20 nm to avoid supercooling problems.

3. Experimental setup

A specific experimental test loop (see Figure 5) has been developed in order to get thermo hydraulic performances in a controlled manner in laminar and turbulent conditions at imposed heat flux with a small volume of fluid (350 ml).

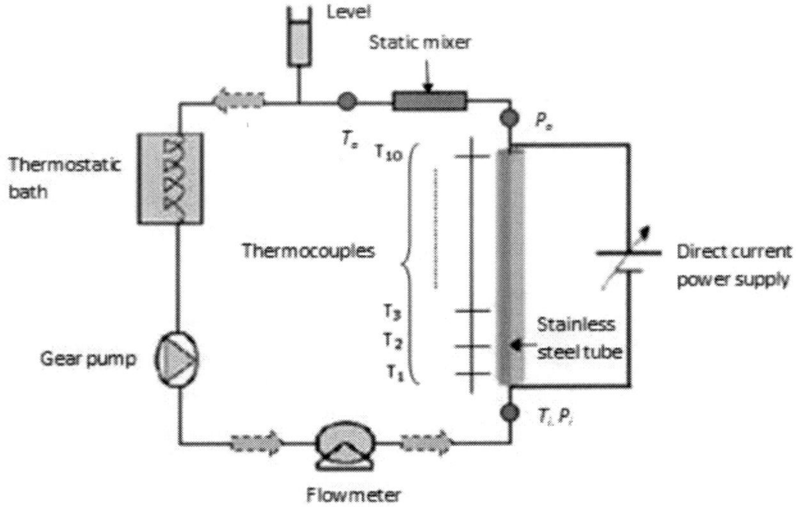

Figure 5. Schematic of the experimental setup

The test section is a circular tube with an imposed flux, with inlet temperatures ranging between 20 and 95°C. The test loop is composed of a fluid circuit, a heater unit, a temperature regulating unit and a control unit and measurement acquisition. The variable parameters are the type of fluid, the Reynolds number, the heat flow and the inlet temperature. The test section is composed of a stainless steel tube of length $L = 0.475$ m, inner diameter di=1.28mm and external diameter de=1.54mm. The test section is heated by Joule effect. The tube is connected to the terminals of a DC generator which generates a maximum power of 625W. Ten micro thermocouples (Tpe) are attached to the outside of the tube in the longitudinal direction, their longitudinal positions (bottom to top of the tube) are given in Table 2. Two platinum sensors: *Pt100*, are located at the inlet and outlet of the tube from within the flow, to measure the inlet temperature "*Ti*" and outlet "*To*". Moreover, pressure sensors are also located before and after the heated section tube to measure the pressure drop Δp caused by fluid flow within the tube.

Table 2. Position of the thermocouples

Thermocouple	T1	T2	T3	T4	T5	T6	T7	T8	T9	T10
Location (x,mm)	8	56	106	160	217	277	323	375	428	462
x/di	6	44	83	125	170	216	252	293	334	361

3.1. Analysis

The analysis of experimental data determines the thermohydraulic properties of the fluid (the heat transfer coefficient, the Nusselt number and the pressure drop). These will then be compared to the usual correlations.

The different dimensionless numbers necessary for analyses are defined below:

$$\mathrm{Re} = \frac{q_m d_i}{\mu S} = \frac{4 q_m}{\mu \pi d_i} \qquad \mathrm{Pr} = \frac{\mu C_p}{k} \qquad h(x) = \frac{q_s}{Tpin(x) - Tf(x)} \qquad Nu(x) = \frac{h(x).di}{k}$$

Where q_m is the mass flow measured with a Coriolis flowmeter, di the inner diameter, $S=\pi di^2/4$ the area of the internal section of the tube, $Tpin$ the temperature of the inner wall at the abscissa x and Tf the fluid temperature at x.

Experimentally we have access only to the outer wall temperatures $Tpe(x)$ at different abscissas x, and Ti and To of the input and output temperature of the fluid. Therefore, it is necessary to establish two equations to determine the temperatures $Tpin$(x) and Tf(x), required in the evaluation of Nu and h, from the measurement of temperatures Ti and $Tpe(x)$. In our case:

$$T_{pin}(x) = \frac{U.I\, d_e^2}{4.\pi.L.\lambda.(d_e^2 - d_i^2)}\left(2\ln\left(\frac{d_i}{d_e}\right) + 1 - \left(\frac{d_i}{d_e}\right)^2\right) + T_{pe}(x) \qquad (11)$$

and

$$Tf(x) = \frac{q_s \pi . d_i}{q_m Cp} .x + Ti \qquad (12)$$

Where UI is the electric power and $q_s = U.I/\pi.d_i.L = q_m.Cp.\Delta T/\pi.d_i.L$ is the surface flux.

3.2. Nusselt number

With the above equations, $Tpin(x)$ and $Tf(x)$, two parameters for determining Nu were defined from temperature measurements Ti, To and $Tpe(x)$. We can now calculate the experimental Nu and compare this to the usual correlations for different flow regimes. These correlations are defined in terms of Reynolds and Prandtl numbers.

3.2.1. Laminar flow

Figure 6 shows the experimental Nusselt (black squares) as a function of x/d for two Re. These results are compared to the classical Shah correlation[10], valid in laminar flow which gives the local Nusselt number :

$$Nu = 1.302\left(x^+/2\right)^{-1/3} - 0.5 \text{ if } x^+ \le 0.003$$

$$Nu = 4.364 + 0.263\left(x^+/2\right)^{-0.506}\exp\left(-41\left(x^+/2\right)\right) \text{ if } x^+ > 0.003\text{, with } x^+ = \left[2(x/d_i)\right]/\text{Re.Pr} \quad (13)$$

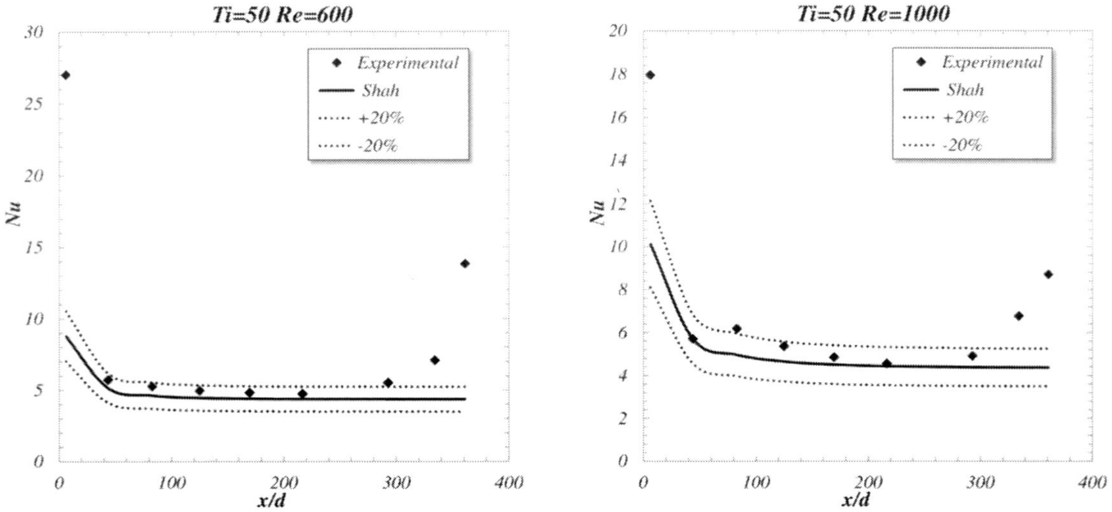

Figure 6. *Nu* as a function of *x/d*.

We see in this figure a good agreement between experimental and theoretical temperatures located in the middle of the section. For thermocouples 1, 9 and 10, there are discrepancies attributed to thermal conduction edge effects that disturb the measurements. Finite element calculations were performed and clearly show the side effect of conduction in the tube, particularly at low Reynolds. Therefore, in laminar flow, only thermocouples 2 to 8 will be used to get correct results.

3.2.2. Turbulent flow

Figure 7 represents the local Nusselt determined experimentally as a function of the dimensionless length *x/d* for different Reynolds. The heat flux is respectively *UI*=219W, 147W and 78W.

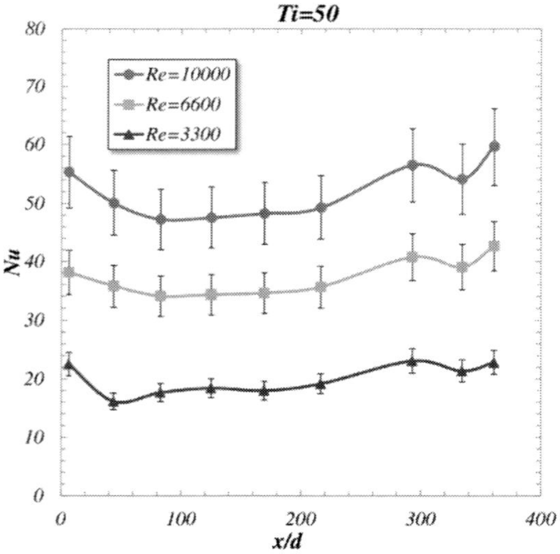

Figure 7. Nusselt number as a function of x/d at different Reynolds.

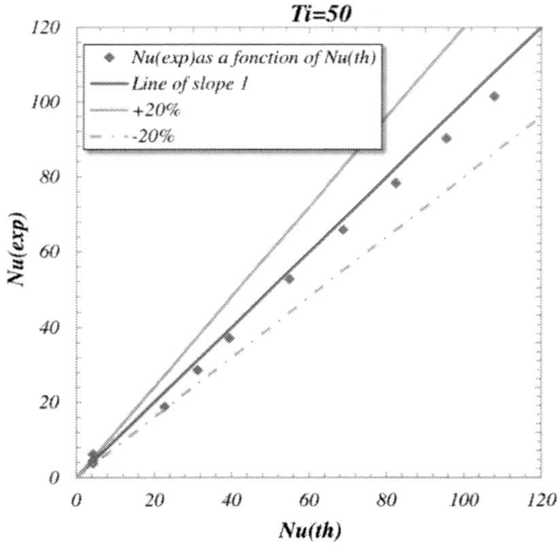

Figure 8. Experimental Nusselt as a function of theoretical Nusselt.

We can see from figure 7 that the edge effects are negligible in the turbulent regime and all thermocouples can be used.

Figure 8 shows the mean Nusselt (thermocouples 1-10) determined experimentally as a function of theoretical *Nu* calculated using Colburn's correlation. For this figure, *Re* is between 650 and 24,000.

The correlation of Colburn: $Nu = 0.023.\mathrm{Re}^{0.8}.\mathrm{Pr}^{1/3}$ is valid for $0.7<Pr<160$ and $10^4<Re<1.2.10^5$ with *Re* and *Pr* calculated film temperature: $T_{film} = (T_{pin} + T_f)/2$.

We observe a good agreement between our experimental results and correlations. The experimental uncertainties were calculated for the case *Re*=3300, *Nu*=22 and *Re*=23000, *Nu*=107, the results are around ± 9% and ± 12.5%. This is reasonably good for this kind of correlation.
In conclusion, we can say that the tests with demineralised water show behaviour consistent with classical correlations for heat transfer.

3.3. Pressure drop

The regular pressure drop in a tube Δp can be written in general:

$$\Delta p = \Lambda.(\frac{L}{d_i}).\rho.\frac{v^2}{2} \quad (14)$$

with $\Lambda = 64/\mathrm{Re}$ for Re ≤ 2300 (laminar flow) and $\Lambda = 0.3164/\mathrm{Re}^{1/4}$ for 2300≤ Re ≤100 000 (Blasius correlation for turbulent flow) [11].
However, these classical correlations have been established in the case of isothermal environments. To take into account the temperature variation in the axial direction of the tube, it is imperative to apply a correction factor:

$$\Lambda_c = \Lambda.(\mu_f/\mu_p)^n \quad (15)$$

With n=0.25 in turbulent flow [11].
Figure 9 shows the experimental Darcy coefficient obtained with demineralised water and the Blasius correlation. The measurement uncertainties are represented by error bars.

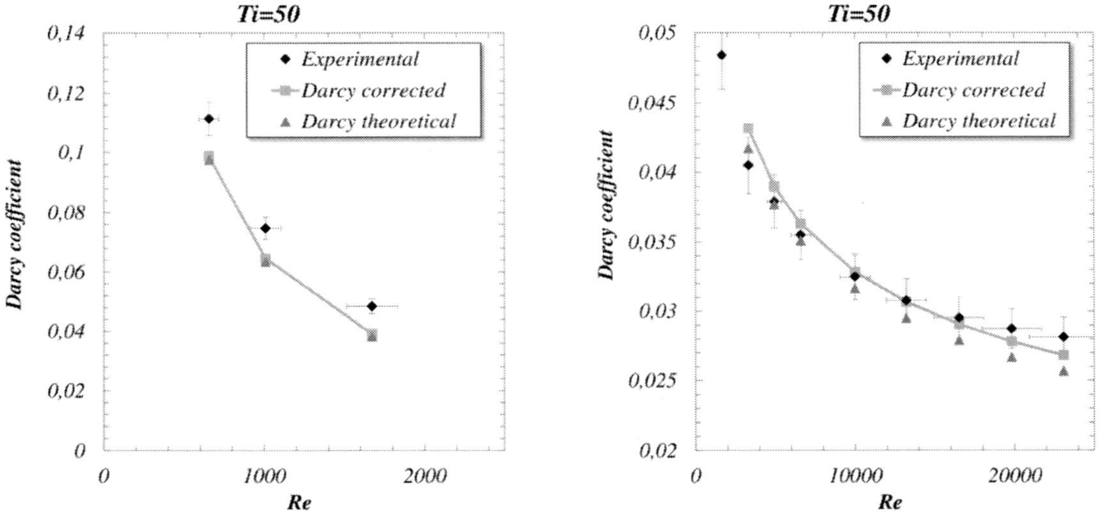

Figure 9. Evolution of the Darcy coefficient as a function of Reynolds number

The conformity of the results of heat exchange and pressure to the demineralized water in the loop allows us to validate the measurement system. We can consider it also allows us to perform the same type of measurements applied to nanofluids with confidence.

4. Conclusions

Different fluid/material *PEC* simulations have led to a choice of a water/metal nanofluid that may increase heat transfer up to 20% at same pressure loss. A summary table is given. This allows to select the right alloy having an adequate melting temperature.
A specific experimental test loop has been developed in order to test thermo hydraulic performance in laminar and turbulent conditions at imposed heat flux with a small volume of fluid (350 ml).
Tests have been performed with base fluid (water) in laminar and turbulent flow on this loop, the classical thermo hydraulic models for heat transfer and pressure loss have been verified thus validating the experimental method.
Note that the experimental results presented in the current paper are only for pure water and not for the proposed NEPCM fluids.
These fluids have a great potential for electronic cooling: for example at same junction temperature we can increase the maximum heat flux or decrease pumping power; it is also possible to increase reliability decreasing the junction temperature with the same cooling system.
Development and tests of some of these fluids are under way in our laboratory. Apart from physical properties and thermohydraulic characterizations, some problems as sedimentation, agglomeration of particles, fouling and so on have to be investigated.

References

[1] Delgado, M. et al. *Review on phase change material emulsions and microencapsulated phase change material slurries: Materials, heat transfer studies and applications* In : Renewable and Sustainable Energy Reviews. 2012, vol. 16, pp. 253-273.
[2] Cingarapu, S. et al. *Nanofluids with encapsulated tin nanoparticles for advanced heat transfer and thermal energy storage* In : International Journal of Energy Research. 1-6-2013, pp. n/a.
[3] Delgado, M. et al. *Experimental analysis of a microencapsulated PCM slurry as thermal storage system and as heat transfer fluid in laminar flow* In : Applied Thermal engineering. 2012, vol. 36, pp. 370-377.
[4] Alvarado, J.L. et al. *Thermal performance of microencapsulated phase change material slurry in turbulent flow under constant heat flux* In : International Journal of Heat and Mass Transfer. 2007, vol. 50, pp. 1938-1952.
[5] Ho, C.J. et al. *Water-based suspensions of Al_2O_3 nanoparticles and MEPCM particles on convection effectiveness in a circular tube* In : International Journal of Thermal Sciences. 2011, vol. 50, pp. 736-748.
[6] Goel, M., Roy, S.K. et Sengupta, S. *Laminar forced convection heat transfer in microcapsulated phase change material suspensions* In : International Journal of Heat and Mass Transfer. 1994, vol. 37, pp. 593-604.
[7] Hamilton, R. et Crosser, O. *Thermal conductivity of heterogenouseous two-component systems* In : Ind.Eng.Chem.Fundamentals. 1962, vol. 1, pp. 187-191.
[8] Einstein, A. *Eine neue Bestimmung der Moleküldimensionen* In : Annalen der Physik. University of Zurich, 1906, vol. 324, pp. 289-306.
[9] http://bourse.lesechos.fr/bourse/matieres_premieres/metaux.jsp
[10] Shah, R. *Correlation for laminar hydrodynamic entry lengh solutions for circular and noncircular ducts* In : Journal of Fluids Engineering, Transactions of the ASME. 1978, vol. 100, pp. 177-179.
[11] Petukhov, B. *Heat Transfer and Friction in Turbulent Pipe Flow with Variable Physical Properties* In : Advances in heat transfer. Academic Press, 1970, vol. 6, pp. 504-564.

Journal of Physics: Conference Series **525** (2014) 012024 doi:10.1088/1742-6596/525/1/012024

Effect of surface wettability on carbon nanotube water–based nanofluid droplet impingement heat transfer

Robert Gordon Jackson[a], Mostafa Kahani[a], Nitin Karwa[a], Alex Wu[b], Robert Lamb[b], Robert Taylor[c], Gary Rosengarten[1,a]

[a] School of Aerospace, Mechanical and Manufacturing Engineering, RMIT University, Melbourne, Australia

[b] Department of Chemistry, The University of Melbourne, Parkville, Australia

[c] School of Mechanical and Manufacturing Engineering, The University of New South Wales, Kensington, Australia

E-mail: gary.rosengarten@rmit.edu.au

Abstract. Recent studies into droplet impingement heat transfer have demonstrated that it has great potential for providing high heat flux cooling in areas such as thermal management of electronics. The wettability of the surface affects the flow dynamics of the impingement process and the resulting heat transfer. In this study, the effect of surface wettability on carbon nanotube water-based nanofluid droplet impingement heat transfer has been studied and compared with water. Superhydrophobic or hydrophilic coatings are applied on one face of monocrystalline silicon wafers (the drop impinges on this face) while the other face is painted matt black to permit infrared thermography. The silicon wafer is preheated to 40 °C and a single droplet impinges normally on the top facing coated surface of the monocrystalline silicon wafer. The inverse heat conduction problem has been solved using the measured black face temperature. For both the water and nanofluid droplets, the convective heat transfer coefficient reduces with the decrease in surface wettability. It is found that the nanofluid produce a significantly higher convective heat transfer coefficient during droplet impingement than water, with the enhancement increasing with increasing wettability.

1. Introduction

There are a large number of applications that require high heat-flux cooling, such as thermal management on high-powered chips, fire suppression and the rapid quenching of steel. Droplet impingement, particularly in the form of spray cooling, has been shown to be an effective means of rapid heat removal. Spray cooling involves heat transfer to multiple droplets impinging simultaneously and in rapid succession over a large area. This method of cooling is capable of providing high heat transfer coefficient like impinging jets, but with a mass flux as much as one order of magnitude less [1,2]. Spray cooling can either occur as non-boiling or evaporative/boiling. Higher heat flux can be achieved through boiling and evaporation, but non-boiling heat transfer systems are more suited for electronics cooling applications as the system is less complex. Karwa et al. [2] reported a non-boiling heat transfer coefficient of 25000 $W/m^2 \cdot K$ for mass flux of 9.94 $kg/m^2 \cdot s$. Wang et al. [3] reached an

[1] To whom any correspondence should be addressed.

Content from this work may be used under the terms of the Creative Commons Attribution 3.0 licence. Any further distribution of this work must maintain attribution to the author(s) and the title of the work, journal citation and DOI.
Published under licence by IOP Publishing Ltd

even higher non-boiling heat transfer coefficient of 66000 $W/m^2{\cdot}K$ for mass flux of 53 $kg/m^2{\cdot}s$ but reported that a thick fluid film remains on the surface.

A nanofluid is a suspension of particles, less than 100 nm in size, within a liquid. The particles can be used to tailor the fluid properties. For example, for better thermal properties than the base liquid, double-walled carbon nanotube nanoparticles ($k{\sim}2000$ W/m·K [4]) are added to water ($k{\sim}0.607$ W/m·K) in concentration of just 0.6% w/w, and the resulting thermal conductivity of the nanofluid is 34% higher than that of water [5]. The thermal conductivity of the nanofluid is a function of the thermal conductivities, particle sizes and relative concentrations of the constituent media. Patel et al. [6] suggested the following equation for the thermal conductivity of carbon nanotube water-based nanofluid

$$k_{nf} = k_f\left[1 + \frac{k_p \varphi r_f}{k_f(1-\varphi)r_p}\right]. \quad (1)$$

The dynamic viscosity can be found using the correlation [7]

$$\mu_{nf} = \mu_f(1 + 22.7814\varphi - 9748.4\varphi^2 + 1\,000\,000\varphi^3). \quad (2)$$

The density and specific heat capacity are calculated as

$$\rho_{nf} = \varphi\rho_p + (1-\varphi)\rho_f, \quad (3)$$

$$C_{p_{nf}} = \varphi C_{p_p} + (1-\varphi)C_{p_f}. \quad (4)$$

The dynamics of droplet impingement depends on various factors, such as the Weber number of the droplet, the Reynolds number of the flow during impact, and the wettability of the surface (the equilibrium contact angle between the droplet and the impingement surface is a measure of the surface wettability) [8]. Rosengarten et al. [9] have shown that the surface contact angle influences the maximum spreading diameter of the droplet, contact time and effectiveness of heat transfer; for single droplet impingement events, it was observed that hydrophilic surfaces (contact angle < 25°) offer a higher cooling effectiveness than both uncoated (contact angle ~ 90°) and superhydrophobic surfaces (contact angle > 160°). If this is also true for a nanofluid, then it will provide an additional means by which heat transfer can be enhanced. Relevant studies on the effect of surface wettability on the nanofluid droplet impingement are not available.

In the present investigation, the effect of surface wettability on heat transfer to a carbon nanotube water-based nanofluid droplet and the enhancement with respect to water is determined.

2. Experimental apparatus and method

Figure 1a shows the schematic layout of the experimental setup. A photograph of the experimental setup is shown in Figure 1b. Wafers of monocrystalline silicon were used as the impingement targets due to the high rigidity and thermal conductivity of silicon. Each wafer was 0.5 mm thick and 100 mm in diameter. The undersides of the wafers were coated with a matte black spray paint allowing temperature measurement by thermal imaging camera. Coatings were applied to the topside of the silicon wafers to modify the surface wettability; one was given a superhydrophobic coating, one was given a hydrophilic coating and the third wafer was left uncoated as the reference surface. The wafer is heated by warm air rising from a ring heater. The ring heater also allows unobstructed view of the bottom surface for thermal imaging.

The superhydrophobic coating was comprised of a mixture of silica nanoparticles (Evonik), dimethylsiloxane polymer (Gelest) and methyltrimethoxysilane (Sigma-Aldrich) as the linking agent. This is produced and applied to the silicon wafers by the methods described in [10]. The hydrophilic coating was prepared in the same way, except that tetraethoxysilane is used as the linking agent with an ethoxylated siloxane polymer. The surface roughness of the superhydrophobic surface was measured to have an RMS value of 308 nm and a roughness ratio of 1.5 [9].

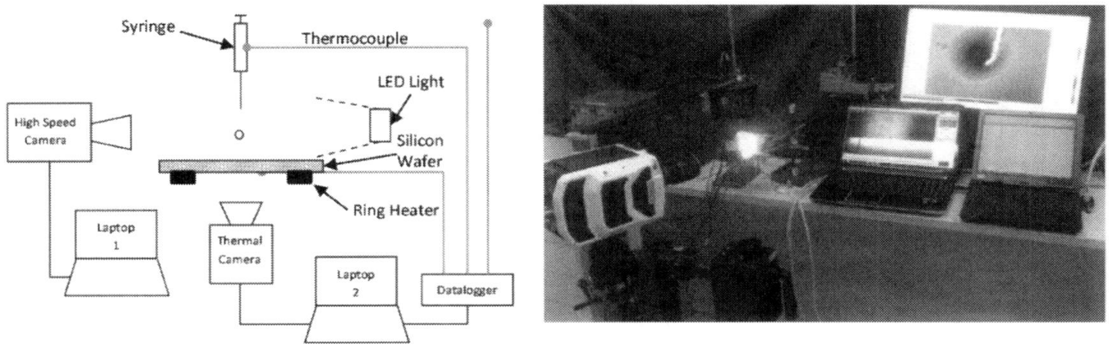

(a) (b)

Figure 1. a) Schematic of the experimental setup, **b)** Photograph of the experimental setup.

Double-walled carbon nanotubes (DWCNT) (Sigma Aldrich) with an average diameter of 3.5 nm and a mean length of 1–10 μm were dispersed in Type 1 water to form DWCNT/water nanofluid. Sodium dodecyl sulphate (SDS) was employed as the dispersant. DWCNT and SDS powders were added to Type 1 water while stirring with a magnetic stirrer for an hour, the suspensions were then sonicated for 35 minutes. Nanofluids were prepared with DWCNT particle concentrations of 0.25, 0.5 and 0.75% w/w. Weight fractions are calculated by dividing the mass measured for the CNT nanoparticles by the final nanofluid mass, using an analytical balance with 0.01 mg resolution (Ohaus Discovery Semi-Micro). Mass fraction can be converted to volume fraction using the following equation:

$$v/v\% = \frac{w/w\%}{w/w\% + (1 + w/w\%)\rho_p/\rho_f} \tag{5}$$

The concentration of surfactant in these nanofluids was 0.25, 0.50 and 0.75% w/w, respectively. The samples were left stationary for 7 days and thereafter were found to be sufficiently stable with no visually observable sedimentation. The thermal properties of the three different concentration nanofluids are calculated using Eq. 1–4 and are shown in Table 1. The predicted enhancement in thermal conductivity of these nanofluids is higher than the range shown in Patel et al. [6] because the diameter of the CNTs used is different. Patel et al. [6] uses CNTs with an average diameter of 30nm, while the CNTs used here have an average diameter of 3.5nm (Sigma Aldrich). Low concentrations of carbon nanotube nanoparticles, such as those used in this experiment, do not cause the surface tension of the nanofluid to differ significantly from that of the base fluid [11], so the surface tension is assumed to be the same as that of an equivalent water/SDS solution. Elworthy and Mysels [12] provide data showing the surface tension of water containing various concentrations of SDS; at a mass concentration of 0.25% the surface tension is 39.5 mN/m, this decreases to 38.5 mN/m as the concentration is increased to 0.75%.

Table 2 shows the wettability of water and nanofluid on the various surfaces. Contact angle measurements were taken from images of the droplets taken using the high-speed camera after the droplet had become sessile. The image of the droplet was processed in Matlab by firstly extracting the pixels representing the droplet from the surrounding image, then drawing a line tangent to the first five pixels that described the curvature of the top of the droplet. The maximum error for this process is for contact angles close to 90°, where the error is ±5.5°. For superhydrophobic (γ>165°) and hydrophilic (γ<15°), the error is less than ±0.4°. As can be seen, the only property of water that significantly changes at these nanoparticle concentrations is the thermal conductivity and the contact angle. Most importantly, the nanofluid is able to wet the hydrophobic surface. This means that the thermohydraulics of nanofluid droplet impingement will be different than a water droplet. Also, it is not known how a nanofluid fundamentally interacts with a superhydrophobic surface.

Table 1. Thermal and transport properties for water, carbon nanotube nanoparticles and nanofluid.

		Water	DWCNT	0.25% w/w		0.50% w/w		0.75% w/w	
				Nanofluid	% Difference	Nanofluid	% Difference	Nanofluid	% Difference
k	W/(m·K)	0.607	2000	0.7427	22.36	0.8778	44.6054	1.0121	66.7421
μ	kg/(m·s)	8.9×10^{-4}	-	0.0009	1.4975	0.0009	1.2591	0.0009	0.2911
ρ	kg/m^3	997	2100	998.31	0.1312	999.61	0.2615	1000.90	0.3908
c_p	J/(kg·K)	4180	530	4175.67	-0.1036	4171.37	-0.2064	4167.11	-0.3084
γ	mN/m	72.8	-	39.5	-45.7	38.8	-46.7	38.5	-47.1

Table 2. Wettability of water and nanofluid on test surfaces.

Surface coating	Surface contact angle (degrees)			
	Water	0.25% w/w	0.50% w/w	0.75% w/w
Hydrophilic	25	11.0	8.4	11.5
Uncoated	43	10.9	7.2	11.5
Superhydrophobic	165	81.8	59.5	58.9

A 30 gauge needle (OD 0.31 mm, ID 0.16 mm) was fixed onto a syringe and its tip was 100 mm above the silicon wafer. A thermocouple was installed in-line between the syringe and needle to measure the temperature of the droplet immediately at the moment of dispensation. A second thermocouple was hung near the apparatus to measure the ambient air temperature. A third thermocouple was attached to the bottom of the silicon wafer to measure the wafer temperature during IR camera calibration. A National Instruments NI9219 analogue input data logger recorded the temperature data at a rate of 10 kHz and passed this to a laptop running LabView.

A FLIR Titanium thermal imaging camera was mounted beneath the silicon wafer to view the black-painted underside. This was connected to a laptop running the FLIR Altair recording and analysis software. The camera was configured to run at 870 fps, with a 1 ms integration time and a resolution of 160×128 pixel^2. The sensor in the camera was calibrated using a thermocouple attached to the wafer over a temperature range of 17–49 °C.

The droplet travel path from the needle to the impingement surface was back illuminated by a high-powered LED light fitted with a diffuser. A Phantom v1610 high-speed camera was used to image the impingement phenomenon from the side. The camera was configured to record at 8700 fps, a multiple of the thermal camera's frame rate.

The wafer was heated until it reached a stable temperature of approximately 40°C. A droplet was dispensed from the syringe onto the heated wafer. The thermal and high-speed cameras simultaneously recorded the droplet impingement process. After each droplet impingement, the surface was rinsed with Type 1 water, allowed to dry and then reheated. This process was repeated three times for each concentration of the nanofluid and for each surface coating.

3. Data Analysis

The high-speed visible images are processed in MATLAB to determine the contact line diameter as it spreads on wafer. Using the Altair thermal camera software, the IR intensity values recorded by the camera were converted to temperature. The temperature data was averaged over a 1 mm diameter circular area around the stagnation point (shown by a blue circle in Fig. 2). From this measured temperature, the impingement face temperature and heat flux are calculated by solving the inverse heat conduction problem. This calculation is performed using the FORTRAN code presented in § 18.5 of [13], which solves the following equations:

$$T(L,t) = T(0,t) + \sum_{n=1}^{\infty} \frac{1}{(2n)!}\left(\frac{L^2}{\alpha}\right)^n \frac{d^n T(0,t)}{dt^n}, \tag{6}$$

$$q(t) = k \sum_{n=1}^{\infty} \frac{L^{2n-1}}{(2n-1)!} \frac{1}{\alpha^n} \frac{d^n T(0,t)}{dt^n}. \tag{7}$$

The instantaneous convective heat transfer coefficient due to droplet impingement can be calculated as

$$h(t) = \frac{q(t)}{T(L,t) - \bar{T}_{drop}}. \tag{8}$$

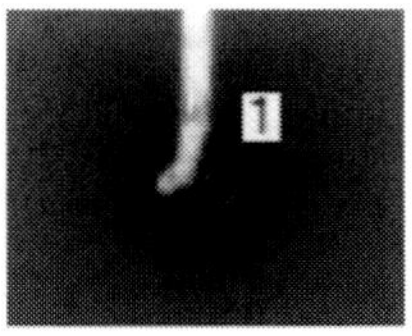

Figure 2. Area-averaged temperature measurement being conducted in Altair. The white object is the thermocouple. The blue circle is 1 mm in diameter and is centered at the thermal footprint on the impinging droplet. The instantaneous wafer temperature for all heat transfer calculations is determined by averaging the temperature over an area enclosed by this circle.

4. Results and discussion

The images in Figure 3 show simultaneous optical and thermal views of the droplet impingement process for water and 0.5% w/w concentration nanofluid droplets on an uncoated silicon surface. The variation of droplet spreading diameter normalized with the initial droplet size is shown in Fig. 4. As the water droplet impacts the wafer surface, its vertical momentum is redirected radially, causing the droplet to flatten and rapidly spread outwards. Surface tension and viscous forces decelerate the radial flow and stop the droplet spreading after approximately 3 ms; this is the maximum diameter to which the droplet spreads, called the equilibrium diameter. Thereafter, the surface contact line remains pinned as the fluid lacks the energy necessary to overcome the adhering forces of the surface but the fluid within the droplet continues to oscillate. First, aided by surface tension pulling the fluid back towards the centre, the internal flow is directed radially inwards. As all of the fluid converges at the centre of impact, an upwash appears at about 9 ms after impact. After reaching the centre the flow reverses, moving the main bulk of the fluid towards the outer edge of the droplet again, although it does not spread beyond the initial maximum diameter. The fluid continues to oscillate between the outer edge and the centre, but the amplitude of the upwash reduces with each oscillation as the flow energy is dissipated by viscous forces. After approximately 30 ms, no fluid motion within the droplet motion is detected.

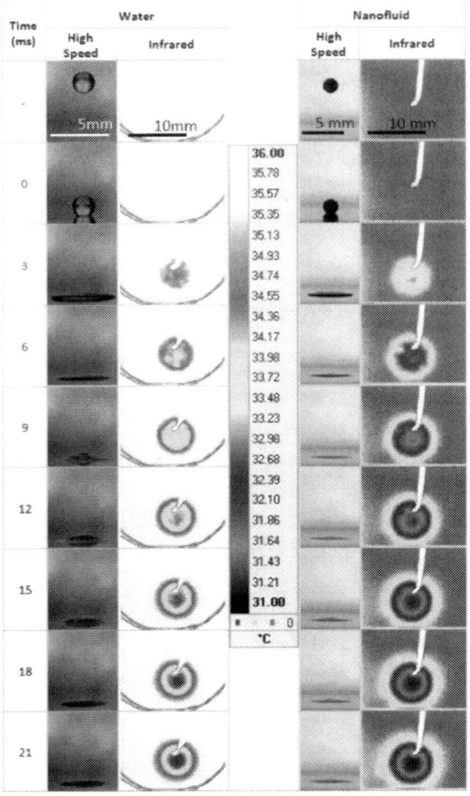

Figure 3. High-speed and infrared images of water (left) and 0.5% w/w concentration nanofluid (right) droplet impingement onto an uncoated silicon surface.

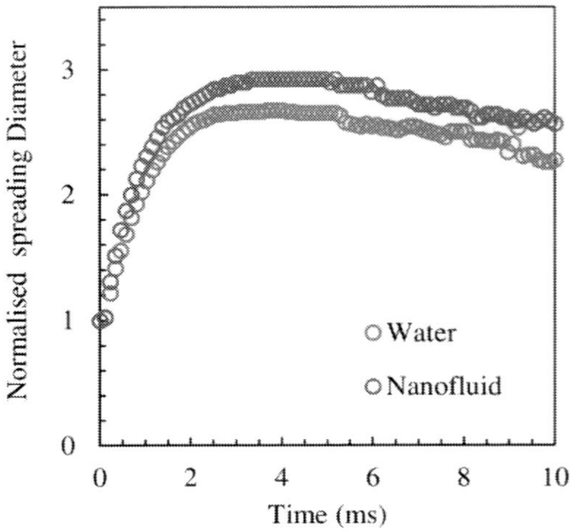

Figure 4. Normalised droplet diameter over time for water and 0.5% w/w concentration nanofluid droplets impinging onto an uncoated surface.

The initial dynamics of the nanofluid droplet impact is similar to those of the water droplet; it spreads rapidly upon coming into contact with the surface, reaching a maximum diameter after 3 ms. However, the nanofluid droplet does not rebound towards the centre as strongly as the water (compare the droplet in Figure 3 at 9 ms). Fluid motion in the nanofluid droplet ceases by 12 ms as compared to 30 ms for the water droplet, which may be due to the lower surface tension of the nanofluid.

Thermal images are shown to the right of the high speed images; these show the temperature of the underside of the wafer as it changes over the course of the droplet impact. Prior to impact, the wafer temperature is uniform. The impingement of the droplet on top of the silicon surface leads to the decrease in surface temperature of a circular area on the bottom face of the wafer; this area is approximately of the same size and shape as the contact area between the droplet and the surface. The cooled area increases in diameter over time, but its growth rate decreases as the droplet spreading velocity decreases. The centre of the circular area is the stagnation point and has the lowest temperature. The wafer temperature increases in the radially outwards direction due to decrease in the convective heat transfer rate.

Journal of Physics: Conference Series **525** (2014) 012024 doi:10.1088/1742-6596/525/1/012024

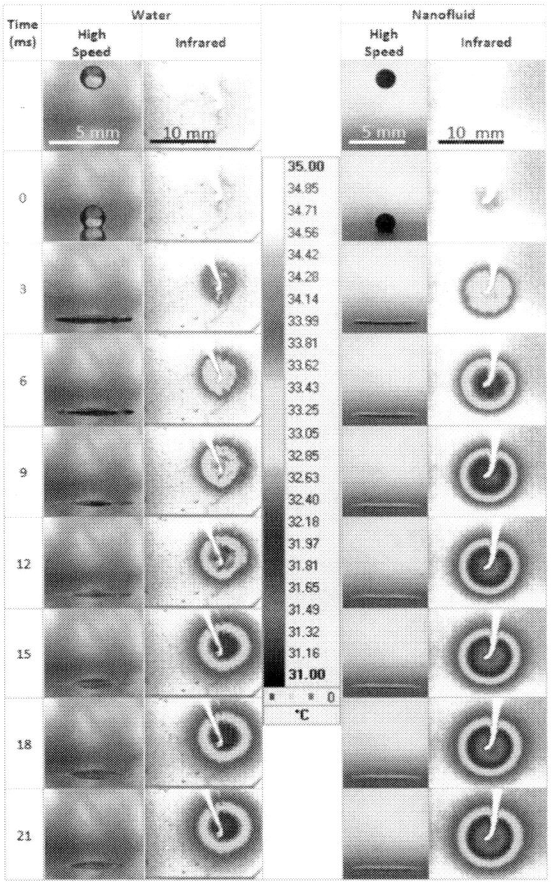

Figure 5. High-speed and infrared images of water (left) and 0.5% w/w concentration nanofluid (right) droplet impingement onto a hydrophilic surface.

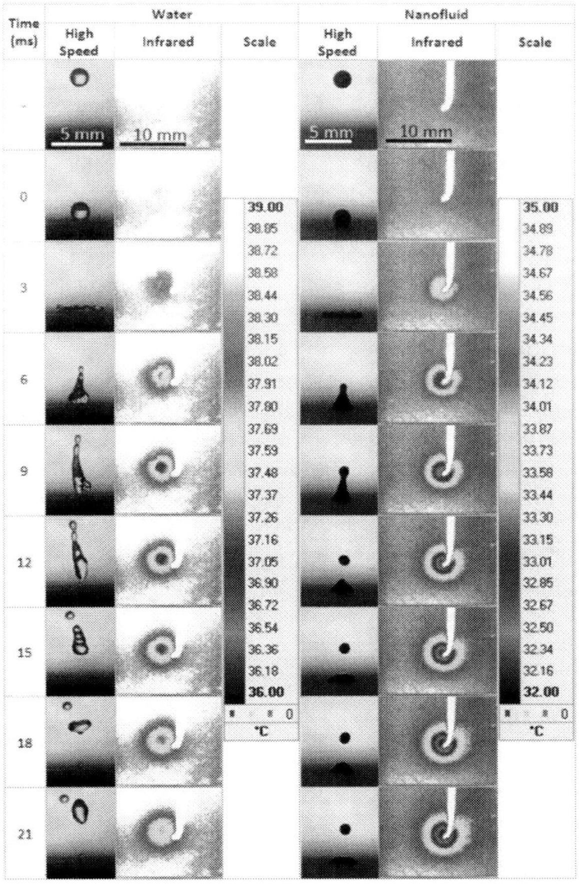

Figure 6. High-speed and infrared images of water (left) and 0.5% w/w concentration nanofluid (right) droplet impingement onto a superhydrophobic surface.

Figure 5 shows water and nanofluid droplet impinging onto hydrophilic surfaces. The area affected by the nanofluid droplet is larger than the size of the area affected by the water droplet. The temperature difference caused by the nanofluid droplet is greater than the temperature difference caused by the water droplet at each point in time. Both of these factors, the size of the area affected and amount of temperature change, indicate that the nanofluid droplet removed more heat from the surface than the water droplet.

Figure 6 shows droplets impinging onto a superhydrophobic surface. After the water droplet has impacted and spread, it rapidly retracts back towards the centre and forms a vertical jet with sufficient momentum to lift the entire droplet off the surface and rebounds. In air, surface tension forces attempt to change its shape into a sphere again. Resultantly, the droplet only stays in contact with the surface for about 9 ms. The thermal image for the water droplet shows that no further cooling is performed after this time and the plate begins to return to its initial temperature as heat from the surrounding solid is conducted towards the cooled spot. The nanofluid droplet impinging onto the superhydrophobic surface does not have as much surface tension as the water droplet, and as such does not rebound as strongly. This results in a significant amount of the nanofluid remaining in contact with the surface, with only a small volume separating as a secondary droplet (visible at 12 ms in Figure 6).

To compare the cooling performance of the two fluids on the three different wafers, the surface temperature is written in a non-dimensional form as

$$\theta = \frac{T_s(0) - T_s(t)}{T_s(0) - T_{drop}(0)} \tag{9}$$

where the surface temperature $T_s(t)$ was the instantaneous average temperature (see in Fig. 2). The variation of θ_0 with time for water and nanofluid for the three surfaces is shown in Fig. 7. For all three surface types, the nanofluid droplet reduces the non-dimensional surface temperature more rapidly than the water droplet does. It can be concluded that nanofluid is more effective at droplet impingement cooling than water. The contact angle between the surface and droplet appears to have a significant effect on the rate at which heat is transferred. Using the uncoated surface as a reference for comparing heat transfer rates, the superhydrophobic surface transfers less heat over time while the hydrophilic surface transfers more heat over time. The low level of heat transfer that results from droplets impinging onto the superhydrophobic surface is due to the small contact area and short contact time. The high-speed images in Figure 6 show the rebounding water droplet loses contact with the surface sometime between 9 and 12 ms, and resultantly the convective heat transfer drops significantly. This can also be seen in the graphs of dimensionless diameter against time shown in Figure 8; by 8 ms the water droplet has returned to a dimensionless diameter of 1, indicating that the droplet is about to lose contact with the surface. The graphs also indicate that the superhydrophobic surface causes a smaller maximum spreading diameter than the other surface types.

The variation of convective heat transfer coefficient over time is shown in Fig. 9. The nanofluid droplets consistently produce a higher convective heat transfer coefficient than water. This is due to the higher thermal conductivity of the nanofluid and the lower surface tension. The peak heat transfer coefficient for nanofluid is particularly high for the hydrophilic surface, approximately three times greater than it is for water. The poor thermal contact between the droplet and superhydrophobic surface results in a peak convective heat transfer coefficient that is significantly lower than for any other surface type. The heat transfer to the water droplet stops about 8 ms after impact because at this point the droplet has left the surface and is therefore no longer cooling it. The nanofluid droplet remains on the surface and continues to cool the surface. Overall, the enhancement in heat transfer rate by nanofluid is least for the hydrophobic surface. The effect of different nanofluid concentration on heat transfer enhancement on an uncoated surface is shown in Figure 10. Note that the surfactant concentration increases in the same proportion as the nanoparticles. While it was expected that increase in the concentration of both would lead to increase in the heat transfer coefficient, the results show that there is no net change in the enhancement factor with the change in nanofluid concentration. The reason for this is not understood at this time and further research in this area is required, with specific emphasis on separating the effect of surfactant and nanoparticle concentration.

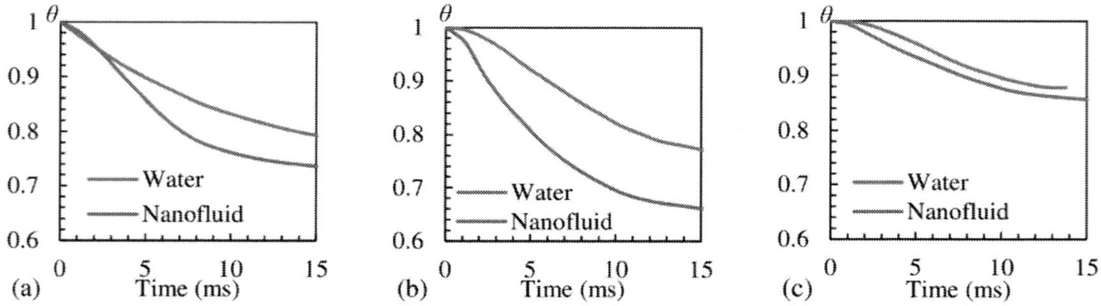

Figure 7. Normalised surface temperature over time for water and 0.5% w/w concentration nanofluid droplets impinging onto (a) uncoated, (b) hydrophilic and (c) superhydrophobic surface.

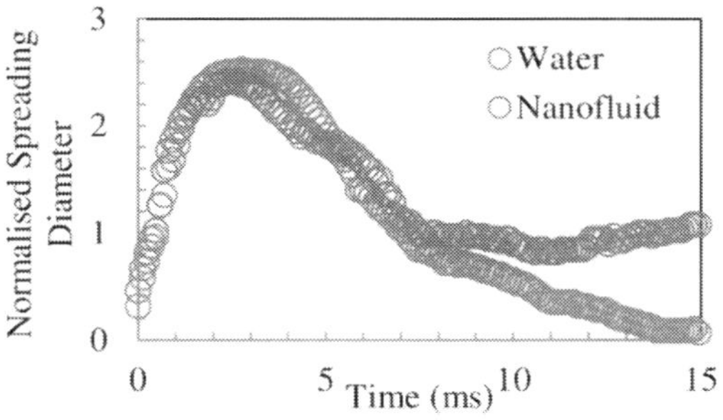

Figure 8. Normalised droplet diameter over time for water and 0.5% w/w concentration nanofluid droplets impinging onto a superhydrophobic surface.

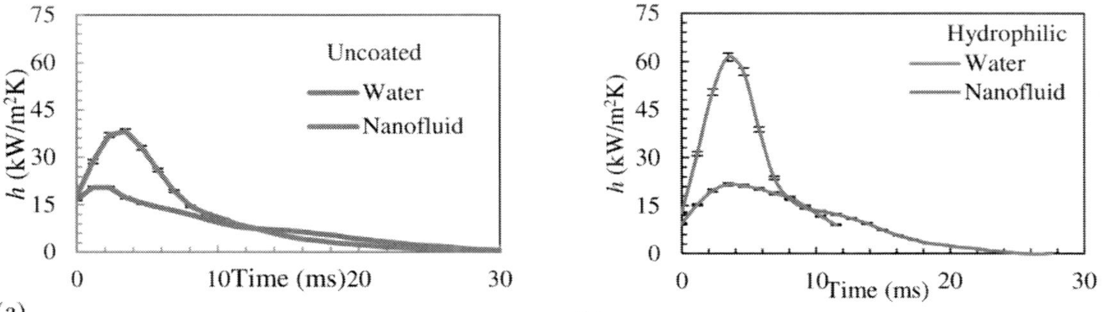

(a) (b)

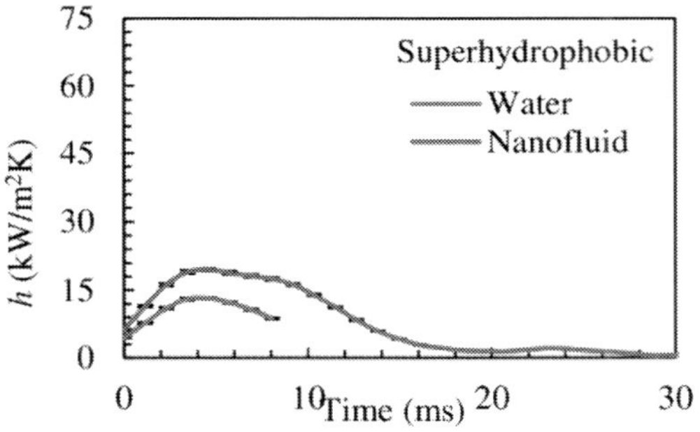

(c)

Figure 9. Convective heat transfer coefficient of water and 0.5% w/w concentration nanofluid droplets impinging onto (a) uncoated, (b) hydrophilic and (c) superhydrophobic surface.

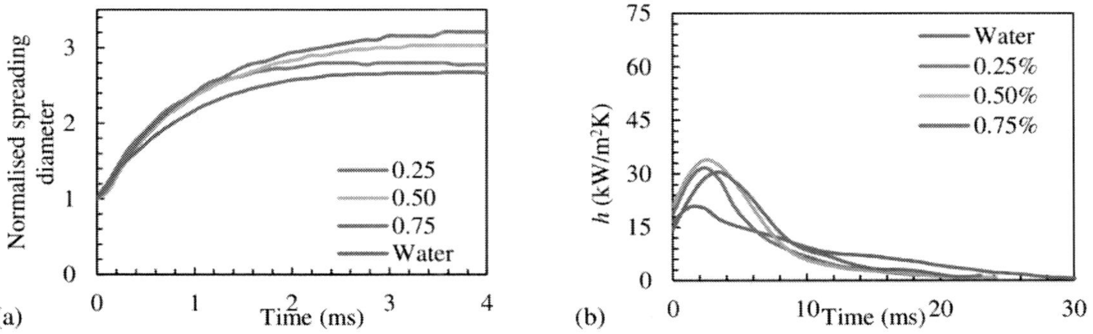

(a) (b)

Figure 10. Effect of nanoparticle concentration on droplet spreading and convective heat transfer coefficient on a uncoated surface.

5. Conclusions

Experiments were performed to study the effect of surface wettability on droplet impingement heat transfer to both water and nanofluid. The key findings from the study are

1. The convective heat transfer coefficient of nanofluid droplets is significantly higher than that of water droplets due to its higher thermal conductivity and lower surface tension.
2. Poor wettability of surface results in reduction of heat transfer rate.

These preliminary studies show that the wettability has a major effect on the heat transfer phenomenon. However, no change in the heat transfer rate was observed with the change in nanoparticle concentration. The reason for this observation is not clear and requires further investigation with specific emphasis on separating the effect of surfactant and nanoparticle concentration. The present heat transfer analysis is only one dimensional and does not account for the radial heat conduction within the wafer. A two-dimensional analysis will be performed in the future to determine the distribution of heat transfer coefficient on the surface. Furthermore, the spreading area on the three surfaces is different and it is worthwhile to determine the total heat removed by a droplet.

6. Nomenclature

		Greek Letters		Subscripts	
c_P	Specific heat capacity	α	Thermal diffusivity	*drop*	Droplet
k	Thermal conductivity	γ	Surface tension	*f*	Continuous fluid
L	Wafer thickness	μ	Dynamic viscosity	*nf*	Nanofluid
q	Heat flux	ρ	Density	*p*	Nanoparticle
r	Particle radius	τ	Dimensionless time	*s*	Top surface of silicon wafer (the impingement face)
T	Temperature	φ	Volume fractions		
		θ	Dimensionless temperature		

7. References

[1] Oliphant K, Webb B and McQuay M 1998 *Exp. Therm Fluid Sci.* **18(1)** 1–10.
[2] Karwa N, Kale S R and Subbarao P M V 2007 *Exp. Therm Fluid Sci.* **32** 571–579.
[3] Wang Y, Liu M, Liu M and Xu K 2011 *Heat Transfer Eng.* **32**(11–12) 1075–1081.
[4] Kahani M, Zeinali Heris S and Mousavi S M 2013, *Ind. Eng. Chem. Res.* **52(36)** 13183–13191.
[5] Assael, M J, Metaxa I N, Arvanitidis J, Christofilos D and Lioutas C 2005 *Int. J. Thermophys.* **26(3)** 647–664.
[6] Patel H E, Anoop K B, Sundararajan T and Das S K 2008 *Bull. Mater. Sci.* **31(3)** 387–390.
[7] Ebrahimnia-Bajestan E and Niazmand H 2011 *Iran J Chem. Eng.* **8**(2) 81–97.
[8] Vadillo D C, Soucemarianadin A, Delattre C and Roux D C D 2009 *Phys. Fluids* **21(12)** 122002.
[9] Rosengarten G, Tetuko A, Li K K, Wu A and Lamb R 2011 *ASME Int. Mechanical Eng. Congress, Denver.*
[10] Jones A, Lamb R N and Zhang H 2004 US 6,743,467 B1.
[11] Tanvir S and Qiao L 2012 *Nanoscale research letters*, **7**(1) 1-10
[12] Elworthy PH and Mysels KJ 1966 *Journal of colloid and interface science*, **21**(3) 331-347
[13] Taler J and Duda P 2010 *Solving Direct and Inverse Heat Conduction Problems* (Springer).

Aerodynamic performance of a vibrating piezoelectric fan under varied operational conditions

J Stafford[1], N Jeffers
Thermal Management Research Group, Bell Labs, Alcatel-Lucent, Dublin, Ireland

E-mail: jason.stafford@alcatel-lucent.com

Abstract. This paper experimentally examines the bulk aerodynamic performance of a vibrating fan operating in the first mode of vibration. The influence of operating condition on the local velocity field has also been investigated to understand the flow distribution at the exit region and determine the stalling condition for vibrating fans. Fan motion has been generated and controlled using a piezoelectric ceramic attached to a stainless steel cantilever. The frequency and amplitude at resonance were 109.4 Hz and 12.5 mm, respectively. A test facility has been developed to measure the pressure-flow characteristics of the vibrating fan and simultaneously conduct local velocity field measurements using particle image velocimetry. The results demonstrate the impact of system characteristics on the local velocity field. High momentum regions generated due to the oscillating motion exist with a component direction that is tangent to the blade at maximum displacement. These high velocity zones are significantly affected by increasing impedance while flow reversal is a dominant feature at maximum pressure rise. The findings outlined provide useful information for design of thermal management solutions that may incorporate this air cooling approach.

1. Introduction

Recent trends in thermal management of electronics using air cooling approaches have demonstrated the need for technologies that provide enhanced heat transfer performance at higher efficiency and reliability than the current technologies in widespread use. This has led to an improvement in conventional rotational fan designs and also the emergence of solid-state air movers such as synthetic jets and vibrating fans [1,2].

In the late 1970's, Toda [3] and Toda and Osaka [4] presented the concept of using a vibrating fan, driven by a piezoelectric polymer, to induce an airflow that could be used for electronics cooling or ventilation. The vibrating fan was compared to a small rotational fan, and it was noted that the solid-state piezoelectric fan required 1/7 of the power input to achieve similar exit velocities. Despite the potential of piezoelectric vibrating fans as an air cooling technology identified by Toda and Osaka [4], it was twenty years later when more intensive research into these air movers began. Yoo et al. [5] presented a number of piezoelectrically driven vibrating fan designs as a solution to remove electromagnetic noise for sensitive electronic systems that require forced cooling. The cantilever parameters of length and material were varied and it was noted that phosphor bronze provided the highest performance over aluminium and brass.

[1] To whom any correspondence should be addressed.

Content from this work may be used under the terms of the Creative Commons Attribution 3.0 licence. Any further distribution of this work must maintain attribution to the author(s) and the title of the work, journal citation and DOI.
Published under licence by IOP Publishing Ltd

Previous research into vibrating flexible cantilevers as air movers suggests that such devices can provide an increase in local heat dissipation [6] and also augment heat transfer due to auxiliary air flows [7]. Kimber and Garimella [6] measured the local heat transfer coefficients on a flat surface cooled by a piezoelectric fan. The vibrating fan was positioned vertically and normal to the flat surface for various gap distances between the surface and fan tip. Heat transfer distributions were shown to vary from a lobed pattern at close gaps to elliptical as the gap distance was increased. Local heat transfer coefficients were of order 100 W/m^2K demonstrating the potential use of vibrating fans for improving discrete cooling applications. Lin [7] considered the use of a vibrating fan in the wake region behind a cylinder in cross flow. The aim of this approach was to investigate the potential for improving the local heat transfer due to the adverse impact of the wake region of a cylinder in cross flow. A twin jet like structure was observed and for a range of operational criteria including fan vibration amplitude and cylinder Reynolds number, an increase in heat transfer was apparent in the wake region. However, it was also noted that variation of these criteria beyond certain limits can increase the wake region and ultimately reduce the local heat transfer performance over the baseline case.

Bulk cooling through the implementation of piezoelectric fans with finned heat sinks has also been analysed both experimentally [8-10] and numerically [8]. Abdullah et al. [8] and Shyu and Syu [10] both examined the potential of integrating piezoelectric fan arrays within the fin gap of parallel fin heat exchangers to reduce the cooling solution volume. In both studies, enhancement in heat transfer over natural convection was found to be above a factor of 1.5 and up to 3.3. Li et al. [9] experimentally examined the flow field and heat transfer due to positioning a piezoelectric fan adjacent to parallel fin heat sink geometries. The study focused on the position of the fan relative to the heat sink, and it was concluded that a horizontally orientated fan should be placed at the entrance to the heat sink for increased thermal performance.

The previous studies [6-10] examined heat transfer characteristics due to piezoelectric fans with the intended application of electronics cooling. The majority of research studies on piezoelectric fans have focused on the velocity and heat transfer performance in an environment free of the global constraints that exist within a practical system. However, the high density of contemporary electronic systems implies that these fans are likely to experience flow resistances that are significantly greater than this type of operating condition. There is limited research in this area, despite the low pressure-flow characteristics of vibrating fans [11] and the higher sensitivity to confinement compared to rotational fan designs. Kimber et al. [11] investigated the pressure-flow characteristics of two piezoelectric fans to determine the relationship between aerodynamic performance and the frequency and amplitude parameters. The bulk measurements were conducted using conventional fan performance characterisation standards and provided a comparison between an axial fan of similar scale. The axial fan provided superior pressure rise and lower efficiency than the piezoelectric fan under consideration. Although this study highlights the typical performance characteristics of piezoelectric fans, the recommended operating region and occurrence of stall effects have yet to be investigated in the literature.

In addition to bulk pressure-flow information, understanding the momentum field generated by air movers under varied operational conditions is important to ensure optimum performance is achieved for a combined fan – heat exchange solution. Egan et al. [12] demonstrated increased heat dissipation of up to 20% by aligning miniature centrifugal fan exit flows with parallel fin configurations. This was achieved using velocity field information for the combined cooling solution. Grimes et al. [13] highlighted the change in swirl characteristics at the exit region of an axial fan air mover with changes in operating pressure. Higher operating pressures result in an increase in the radial velocity component, and therefore alter the flow distribution for a convective heat exchange system. Considering both of these studies, an optimum fan - heat exchanger arrangement for maximal heat dissipation is dependent on knowledge of the air mover response to the system characteristics, or operating condition.

The aim of the current study is to determine the effect of operating conditions on the exit flow field of a vibrating piezoelectric fan. Using this information, the occurrence of stall for this type of air moving device will be examined. The stall region has been defined as the region along the pressure-flow characteristic where flow reversal into the blade passage occurs. Combined with the bulk aerodynamic performance characteristics, this information will be used to determine the recommended operating region for vibrating fans. Therefore, the results presented in the current study will assist in the optimisation of thermal management solutions incorporating vibrating fans subjected to practical system conditions.

2. Experimentation

In this section the experimental testing facility designed and implemented for the measurement of bulk pressure-flow characteristics and velocity fields in the fan exit region is presented. Geometric and material information is also provided on the vibrating fan used in the current investigation.

2.1 Pressure-flow characteristics

The measurement of pressure-flow characteristics was achieved by developing an experimental facility based on BS 848 [14] standards outlined for performance testing of fans. This facility is shown in figure 1 a) and was designed to span a wide range of fan performance characteristics that can range from 0 – 1000 Pa in static pressure rise, and 0 – 0.04 m^3/s in volumetric flow rate. In addition to the design guidelines recommended by the standards [14], a refinement was carried out using computational fluid dynamics to ensure that a suitable auxiliary fan was selected and flow conditioning in the test chamber was acceptable for the inflow to the test air moving device.

Measurement of static pressure rise was achieved through settling chamber pressure taps, and volumetric flow rate was indirectly determined by measuring differential pressure across an orifice plate and using a pre-calibrated relationship between the orifice pressure drop and flow rate. Accuracy in the measurement of differential pressure across a wide range was achieved using three pressure transducers which spanned three different scales. A Novasina PascalSwitch transducer was used for 0 – 20 Pa range with uncertainty of +/-0.05 Pa for < 4 Pa and +/- 1.5% of measured value for > 4 Pa. For 20 – 200 Pa, a Novasina Pascal-STV200 Z instrument was used with uncertainty of +/-0.2% of full scale. Finally, three Freescale semiconductor MPXV5004DP transducers were used for measurements > 200 Pa and calibrated to +/- 10 Pa. For the vibrating fan performance characteristic presented in this study, the maximum uncertainty in static pressure rise and volumetric flow rate are +/-0.068 Pa and +/- 0.6 x 10^{-5} m^3/s. For all experiments, the ambient air temperature was constant at 21.5°C +/- 0.1°C.

An enhanced view of the air moving device arrangement is illustrated in figure 1 b). The measurement facility is based on an inlet-side design, where air is moved from the settling chamber and expelled to the environment. The vibrating piezoelectric fan was positioned within the settling chamber and anchored to a support structure. This design maintained rigidity and ensured the electrical energy input to the piezoelectric ceramic was used solely for the mechanical vibration of the cantilever fan. The support structure had cross-sectional dimensions 11mm x 3mm and the edge of the cantilever was 15mm from the support leg. The structure dimensions were kept to a minimum and the distance of both legs from the fan was maximized to remove confinement effects that could be detrimental to the fan aerodynamics. Table 1 contains the geometric and material properties of the vibrating piezoelectric fan used for the current study. The fan was operated at the first vibration mode which resulted in maximum displacement amplitude, or tip deflection. A TTi TG1000 function generator supplied a sinusoidal waveform which was amplified to provide an electrical input to the piezoelectric ceramic of 50VRMS. The outflow region, defined by dimensions w and h in figure 1 b), was larger than the fan width and displacement amplitude by 0.5 mm to allow clearance for the fan motion. The free end of the cantilever was centred and coplanar with the outflow region when in the unexcited state (zero electrical power to piezoelectric ceramic).

It is anticipated that the outlet geometrical configuration described in figure 1 b) will also have some impact on the performance of the vibrating fan compared to a fan in free space, or due to other geometrical confinements. This was also noted by Kimber et al. [11] who measured bulk performance

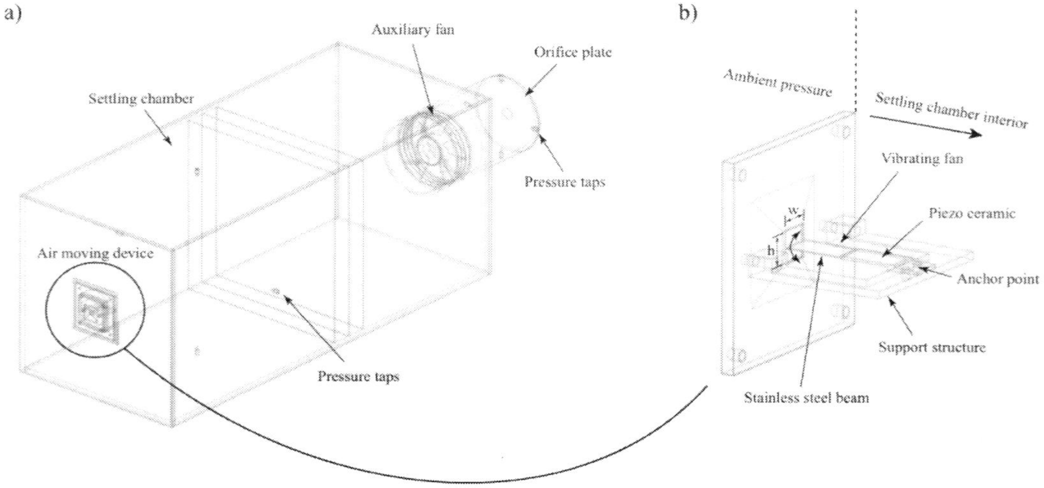

Figure 1. a) Apparatus for the characterisation of air moving devices and b) a detailed view of the vibrating fan installation.

Table 1. Details of the vibrating fan.

Vibrating fan parameters	
Length	48 mm
Height	12.5 mm
Blade material	Stainless steel
Blade thickness	75 μm
PZT patch length	24 mm
Input waveform	Sine wave
Frequency	109.4 Hz
Displacement amplitude	12.5 mm

using an outlet-side characterisation method. The dependence of performance on installation is a difficulty if the aim is to determine a generic fan performance characteristic for vibrating fans or one which is as robust in practice as for rotational fan designs. However, by examining the performance of different air movers in a similar test environment [14] relative comparisons can be useful for thermal design decisions.

2.2 Local velocity field

Local velocity field measurements were recorded using particle image velocimetry (PIV) on the exit flow field simultaneously with the bulk aerodynamic performance characteristics described previously in section 2.1. The experimental configuration is shown in figure 2. A laser sheet with 532nm wavelength and 7.5Hz pulse frequency was generated using a Quantel Big Sky Laser. The laser sheet illuminated glycol-based tracer particles < 10μm in diameter that were introduced into the test environment using a Magnum 2500 hazer. A homogeneous particle distribution was achieved after approximately 10 minutes of continuous issuing of tracer particles. This period of time was necessary to ensure sufficient seeding was achieved both internally and externally to the settling chamber. Once

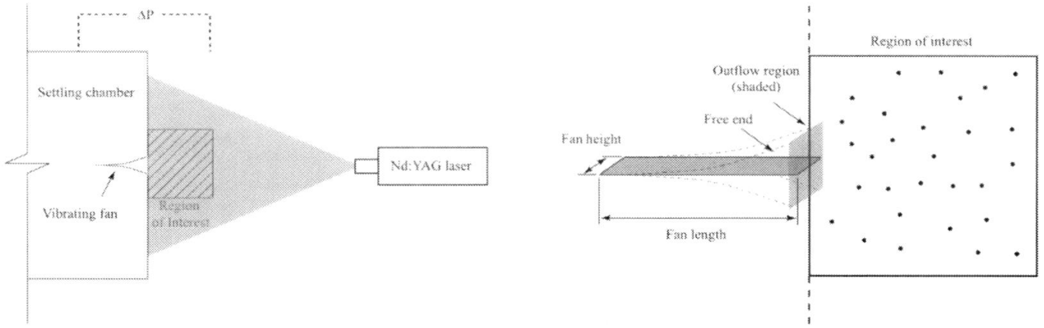

Figure 2. Region of interest for local velocity field analysis.

sufficient tracer particle density was achieved, the hazer was turned off for 1 minute prior to recording velocity field data. This approach ensured that air currents unrelated to the vibrating fan motion in the test environment were alleviated.

The laser illuminated the tracer particles twice in succession for the plane of interest. A single plane was considered for the purpose of investigating the impact of operating conditions. This plane was located at the centre of the blade height in a region where the fluid motion exiting the fan is dominant. Timing between pulses was adjusted based on flow velocities at each operating point considered. A TSi PowerView Plus 4MP CCD camera was positioned perpendicular to the laser sheet plane to record the scattered light from the particles on two separate frames for the region of interest illustrated in figure 2. The camera and laser were synchronized using a TSi laser pulse synchronizer. Each image pair was split into interrogation regions that were 32 x 32 pixels and processed using TSi Insight 4G software. The resulting spatial resolution of the vector field was 245μm.

Fixed errors for the PIV system were estimated to produce a relative uncertainty in velocity magnitude of < 2% [15]. A total of 2000 image pairs were recorded for each operating condition of interest. Based on the vibration frequency of the air mover and the acquisition rate of the PIV system, a random sampling assumption is valid. The method presented by Stafford et al. [15] was used to analyze convergence of full field measurement statistics. It was determined that the ensemble average velocity field is sufficiently converged towards the time-average within 3.24% using this sample size.

The vibrating fan is a composite of PZT and stainless steel materials with epoxy bond layers. Vibration is induced at the lower section of the cantilever, where the rigidity differs from the more flexible stainless steel free end of the fan. Therefore, the motion in space and time is dependent on the material and geometric properties of the cantilever. Due to this complexity and importance for the fan aerodynamics, the shape of the vibrating cantilever has been measured over the cycle period. A TSi PowerView Plus 2MP camera was used to measure the shape of the flexible cantilever at the maximum vibration amplitude to assist in discussing the resultant vector field. This test was conducted using ambient lighting and the exposure of the camera was adjusted to ensure the shape of the fan path was captured on a single frame.

3. Results and discussion

The pressure-flow characteristic for the vibrating piezoelectric fan is presented in figure 3. The data highlights the low pressure rise and flow rate that is typically achieved using vibrating piezoelectric fans, particularly when compared to small scale rotational fans [16,17]. Volumetric flow rate monotonically increases from the point of maximum static pressure rise to zero static pressure rise. The shape of the characteristic curve is similar to that observed by Kimber et al. [11] for other piezoelectrically actuated vibrating fans with different material and geometric properties. Figure 3 has no distinctive stalling dip that usually occurs in static pressure curves for rotating axial fans. This is

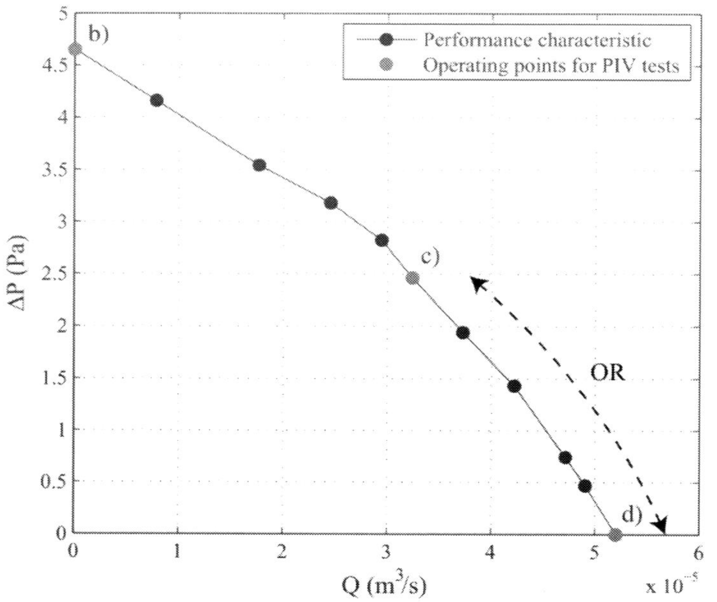

Figure 3. Performance characteristic for a vibrating fan operating at a resonant frequency of 109.4 Hz and amplitude of 12.5 mm. Operating region (OR) has been outlined based on the flow field analysis.

expected, as the static pressure rise generated by a vibrating piezoelectric fan is not produced by airfoil lift as in rotating axial fans. Air is moved by the oscillation of a simple blade structure, and comparisons are better suited with the aerodynamic characteristics of centrifugal-type fans. Static pressure for centrifugal fans is generated through the centrifugal action, and various blade designs produce a pressure-flow characteristic that is void of an obvious stalling point [18]. However, these air movers do experience similar stall symptoms as there is a static pressure region which results in flow separation and reversal in the blade passages of the fan. In the current study, the characteristics leading to stall for vibrating fans will be discussed using the simultaneous local velocity field measurements. The occurrence of stall has been defined as the point when flow is reversed and the vibrating fan begins to entrain air from the fan exit region.

Three points along the characteristic curve of figure 3 have been highlighted and labelled as b), c) and d). These points correspond to the vibrating fan operating at maximum, medium and zero pressure rise conditions. The local velocity magnitude and vector field at the exit region of the vibrating fan is presented in figure 4 for each of these operating points.

Figure 4 a) is a measurement of the path taken by the vibrating fan over a full cycle, scaled to the velocity field axes. The piezoelectric ceramic patch and stainless steel regions of the cantilever are defined and the influence of this anisotropic design is evident in the shape at maximum amplitude. The ceramic is displaced almost linearly from the anchor point, whereas displacement of the flexible stainless steel region is characteristic to the motion of an isotropic cantilever beam. The combination of figure 4 a) and b) provide an insight into the relationship between the fan motion and the resultant velocity field at the exit. High velocity zones correspond to the maximum displacement of the vibrating fan and are apparent for all operating points in figure 4. In this measurement plane, these zones form a time-averaged v-shaped flow structure.

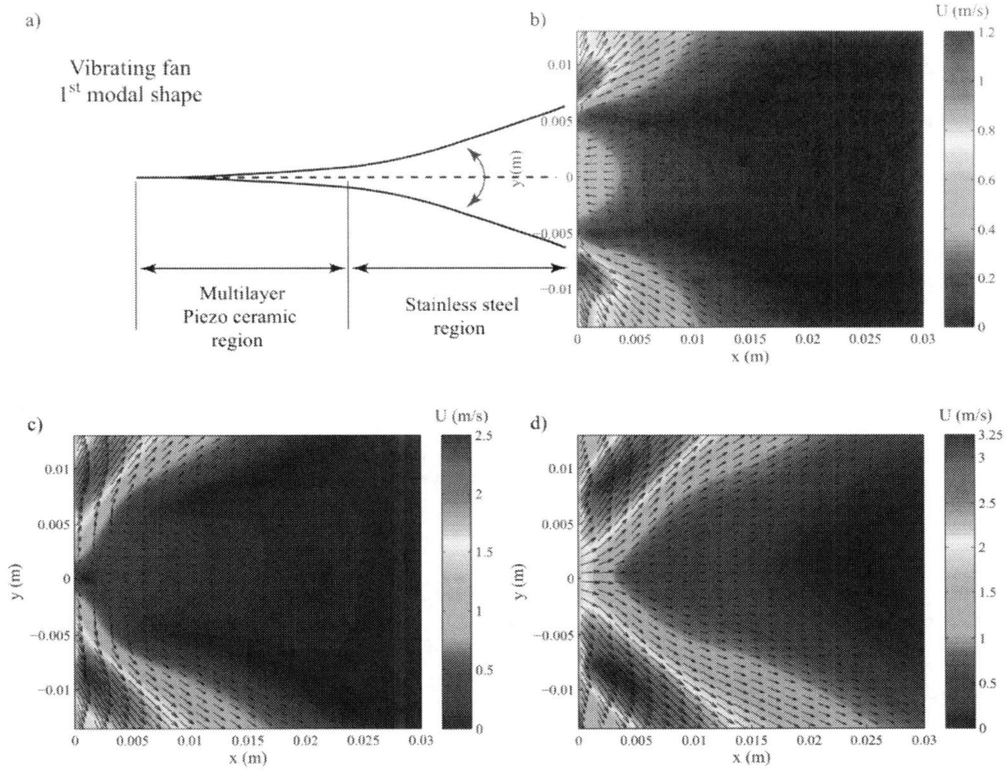

Figure 4. Measured fan modal shape a) and ensemble averaged velocity fields at the exit region for various operating conditions b), c), d) spanning the performance characteristic shown in figure 3.

The time-averaged distributions in figure 4 are due to the oscillating behaviour of the piezoelectric fan. During this motion, a pressure differential is generated across the fan surfaces, producing two counter-rotating vortices that roll over on the suction side of the fan and has been discussed previously in the literature by Lin [19]. Simultaneously, a negative pressure gradient exists between blade tip to root due to the difference in blade tip and root velocity. This static pressure gradient has been confirmed by Choi et al. [20]. Collectively, this provides the formation and development of a tip vortex which is forced to detach when the fan motion goes to zero velocity at maximum displacement [20]. It is this release of high momentum fluid at the maximum displacement which produces the time-averaged behaviour observed in figure 4 and the stable pressure-flow curve in figure 3. This is confirmed by examining the velocity vectors in the high momentum regions which are tangent to the fan blade profile at maximum displacement. A recent experimental study by the authors on the three-dimensional velocity field surrounding a vibrating piezoelectric fan in a free field also provides further detail on the local flow behaviour in the blade passage region [21].

Although the location of the high velocity zones is similar for figure 4 b) – d), there are discernible differences in the flow fields due to varied operating conditions experienced by the vibrating fan. At zero pressure rise and maximum flow rate (d), air exits radially outwards with two high velocity zones as previously discussed. Utilization of this distribution of momentum is important, as it influences the location and geometry of heat exchanger structures in the design for maximal heat dissipation. For example, heat sink structures could be aligned with the flow as this approach has been found to increase heat dissipation in forced convection heat sinks at miniature scales [12].

As static pressure is increased to just above 50% of ΔP_{max} (c), two weak recirculation zones emerge in the low velocity region (< 0.7 m/s) of the vector map. These are formed by the interaction between the high velocity flow exiting and the entrainment of ambient fluid from outside the test chamber due to flow reversal. At this point the onset of stall begins, and the vibrating fan is able to provide approximately 60% of Q_{max}. The increase in static pressure also alters the shape and increases the angle between the high velocity zones when comparing figure 4 c) and d). This may be a similar effect to that observed for conventional rotational fans [13] however further investigation is necessary to confirm.

In figure 4 b), the vector field resultant from operating at maximum pressure rise and zero flow rate is presented. At this operating point, the vibrating fan is fully stalled and flow reversal dominates the flow field. The net volumetric flow rate is zero; therefore the original exit flow region is performing two functions as both an exit and inlet. Air is entrained from the external environment at $x = 0$ and -0.005 m $< y <$ 0.005 m and then expelled at the fan tip regions. This also produces two recirculation zones that are centered in close proximity to the fan tip, compared to that observed for the point when stall is initiated in figure 4 c). In addition to a significant reduction in volumetric flow rate, the flow reversal associated with operating a vibrating fan in the stall region is of considerable importance for thermal management applications. During stalling, there is a risk that high temperature fluid could be entrained from downstream heated bodies and circulated within the electronic system to heat nearby sensitive components. However, this flow reversal effect could also be used to benefit the thermal system if applied correctly at the initial design stages.

Both the bulk aerodynamic performance characteristics and local velocity field analysis have been used to determine the initiation of stall. This has been used to provide a recommended operating region (OR) for vibrating fans. The operating region is defined in figure 3 and extends to 50% of the static pressure rise. Operating in this region of the pressure-flow characteristic curve ensures that the vibrating fan avoids stalling effects.

4. Conclusions

The aerodynamic performance of a vibrating piezoelectric fan has been experimentally investigated to understand the impact of operating conditions on the exit flow field and determine the stalling characteristics of this type of air moving device. A custom fan characterisation apparatus has been designed for the measurement of the vibrating fan pressure-flow rate performance characteristics. This apparatus also enabled simultaneous measurement of the velocity field at the exit flow region of the vibrating fan using particle image velocimetry. The bulk aerodynamic performance assessment demonstrated the stable pressure-flow characteristic that vibrating fans generate. However, time-averaged exit flow fields have been shown to alter from the zero to maximum static pressure rise performance range. At maximum static pressure, flow reversal dominates the flow field. The initiation of flow reversal at the fan exit, classified as the onset of stall, was found to occur at approximately 50% of the maximum static pressure. The local velocity field observations have been used to outline a recommended operating range and also provide valuable information for the design of thermal management solutions incorporating vibrating fans.

5. Acknowledgments

Bell Labs would like to thank the Industrial Development Agency (IDA) Ireland for their continued support.

References

[1] Chaudhari M, Puranik B and Agrawal A 2011 *Int. J. Heat Mass Transf.* **54** 2056-65
[2] Kimber M, Garimella S V and Raman A 2007 *J. Heat Transf.* **129** 1168-76
[3] Toda M 1979 *Ferroelectrics* **22** 911-18
[4] Toda M and Osaka S 1979 *Proc. IEEE* **67**(8) 1171-3
[5] Yoo J H, Hong J I and Cao W 2000 *Sens. Actuators* **79** 8-12

[6] Kimber M and Garimella S V 2009 *Int. J. Heat Mass Transf.* **52** 4470-8
[7] Lin C N 2013 *Int. J. Heat Mass Transf.* **60** 296-308
[8] Abdullah M K, Ismail N C, Abdul Mujeebu M, Abdullah M Z, Ahmad K A, Husaini M and Hamid M N A 2012 *Int. J. Heat Mass Transf.* **55** 5514-25
[9] Li H Y, Chao S M, Chen J W and Yang J T 2013 *Int. J. Heat Mass Transf.* **57** 722-32
[10] Shyu J C and Syu J Z 2014 *Appl. Therm. Eng.* **62** 573-80
[11] Kimber M, Suzuki K, Kitsunai N, Seki K and Garimella S V 2009 *IEEE. Trans. Comp. Pack. Technol.* **32**(4) 766-75
[12] Egan V, Stafford J, Walsh P and Walsh E 2009 *J. Heat Transf.* **130**(7) 071402
[13] Grimes R, Davies M, Punch J, Dalton T and Cole R 2001 *J. Elect. Pack.* **123** 112-9
[14] BSi 2007 *BS848-1:2007*
[15] Stafford J, Walsh E and Egan V 2012 *Flow Meas. Instr.* **26** 1-9
[16] Walsh P, Egan V, Grimes R and Walsh E 2009 *Heat Transf. Eng.* **30**(1) 130-7
[17] Stafford J and Fortune F 2014 *J. Elec. Pack.* **136** 011008
[18] Bleier F P 1998 *Fan Handbook: Selection, Application, and Design* (New York: McGraw-Hill)
[19] Lin C N 2013 *Appl. Therm. Eng.* **50** 693-703
[20] Choi M, Cierpka C, and Kim Y H 2012 *J. Fluids Struct.* **31** 67-78
[21] Jeffers N, Nolan K, Stafford J and Donnelly B 2014 *Eurotherm Seminar 102, Limerick, Ireland.*

Journal of Physics: Conference Series **525** (2014) 012026 doi:10.1088/1742-6596/525/1/012026

Feasibility and design study of a frictionless air mover for thermal management of electronics

R. Schacht[1], A. Hausdorf[1], B. Wunderle[2]
[1] Brandenburg University of Technology Cottbus-Senftenberg
Grossenhainerstrasse 57, Senftenberg , D-01968 , Germany
[2] Chemnitz University of Technology
Reichenhainerstrasse 70, Chemnitz , D-09126 , Germany

E-mail: ralph.schacht@b-tu.de

Abstract. A frictionless air mover concept is introduced in this paper. As opposed to a piezoelectric driven fan, the air mover is based on a flexible blade whose vibration is driven by means of a magnetic field. The blade is based on a polymer material. The paper presents the results of a feasibility analysis and an on-going comprehensive design study. The performance of the prototype amounted to 65% of a comparable piezoelectric fan. To enhance the performance, two different blade materials were investigated, as well as the influence of the coil shape and value. A further goal is to reduce the size and to investigate the influence of a casing. The design study resulted in a prototype of size of 50 x 14 x 35 mm² including a casing. The performance could be doubled, to attain a volumetric flow rate of $\mathring{V}$ ~14 l/min and a static pressure of ρ_{stat} = 3 Pa.

1. Introduction

Reliable miniaturized air movers (Mean Time To Failure, MTTF > 100 000 h) with a small form factor are of interest as they can enhance heat transfer by increasing the fluid circulation in regions which are otherwise thermally not manageable. Commercially available miniaturized axial or radial fans use ball bearings that degrade because of friction and have an MTTF < 14 000 h [7]. Frictionless fans like piezoelectric fans or synthetic jet drivers should have lower degradation but, in case of the piezoelectric blade structure, its MTTF is given as > 4 600 h [7], because of the degradation of the adhesive layer between the polymer blade sheet and the piezoelectric material. Additionally, it needs a driving (high) voltage of > 40 V and its price, compared to an axial or synthetic jet, is approximately six times higher (~ USD 180) [2][7]. Synthetic jet fans have an MTTF > 100 000 h, but their form factor and weight (Ø 60 mm, 125 g), compared to a piezoelectric fan (60x 15x 30 mm², 10 g), are larger [4][9].

The approach of this paper is to combine the advantages of a blade structure from a piezoelectric fan and the magnetic field excitation mechanism from a synthetic jet, resulting in a frictionless blade air mover concept that is more wear-resistant.

A first feasibility study has shown good potential for low form factor, low weight and low cost [1]. A comparison between a piezoelectric driven fan [3] and the magnetic field driven air mover introduced in [1] has shown that for the equal blade dimensions (length L = 70 mm, height h = 12 mm, thickness t = 0.1 mm) the magnetic field driven air mover had a 65% lower volume flow rate ($\mathring{V}$ = 10.5 l/min)

Content from this work may be used under the terms of the Creative Commons Attribution 3.0 licence. Any further distribution of this work must maintain attribution to the author(s) and the title of the work, journal citation and DOI.
Published under licence by IOP Publishing Ltd

than the piezoelectric driven fan ($\mathring{V}$ = 27.5 l/min), and the energy consumption was ~ 97% higher (P_V = 1 W in comparison with 0.03 W).

The overall aim of this design study is to reduce the form factor towards the size of a matchbox (52 x 15 x 35 mm^3). Further objectives are to increase the volumetric flow rate and the static pressure, and to reduce the power loss.

To achieve those goals, this paper will mainly target the following variables:

- Volumetric flow rate, $\mathring{V}$
- Static pressure, ρ_{stat}
- Power loss, P_V
- Form factor / cabinet size.

First, this paper introduces the frictionless air mover design and its working principle. Secondly, two experimental set-ups are described, which were used to gauge the air mover variables and to demonstrate its cooling performance. Finally, the experimental details, the results, an optimized prototype and a discussion are given.

2. Frictionless blade air mover design

The blade air mover introduced in [1] consists of a polymer blade, which is fixed at one end. A magnetic steel clip is plugged on the blade, as shown in Figure 1a. Two magnetic rod core shaped coils are assembled on a base plate in position of the steel clip (see Figure 2).

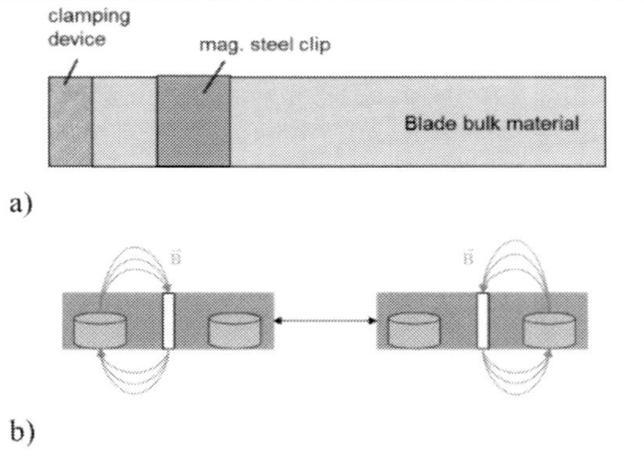

Figure 1. a) Blade with steel clip b) Blade excitation by means of a magnetic field

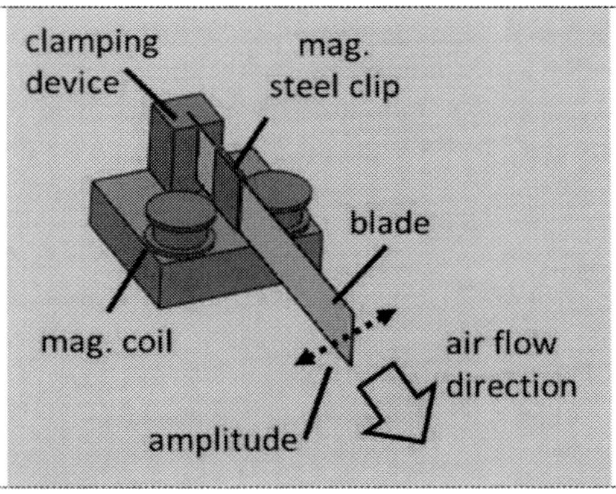

Figure 2. Blade air mover assembly and its air flow direction

The blade is stimulated by means of a magnetic field in 180° phase shift generated by the coils (see Figure 1b). Thus, air flow streams are generated in the direction to the movable end of the blade.

3. Experimental details

To enhance the air mover efficiency and, in addition, to reduce the air mover size, at least four target variables were identified (Table 1).

Table 1. Target variables under investigation.

Variable	Values
Mag. excitation material	steel clip / spring steel sheet (1.4310)
Coil core shape	rod core / U-core
Coil inductance	330μF, 1000μF, 10000μF
Notch	with, without, notch
Blade bulk material	PVC-U / MYLAR

3.1. Magnetic excitation material

The steel clip (0.5 mm) used in [1] was bulky and slowed down the blade movement because of its mass. To reduce its mass, the clip was substituted with two spring steel sheets, which were glued on the blade surface in position of the coils. Two different thicknesses (10μm, 20 μm) were investigated. The effect on the blade movement and volumetric flow rate, using the spring steel sheets instead of the steel clip, were examined.

3.2. Coil shape

The magnetic field flux of a rod core shaped coil surrounded the coil in a circular manner. To concentrate the magnetic field flux in the direction to the ferromagnetic material, the influence of a U-core shaped coil was investigated.

3.3. Coil inductance value

The magnetic force F_{mag} , which is the driving force for the blade movement, is based on Eq. 1:

$$F_{mag} = \frac{I^2 \cdot L_{ind}}{2 \cdot s} \tag{1}$$

where s is the distance between the coil and the ferromagnetic spring steel sheet, I is the current through coil and L_{ind} the coil inductance. The magnetic force, F_{mag}, increases in direct proportion with the coil inductance L_{ind}. To confirm this, three different coil inductances were also studied.

3.4. Notch at the vibration node

To check if a notch, located at the vibration node, leads to an additional acceleration at the blade tip, and thus could enhance the efficiency, the blade structure and its dimensions shown in Figure 3 was examined. In the first step, only a notch height of 2 mm was investigated.

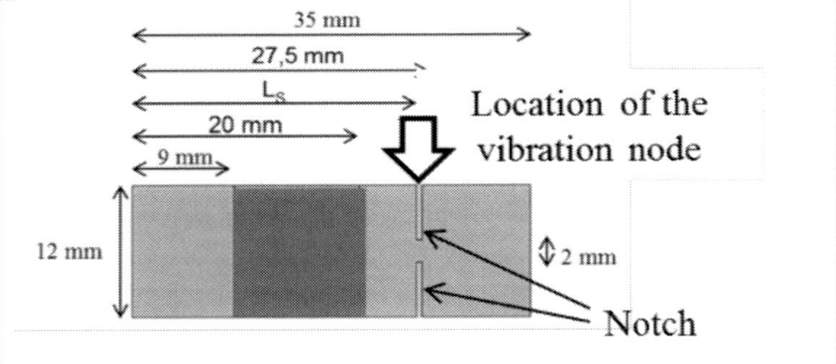

Figure 3. Schematic of the blade and the investigated dimensions

3.5. Blade bulk material

In [1], a simple overhead projector PVC-U (Poly-vinyl chloride) sheet as blade bulk material was used. To test the importance of the blade bulk material on the air mover performance, a spring steel sheet and a Mylar (Poly-ethylene terephthalate) sheet, known from [3], were investigated as bulk materials. A spring steel sheet should have the best long-term in comparison with polymers but, due to the higher stiffness of the steel, a larger current through the coils is necessary. In addition, its movement becomes very noisy. Consequently, it was excluded from this examination. Mylar (from DUPONT) is wear- and chemical-resistant as well as isotropic, has a wide operating temperature range of -70 °C to +150 °C, and is available in various thicknesses: stepwise from 0.012 mm to 0.13 mm. To make a direct comparison between the PVC-U and the Mylar material, equal sheet thicknesses of t = 0.1 mm were investigated.

4. Experimental set-ups

Figure 4 shows the experimental set-up to determine the fan performance curve, according to the AMSE/ANSI Standard 210-07 [6]. It includes a flow chamber, a hot-wire anemometer and a manual inclined tube manometer. The flow chamber has a movable shutter at the air flow outlet to record a fan characteristic. The air flow inlet is located in the centre at the front plane of the flow chamber. Its opening height was set to 14 mm, which takes into account the investigated blade height of 12 mm. Its opening width was set to 26 mm, which considers the maximum blade amplitude at the tip.
Volumetric flow rate $\mathring{V}$ was determined from Eq. 2:

$$\dot{V} = 60000 \cdot w \cdot h \cdot v \quad [l/min] \tag{2}$$

where w [m] and h [m] are the flow chamber outlet opening channel dimensions, and v [m/s] the air speed. The air speed was measured with a hot wire-anemometer from KIMO CTV210 (resolution limit Δv= 0.01 m/s). The static pressure, ρ_{stat}, was measured with a manual inclined tube manometer from DOSCH, Type D580 (resolution limit $\Delta\rho_{stat}$ = 0.25 Pa). The fan power loss, P_V, was determined by the product of the current I through the coils and the voltage V across the coil terminals.

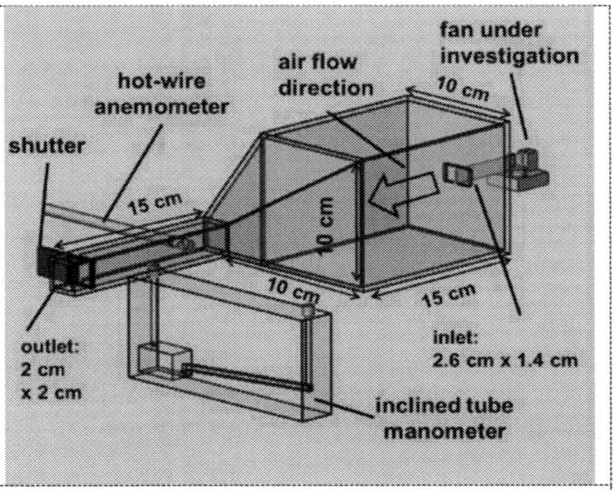

Figure 4. Experimental set-up to determine the air flow rate $\mathring{V}$ and the static pressure ρ_{stat}.

The static fan efficiency η was determined by Eq. 3:

$$\eta = \frac{\rho_{stat} \cdot \dot{V}}{P_V} \tag{3}$$

In Figures 6 and 7, a test set-up is shown, that was used to demonstrate the cooling performance for microelectronic assemblies. As a representative microelectronic component, a thermal test chip (TTC) was chosen. The TTC consisted of a 3 by 3 thermal test die. The TTC was mounted using flip-chip technology on a FR-4 board (chip area: 11.8 x 11.8 mm², substrate-area: 25 x 20 mm²) (Figure 5a). With this TTC, it was possible to have a small-outline heat source. The heat source could be easily controlled using internal resistances. The junction temperature could be measured using an internal temperature-sensitive diode Figure 5b).

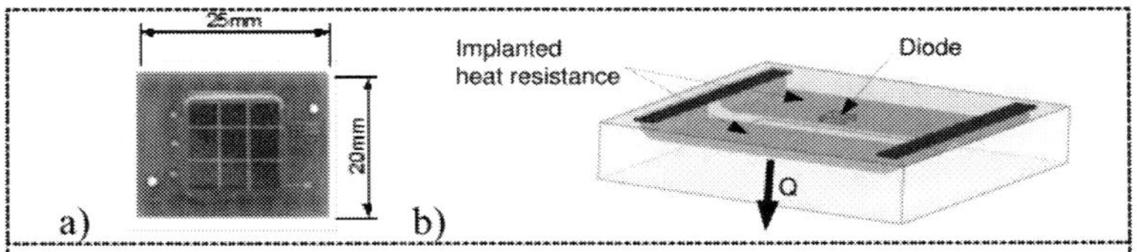

Figure 5. a) TTC assembly b) thermal test die structure.

The air mover was placed in front of the TTC, so that the forced air stream flowed directly over the active TTC surface. The transient junction temperature behaviour of the TTC surface was additional recorded by an IR-camera. The ambient temperature influence was observed by a NiCr thermocouple. The overall set-up was enclosed by a box to avoid external ambient influences.

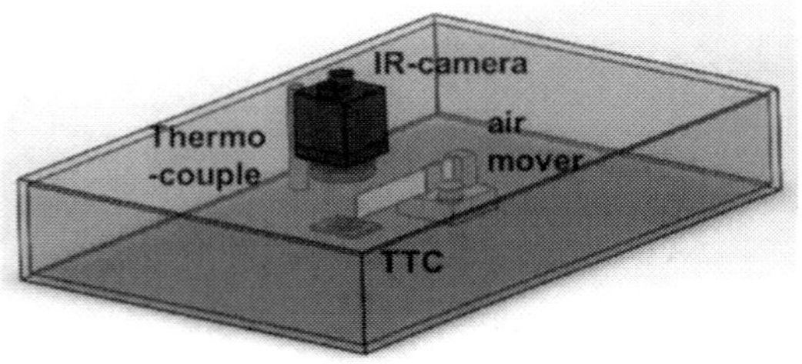

Figure 6. Schematic of test set-up to demonstrate cooling potential

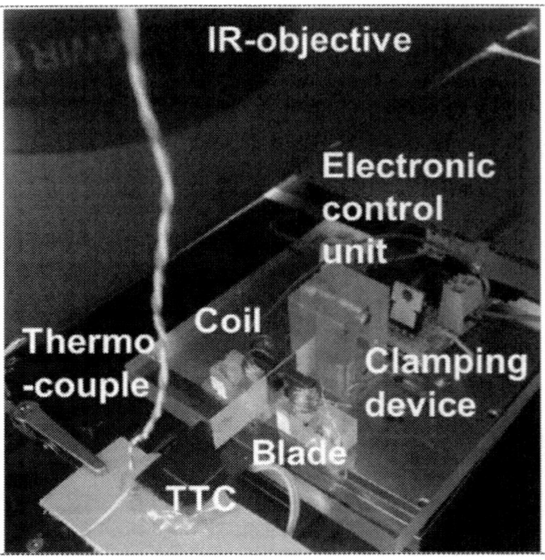

Figure 7. Photograph of the test set-up.

5. Results and discussion

Firstly, the experimental results for the target variables identified in Table 1 are presented. Secondly, a designed prototype, based on the experimental results, is presented, followed by an evaluation regarding its cooling performance for an electronic component. Finally, the classification of the introduced prototype in relation to other small scale fans is discussed.

The following results are based on the blade dimensions and the ferromagnetic material position given in Figure 8. To have comparable results with [1], the previously tested PVC-U was used as the blade material. If not stated otherwise, two rod core shaped coils with a coil inductance of L_{ind} = 1000μH each and a total power loss of P_V = 1W were used.

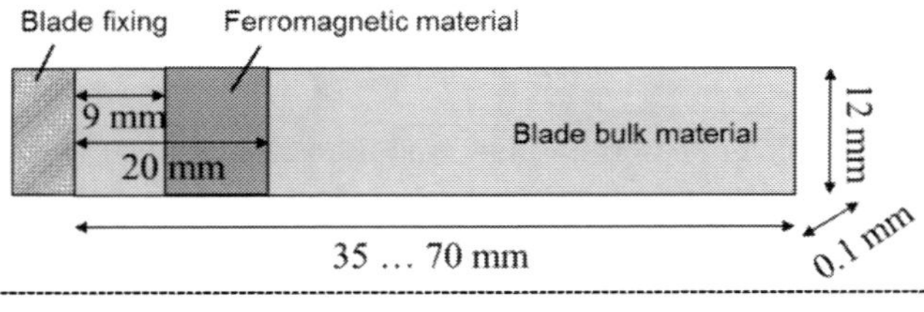

Figure 8. Blade dimensions and ferromagnetic material position.

The volumetric flow rate was determined with the shutter in fully opened position (Figure 4) and, for the static pressure, the shutter was fully closed.

5.1. Magnetic excitation material

In Figures 9 and 10, the blade length dependent volumetric flow rate and the static pressure are presented for the each excitation material, the steel clip and the spring steel sheet. Two different thicknesses (10 μm and 20 μm) of the spring steel sheet were investigated.

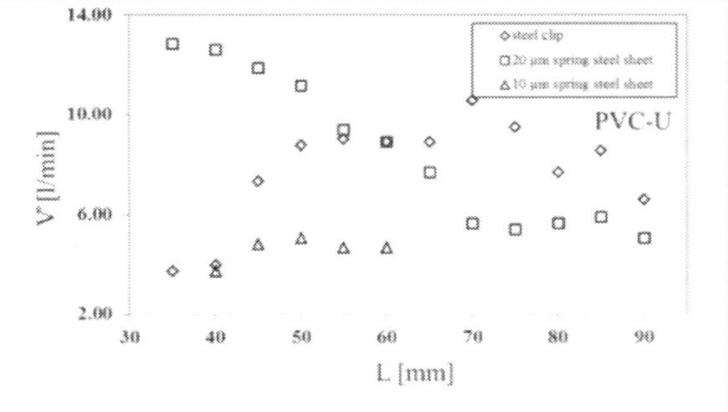

Figure 9. Volumetric flow rate $\mathring{V}$ as function of the blade length using a ferromagnetic spring steel sheet of two different thicknesses and a steel clip.

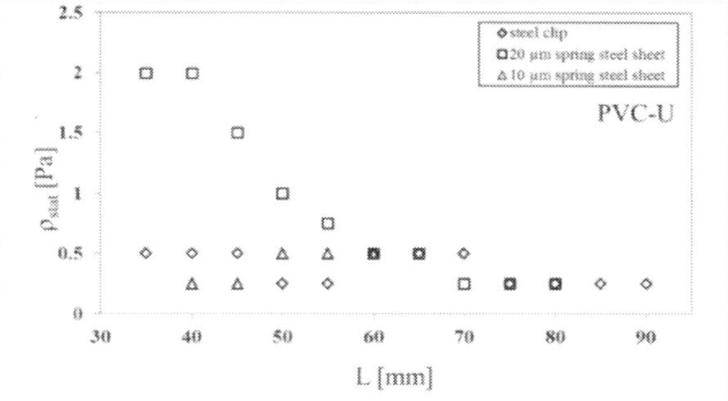

Figure 10. Static pressure ρ_{stat} as function of the blade length using a ferromagnetic spring steel sheet of two different thicknesses and a steel clip.

In Figure 9, it can be observed that the volumetric flow rate shows a linearly decreasing behaviour for the blade with the glued spring steel sheet (thickness 20 µm) compared to the blade with the assembled steel clip. The 10 µm thick spring steel sheet shows only a low response, probably due to a decreasing magnetic response of the magnetic flux.
Figure 10 presents the static pressure behaviour as a function of the blade length. For the 20 µm thick spring steel sheet, the static pressure demonstrates a decaying behaviour. For the steel clip and the 10 µm thick spring steel sheet, a maximum static pressure of only ~ 0.5 Pa could be observed. A maximum for both variables could be found at a blade length of ~ 40 mm using the 20 µm thick spring steel sheet.

5.2. Notch at the vibration node and coil inductance value
In Figures 11 and 12 it can be seen that a notch at the vibration node has a positive influence on the volumetric flow rate and the static pressure.

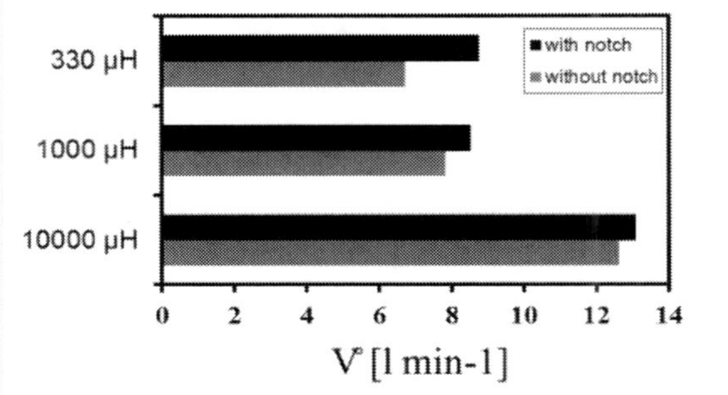

Figure 11. Volumetric flow rate $\mathring{V}$ as function of the coil inductance value as well as with or without a notch at the vibration node.

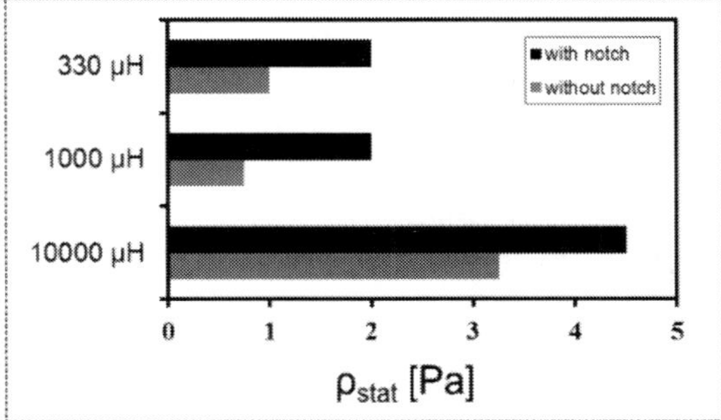

Figure 12. Static pressure ρ_{stat} as function of the coil inductance value as well as with or without a notch at the vibration node.

It can also be seen that, in this investigated case, a notch could help to increase the volumetric flow rate and to double the static pressure. In addition, it can be noted that increasing the coil inductance value by up to one order of magnitude, both variable values could only nearly be doubled. In this investigation commercially-available rod core shaped coils were used.

5.3. Coil shape
Commercial small scale U-Core shaped coils were not available, so a custom wire winding had to be done. Because of the small scale U-core form factor (9.9 x 8.2 x 2.82 mm^3), a maximum practical wire

turn ratio could only be realized for an inductance value of L_{ind} = 470 μH, which will be investigated further.

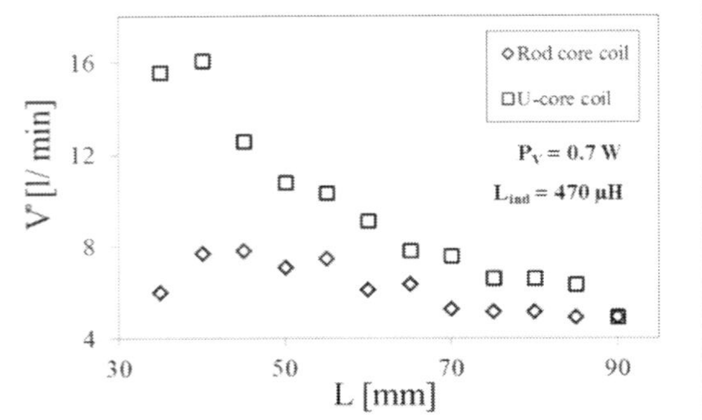

Figure 13. Volumetric flow rate $\mathring{V}$ as function of the blade length comparing U-core shaped coils instead of rod core ones.

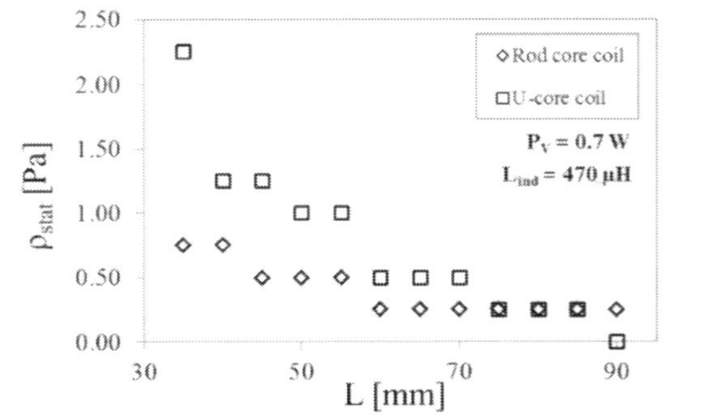

Figure 14. Static pressure ρ_{stat} as function of the blade length comparing U-core shaped coils instead of rod core ones.

In Figures 13 and 14, it can be observed that the volumetric flow rate and the static pressure can be increased by decreasing the blade length and using U-core shaped coils.

5.4. Blade material

The result presented in Figures 15 and 16 are based on the use of a U-core shaped coil with an inductance value of L_{ind} = 470 μH.

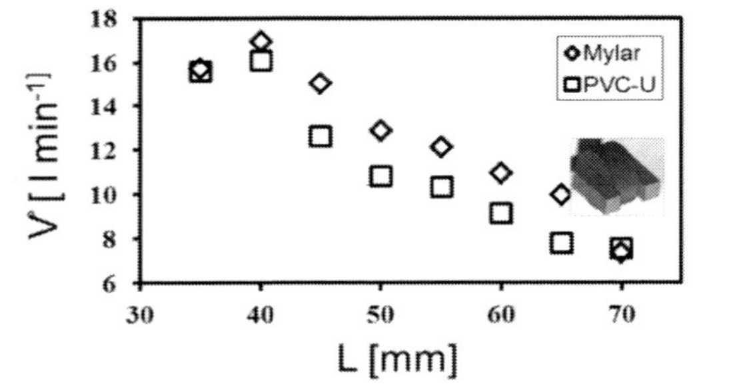

Figure 15. Volumetric flow rate $\mathring{V}$ as function of blade length comparing PVC-U and Mylar as blade material.

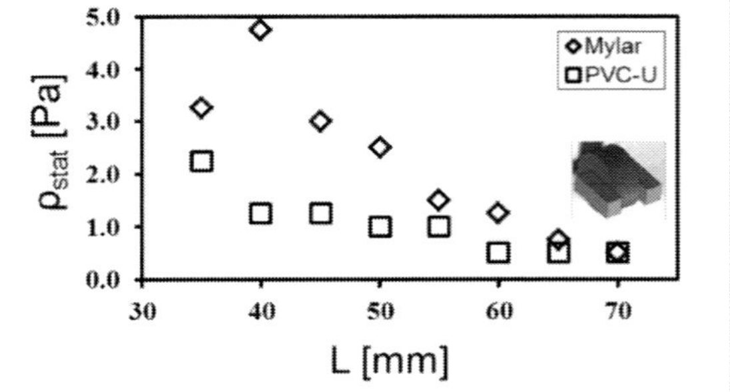

Figure 16. Static pressure ρ_{stat} as function of blade length comparing PVC-U and Mylar as blade material.

A maximum of both variables can be observed with Mylar as blade bulk material at a blade length of L = 40 mm. Due to the fact that the Young's modulus from Mylar (~ 3.6 GPa) is 1 GPa lower than that for PVC-U (~ 4.6 GPa), the blade amplitude *y* increases and the static pressure is increased, from 1 Pa to 4.8 Pa. The volumetric flow rate is also increased from 16 to 17 l/min.

5.5. Prototype design

Based on the experimental results, it was possible to reduce the length of the blade from 70 mm down to 40 mm, in order to achieve a 'matchbox' size. To this end, the blade bulk material was changed to MYLAR and, instead of the steel-clip, two magnetic steel sheets of 20 μm thickness were used. In addition, the two rod-core shaped coils were replaced by two U-core ones. As mentioned above, only a maximum coil inductance of L_{ind} = 470μH for the U-core shaped coils was available. With these settings, the power loss could be reduced from P_V = 1 W to 0.24 W (U_{ind} = 1.2 V, I_{ind} = 200 mA) and a maximum volumetric flow rate of $\mathring{V}_{max}$ = 14 l/min (ρ_{stat} = 0 Pa) as well as a maximum static pressure of $\rho_{stat,\ max} = 3$ Pa ($\mathring{V} = 0$ l/min) could be observed. In Figure 17, an initial fan performance curve is presented, based on different shutter positions.

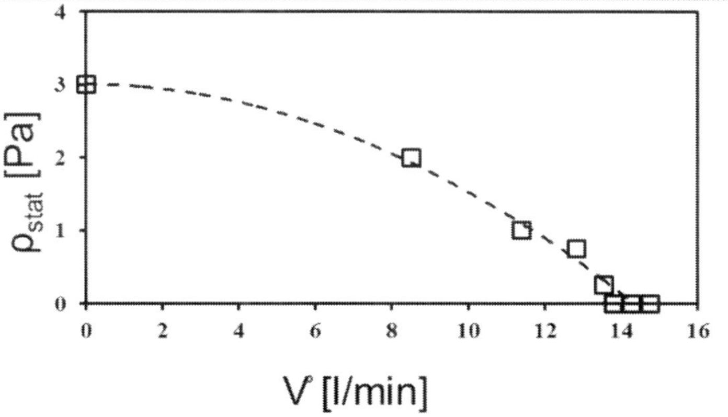

Figure 17. Fan performance curve of the prototype

Figure 18 and 19 depicts a schematic and photograph of the prototype and its geometrical dimensions.

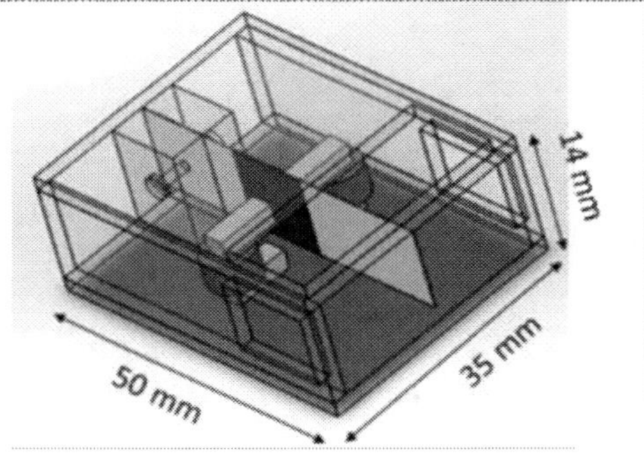

Figure 18. Schematic of the prototype and its geometrical dimensions.

Figure 19. Photograph of the prototype in comparison to a matchbox.

Due to the shortened blade length, the operating frequency shifted, from 18 Hz [1] to 125 Hz, into the audible range.

5.6. Cooling performance for electronic components

In this investigation, a Mylar foil as blade bulk material with a blade length of L = 40 mm, two 20 µm thick spring steel foils, a notch at the vibration node, two U-core coils with an inductance value of $L_{ind} = 470\mu H$ were used. The power loss was reduced to $P_V = 0.7$ W. In Figure 16, the transient junction temperature, T_J, behaviour, after switching on the heated TTC ($t = 0$ s), for three constant driven power losses $P_{V,TTC}$ are presented.

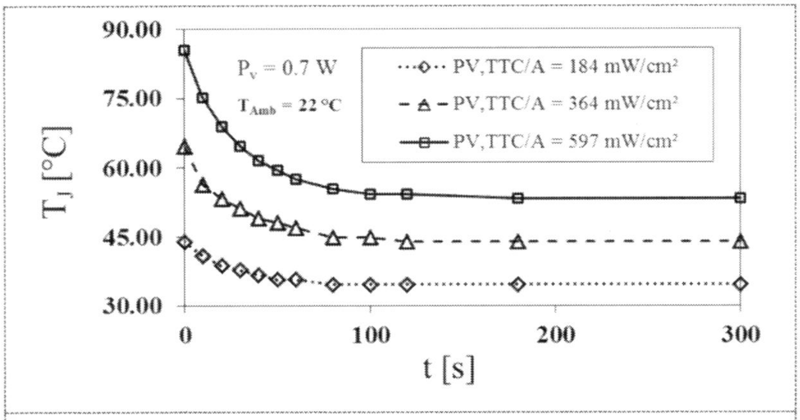

Figure 16. Junction temperature T_J behaviour of TTC after switching on the blade fan at different constant driven TTC power losses.

As an illustrative result, the steady state junction temperature, T_J, of the TTC could be reduced: e.g. for $\Delta T \sim 30$ K, from $T_J = 85$ °C ($P_{V,TTC}/A = 597$ mW/cm²), in case of natural convection (blade air mover turned off) , down to 56°C when the blade air mover was turned on.

5.7. Classification

Several frictional and frictionless fan solutions are currently on the market. In order to classify the magnetic field driven prototype presented in this paper, Table 2 compares a selection of small scale radial and axial fans [5][7] as well as the piezoelectric driven fan given in [3].

Table 2. Comparison between an axial mini fan, a small scale radial fan, a piezoelectric driven fan and the magnetic field driven prototype.

	SUNON mighty radial mini fan	Small scale axial fan	Piezoelectric driven fan	Magnetic field driven prototype
L x W x H [mm³]	9 x 9 x 3	25 x 25 x 10	69 x 30 x 15	50 x 35 x 14
P_V [mW]	185	n. a.	30	240
U_{Supply} [V]	3	n. a.	> 40	1.2
$p_{stat,max}$ [Pa] (at $\dot{V}$ = 0 l/min)	4.44	43	3.5	3
$\dot{V}_{max}$ [l/min] (at p_{stat} = 0 Pa)	1.17	13	27.5	14
η [%]	0.047	n. a.	5.35	0.249

In case of the static pressure, the frictional fans could reach a one order of magnitude higher maximum static pressure than the frictionless fans. In case of the volumetric flow rate, it can be noticed that the piezoelectric driven fan, due to the design concept, shows a two times better maximum flow rate. Comparing the power loss and the efficiency, the piezoelectric driven fan shows the lowest power consumption and the best fan efficiency, but a much higher supply voltage (of > 40 V) must be provided. In the case of the geometrical dimensions, the frictional fans have a smaller form factor than the frictionless ones.

The magnetic field driven prototype can be classified in between the frictional and frictionless fans. It has a comparable volumetric flow rate to the radial fan and is twice as large as the radial fan. Compared to the piezoelectric driven fan, it has similar maximum static pressure and comparable geometrical dimensions. Only the power loss and the voltage supply behave inversely. Because the magnetic field driven concept was planned as a more wear-resistant solution, it shows a good potential to enhance the heat transfer by increasing the fluid circulation in regions which are otherwise thermally not manageable.

6. Conclusions

This paper has presented the results of an on-going design study to reduce power loss and size of a frictionless air mover, as well as to enhance the volumetric flow rate and static pressure. This overall goal was reached. An optimized prototype with a size of 50 x 14 x 35 mm^3 (including casing) was realized. The maximum volumetric flow rate was enhanced from $\mathring{V}_{max}$ = 10 l/min to 14 l/min (at ρ_{stat} = 0 Pa), and the maximum static pressure was enhanced from $\rho_{stat\text{-}max}$ = 0.5 Pa to 3 Pa (at $\mathring{V}$ = 0 l/min). In addition, the power loss P_V was lowered from 1W to 0.24 W. In addition, the fan efficiency was more than doubled (0.1 % to 0.25 %). The air mover is low weight (< 50 g) and could be manufactured at low cost (material costs < USD 5). Depending on the fan system working point, the volumetric flow rate can be adjusted from $\mathring{V}$ ~ 5 – 12 l/min at a static pressure range of ρ_{stat} ~ 1 – 2.7 Pa. An examination of the thermal performance has shown that a representative active electronic component could be cooled down adequately and is therefore thermally manageable. The next steps are to conduct long term stability examinations and to aim to double the current volume flow rate.

7. References

[1] R. Schacht, A. Hausdorf, B. Wunderle, 2012, "Frictionless Air Flow Blade Fan for Thermal Management of Electronics,"13th IEEE ITHERM Conference, San Diego, USA, 978-1-4244-9532-0/12/$31.00 ©2012 IEEE
[2] Mark Kimber, Kazuhiko Suzuki, Nobutaka Kitsunai, Kenichi Seki, and Suresh V. Garimella, 2008, "Quantification of Piezoelectric Fan Flow Rate Performance and Experimental Identification of Installation Effects" 978-1-4244-1701-8/08/$25.00 ©2008 IEEE.
[3] Mark Kimber, Kazuhiko Suzuki, Nobutaka Kitsunai, Kenichi Seki, and Suresh V. Garimella, 2009, "Pressure and Flow Rate Performance of Piezoelectric Fans" IEEE Transactions on Components and Packaging Technologies, Vol. 32, No. 4, December 2009.
[4] Kercher, Lee, Brand, Allen, and Glezer, 2003, "Microjet Cooling Devices for thermal Management of Electronics", IEEE Transactions on Components and Packaging Technologies, vol. Vol. 26, pp. 359-366, Juni 2003
[5] Jason Stafford, Ed Walsh, Vanessa Egan, "Local heat transfer performance and exit flow characteristics of a miniature axial fan", International Journal of Heat and Fluid Flow 31 (2010) pp. 952 – 960.
[6] Laboratory Methods of Testing Fans for Certified Aerodynamic Performance Rating, ANSI/AMCA Standard 210-07.
[7] Radial fan, www. Titan-cd.com
[8] Piezo fan, www.piezo.com (Part Number: RFN1-LV-02)
[9] Membran fan, SynJet® Low Profile Cooler. Co. Nuventix (www.nuventix.com)

Experimental Optimisation of the Thermal Performance of Impinging Synthetic Jet Heat Sinks

Craig Marron and Tim Persoons

School of Engineering, Department of Mechanical & Manufacturing Engineering, Parsons Building, Trinity College, Dublin 2, Ireland

marroncr@tcd.ie, tim.persoons@tcd.ie

Abstract. Zero-net-mass flow synthetic jet devices offer a potential solution for energy-efficient cooling of medium power density electronic components. There remains an incomplete understanding of the interaction of these flows with extended surfaces, which prevents the wider implementation of these devices in the field. This study examines the effect of the main operating parameters on the heat transfer rate and electrical power consumption for a synthetic jet cooled heat sink. Three different heat sink geometries are tested. The results find that a modified sink with a 14x14 pin array with the central 6x6 pins removed provides superior cooling to either a fully pinned sink or flat plate. Furthermore each heat sink is found to have its own optimum jet orifice-to-sink spacing for heat transfer independent of flow conditions. The optimum heat transfer for the modified sink is H = 34 jet diameters. The effect of frequency on heat transfer is also studied. It is shown that heat transfer increases superlinearly with frequency at higher stroke lengths. The orientation of the impingement surface with respect to gravity has no effect on the heat transfer capabilities of the tested device. These tests are the starting point for further investigation into enhanced synthetic jet impingement surfaces. The equivalent axial fan cooled pinned heat sink (Malico Inc. MFP40-18) has a thermal resistance of 1.93K/W at a fan power consumption of 0.12W. With the modified pinned heat sink, a synthetic jet at Re = 911, $L_0/D = 10$, H/D = 30 provides a thermal resistance of 2.5K/W at the same power consumption.

1. Introduction

Recent developments of the ICT industry have led to increasing power density of electronic and photonic packages. The desire to decrease the size of computers as well as increase power means that the direction of the industry is towards greater power from smaller devices. A problem inherent in increasing the power is the heat generated from the chip unit [1]. The power increases and size reductions of electronic devices means that existing methods of cooling are becoming insufficient to meet demands. Much research is currently being carried out in the field of electronics cooling in an effort to find a cooling technology which can fulfil the cooling needs. A large degree of what is being researched and investigated involves jet flow impingement in various forms.

Synthetic jet flow is named so as the fluid motion is 'synthesized from the ambient fluid. As such the system is also known as a Zero-Net-Mass flux device. Synthetic jets have been studied in depth for many decades, in particular as flow alteration devices, such as for stall control [2]. More recently in response to the increasing power density of chips the devices have been investigated as heat dissipation devices.

Content from this work may be used under the terms of the Creative Commons Attribution 3.0 licence. Any further distribution of this work must maintain attribution to the author(s) and the title of the work, journal citation and DOI.
Published under licence by IOP Publishing Ltd

The basis of operation is to create vortical flow which will dissipate the heat from the heated surface to the ambient. The vortical flow is a series of repeated vortices ejected individually through an orifice. The flow is typically generated suction and ejection of fluid from a small cavity using a diaphragm [3]. The pulsating nature of synthetic jet flows increases entrainment of ambient fluid in comparison to a continuous jet, which in turn provides a cooling performance, comparable to continuous impinging jets, yet without the need for a pressurized fluid supply [4]. A major advantage of synthetic jets over other jet flows is there is no need for external pressurized fluid making such devices more spatially efficient and less expensive. Defining parameters of synthetic jets are the Reynolds number Re = U_0D/ν and the stroke length L_0/D. The stroke length is the displacement of a slug of fluid from the orifice during the ejection phase of the cycle [5], in this case made non-dimensional by the diameter of the hole. The velocity of fluid through the system is taken as the frequency at which the stroke length is repeated such that

$$U_0 = f.L_0 = f.\int_{t=0}^{\frac{1}{2f}} u_0(t)dt \quad (1)$$

where $u_0(t)$ is the area-averaged orifice velocity. The stroke length determines the flow characteristics, particularly at low values of L_0/D. Vortex detachment from an axisymmetric sharp-edged orifice will not occur unless $L_0/D > 0.5$[5]. The flow field of impinging synthetic jets is further characterised by the orifice-to-surface distance H. This spacing parameter is typically assigned a non-dimensional form relative to the orifice diameter H/D. The flow characteristics of impingement are heavily dependent on the spacing and studies have shown an optimal value exists within each system in order to achieve maximum heat transfer [6-9].

Although the potential for synthetic jets as cooling devices has been researched there exist few practical applications in existence today. The reason behind this is due to a lack of complete understanding of the best operating conditions of a synthetic jet system. Research by Gillespie et al. [9] showed substantial enhancement in heat transfer using a zero net mass transfer device due to the strong mixing characteristics. Furthermore, studies have shown a wide range of spacing ($4< H/D <11$) at which the heat transfer is maximised [9, 10]. However, the long-term goal of implementing synthetic jets in small electronic devices is to further decrease the size of devices. In this light research such as that carried out by Valiorgue et al. [11] is prevalent as it investigates heat transfer characteristics in small spacing. Particle image velocimetry was used to show how varying parameters such as the stroke length and the Reynolds number dramatically affected the flow and heat transfer characteristics. This study highlights how the ability for a synthetic jet actuator to dissipate heat is largely dependent on the spacing (H/D). Furthermore it provides visual data which contrasts well with data provided ten years earlier by Gharib et al. [12] to show the difference in fully develop vortices with entrainment and vortices which recirculate due to lack of entrainment. A significant flaw of synthetic jets is highlighted in this research, which is recirculation. The suction of previously ejected fluid means that the temperature of air within the cavity of the actuator will increase and thus have inferior cooling potential.

Persoons [13] developed a model to design and operate synthetic jet actuators by coupling the mechanical and fluidic model. Experimentation within the same paper verified that for rectangular and circular orifices the model holds true.

Evidently the heat transfer capabilities of a synthetic jet system are heavily dependent on the geometric parameters which define it. Chaudhari et al. [7] investigated the effect of geometry by empirical means to qualify how the flow produced by different shaped orifices affected the heat transfer away from a heated surface. Research obtained interesting results which clarified that at larger spacing a circular orifice is preferred whereas at smaller spacing a rectangular orifice provides superior cooling. This mirrors studies which visualise flow from slot orifices and show that the long vortices generated along the length of the slot curl-up and collapse on themselves[4, 14]. The reason for this transition to turbulence lies in the generation of counter rotating stream-wise vortex pairs that wrap around the primary vortices. The onset of the counter rotating vortices is caused by instability and small scale motion of the primary vortices[4]. The key to maximising heat transfer using synthetic jets is to correctly design the orifice and system which produces a flow which achieves strong entrainment of ambient fluid and has vigorous mixing near the impingement surface[15].

Although much research has been carried out on the effect of orifice geometry, the understanding of the effect of heat sink geometry on heat transfer characteristics is incomplete. The current study aims to address this lack of knowledge and quantifiably compare the heat transfer achievable with different heat sink configurations. The research investigates the heat transfer as a function of spacing (H/D) with each sink. The study also looks at the how the heat transfer is affected by the frequency under the circumstances of constant Reynolds number and constant stroke length.

2. Experimentation

2.1. Test Facility

The design of the system is inspired by the set-up and testing carried out by Chaudhari et al. [6, 7, 16]. The experimental rig allows easy spacing adjustment of the jet actuator using a slider mechanism. The jet actuation is produced by the oscillatory motion of the diaphragm of a Visaton BF32-8Ω speaker. The cavity is cylindrical aluminium which fits around the perimeter of the diaphragm with 0.5mm clearance to allow unrestricted motion of the diaphragm. The cavity wall has a height of 10mm resulting in a total cavity volume of 7068mm^3. The orifice plate is located directly opposite the speaker. The cylindrical orifice is constructed from nylon of 5mm thickness. The orifice hole is 2mm diameter. For the purposes of comparison tests are carried out with a fan cooling device impinging on the same heat sinks. The fan is a Sunon Maglev GM0504PEV1-8, mounted directly on top of the sink being tested (H = 0).

The jet produced impinges onto a pin fin heat sink (Malico Inc. MPF40-18). This sink is an array of 14x14 fins each 8mm high, diameter 1mm. The sink is anodized aluminium. The system is designed to be modular for easy changing of heat sinks. The heating apparatus is displayed in figure 1. A heater plate is adhered to the underside of a copper block of 7mm thickness. The copper block sits within an insulation block which is insulated on all but one side. The exposed side, measuring 40mmx40mm, faces the synthetic jet. On the exposed side a heat sink of selected geometry will be placed. To ensure good thermal contact a thermal paste is used which minimizes any contact resistance. All heat conducted through the copper block to the heat sink side is assumed to dissipate from the heat sink surface (i.e. conduction resistance of heat sink base is negligible). A constant power supply to the insulated face of the copper block is maintained at 2.6W.

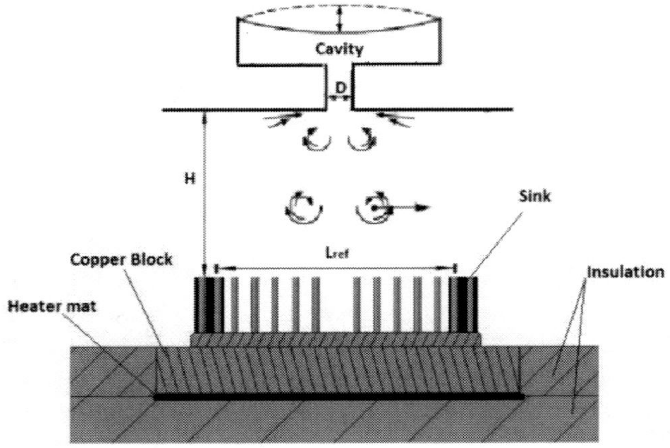

Figure 1. Schematic diagram of the test apparatus

2.2. Synthetic jet operating point

The focus of this study is primarily on the heat transfer characteristics of the system. As such the measurement devices in operation are focused on temperature monitoring more so than flow characterisation or visualisation. The synthetic jet is operated using LabView interface and National Instruments CompactDAQ System designed according to a model derived by Persoons [13,17]. A G.R.A.S. 40BH microphone and amplifier are used for pressure measurement in the cavity. Having set the operating frequency and amplitude the program takes in the pressure measurement and by accounting for the size of the cavity and nature of the orifice exit calculates critical flow data such as the stroke length, area-averaged orifice velocity and Reynolds number. The calculations carried out by the program implemented are those derived by Persoons and O'Donovan to estimate jet velocity from the cavity pressure [17].

2.3. Convective heat transfer measurements

The method of calculating the heat transferred from the surface of the heat sink is described belown. The heat input to the system is equal to the heat dissipated through conduction, radiation and the convective heat transfer away from the surface is obtainable. The heat dissipated by convection is taken as the product of the heat transfer coefficient and the difference in temperature between the surface and ambient. In the field of synthetic jets it is more appropriate to use the difference between the surface and the cavity temperature:

$$q_{conv} = q_{gen} - q_{cond} - q_{rad} = h(T_{surface} - T_{jet}) \qquad (2)$$

As different heat sink geometries will result in varying natural convection capabilities it is necessary to distinguish the forced convection from the total achievable convection. It is assumed that the fin efficiency is 100% so that the base plate temperature is an accurate approximation of the true surface temperature, to appropriately and quantitatively compare how effectively a synthetic jet transfers heat away from a sink due to forced convection it is necessary to calculate the heat transfer from each sink through forced convection only. The Nusselt number for forced convection is found by separating the Nusselt number due to natural convection from the mixed convection Nusselt number [18]:

$$Nu_{forced} = \left(Nu_{mix}{}^{3} - Nu_{nat}{}^{3}\right)^{1/3} \qquad (3)$$

where both Nu_{mix} and Nu_{nat} are found empirically. Testing for natural convection is carried out with the jet actuator in position turned off at a variety of heights. The need to carry out testing in this manner is due to the effect spacing has on radiation and possible restriction of flow away from the surface in close spacing. By obtaining a curve for Nu_{nat} against Ra and H/D equation (3) can be used. The correlation of natural Nusselt number to Rayleigh number is found as;

Case A (Flat Surface)	$Nu = (2.547 - 1.538/(H/D))\, Ra^{1/4}$
Case B (14x14 Pinned Sink)	$Nu = (2.886 - 1.029/(H/D))\, Ra^{1/4}$
Case C (14x14 – 6x6 Pinned Sink)	$Nu = (2.547 - 1.538/(H/D))\, Ra^{1/4}$

The length scale used for Nu_{mix} and Nu_{nat} is the length of the heat sink base. However, for analysis of synthetic jets a more fitting length scale is the orifice geometry. In order to convert Nu_{forced} to Nu_{jet} the following equation must be implemented:

$$Nu = Nu_{forced} \frac{D}{L_{ref}} \frac{(T_{Surface} - T_{ambient})}{(T_{surface} - T_{jet})} \qquad (4)$$

where L_{ref} is the length of the side of the heat sink base. This can now be used to find a synthetic jet heat transfer coefficient.

$$h_{jet} = \frac{Nu.k_{air}}{D} \quad (5)$$

The corresponding thermal resistance is

$$R_{th} = \frac{1}{hA} \quad (6)$$

where A is the heat sink base area.

2.4. *Energy efficiency*

Efficiency of synthetic jets is still undefined due to the different applications in different fields. For the purposes of fluid motion the efficiency is typically taken as the ratio of the fluidic power to the electrical actuation power [13]. However, for the purposes of this study it is more appropriate to compare the heat transfer to the power input. This quantity is known as the effectiveness, ε, such that

$$\varepsilon = \frac{Q_{forced}}{P_{electric}} \quad (7)$$

3. Results & Discussion

3.1. *Effect of jet orifice-to-impingement surface spacing*

Figure 2 shows how heat transfer changes with jet to surface spacing at a given jet operating condition (Re, L_0/D). The heat transfer coefficient has positive correlation with the non-dimensional spacing parameter such that a maximum heat transfer coefficient is reached at a given spacing H/D. Although the optimum height for heat transfer varies depending on the heat sink geometry it is clear that an optimum spacing exists for each sink. In addition to this the results show that the optimum spacing is a function of sink geometry independent of Reynolds number. This is evidenced by comparison of figure 2a and figure 2b. Although the Nusselt number is drastically reduced with reduced Reynolds number the optimum heat transfer relative to spacing occurs at the same point for each sink. Ignoring minor discrepancies which exist between the two data the trend remains very similar. The pinned sinks achieve optimal heat transfer at a larger spacing than the flat plate. In both graphs the Nusselt number of all three heat sinks is shown to increase to a point where the spacing has less effect as it changes. As stated previously there exists an optimum operating spacing for maximising heat transfer.

Interestingly far superior heat transfer is achieved from a heat sink which incorporates pins in its geometry. The figures shown above consist of forced convection only which indicates that the flow interaction between vortices and pins achieves cooling better than vortices impinging on a flat plate. The height, H is measured from the point of flow exit to the point of first impingement. In the case of a pinned sink this is from the top of the pins to the bottom of the orifice plate.

Optimum heat transfer is achieved at a much smaller spacing for flat plate (H/D = 20) than the optimum spacing for a pinned heat sink (H/D = 34). A modified sink was tested in conjunction with the aforementioned sinks. This sink consisted of an array of 14x14 pins with the central 6x6 pins removed. This sink performed better over the entire range than either the flat plate or the fully pinned heat sink. It can be postulated that the existence of pins in the central portion of the plate reduce the flow momentum and impingement. Furthermore, it is possible that the fluid-pin interaction reduced the flow significantly that recirculation occurs for small spacing. For both Reynolds number tests the Nusselt number increases dramatically over the range $15 < H/D < 20$, for the pinned heat sink. Evidenced by the fact that the flat plate outperforms the fully pinned sink at smaller spacing before a drastic increase, it appears that recirculation occurs. When these pins are removed the best of both is achieved. Heat transfer occurs on impingement and as the flow spreads outwards the interaction provides further cooling which is not obtainable with a solely flat plate.

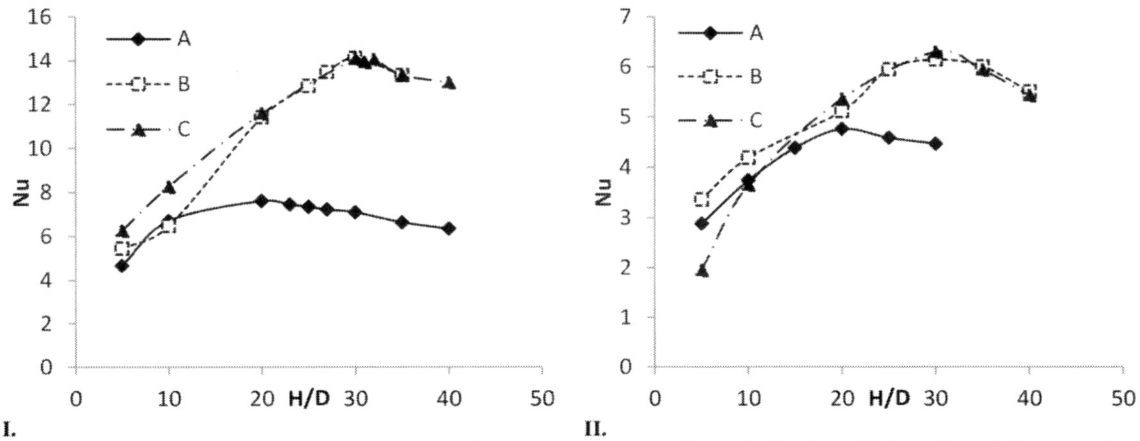

Figure 2. Average Nusselt number as a funcion of nozzle-to-surface spacing H/D for three different heat sink geometries: (A) flat surface, (B) 14x14 Pins, (C) 14x14 array with central 6x6 pins removed (I. Re = 1600, L_0/D = 12 and II. Re = 800, L_0/D = 6)

Case	Sink	Re	Optimum H/D	Nu	h [W/(m^{2}K)]	R_{th} [K/W]
A	Flat Plate	1600	20	7.62	86.6	7.20
		1200	15	6.37	77.4	8.09
		800	20	4.76	60.5	10.36
B	Pinned	1600	30	14.13	151.8	4.10
		1200	30	10.17	115.3	5.37
		800	30	6.15	75.3	8.32
C	Modified	1600	32	14.10	147.7	4.23
		1200	35	10.70	117.1	5.30
		800	30	6.31	76.1	8.20

Table 1. Heat transfer at optimum spacing for three different sink geometries at three different Reynolds numbers

Case	Sink	Nu	R^2
A	Flat Plate	$3.13\dfrac{\left(1+3.069\left(\frac{H/D}{8.625}\right)^2\right)}{\left(1+\left(\frac{H/D}{8.625}\right)^{2.249}\right)}$	0.95
B	Pinned	$5.04\dfrac{\left(1+4.343\left(\frac{H/D}{34.641}\right)^2\right)}{\left(1+\left(\frac{H/D}{34.641}\right)^{4.254}\right)}$	0.991
C	Modified	$5.935\dfrac{\left(1+3.69\left(\frac{H/D}{29.808}\right)^2\right)}{\left(1+\left(\frac{H/D}{29.808}\right)^{3.043}\right)}$	0.981

Table 2. Average Nusselt number relationship with spacing for three diferent sink geometries (Re = 1600, L_0/D =12)

3.2. Effect of synthetic jet actuation frequency

Testing carried out in the course of this research examined how heat transfer corresponds with changes in frequency. A relationship is found which correlates the the Nusselt number with the Reynolds number. In this case the stroke length is kept constant at a specified number and the Nusselt number is normalised by the Reynolds number. The results show the manner in which the heat transfer increases with frequency. At a smaller stroke length the heat transfer is less susceptible to large change with frequency and thus the correlation between the heat transfer and frequency is linear. At higher stroke lengths each change in frequency affects the flow to a greater extent and the heat transfer increases exponentially with increase in frequency. This increase suggests that to maximise heat transfer in any system frequency should be maximised to a point where the power supplied remains safe for the actuator device.

The increase in frequency results in improved heat transfer. The improvement brought about by the increased frequency is dependent on the stroke length at which the system is operating.

Case	Sink	L_0/D	H/D	Nu	R^2
A	Flat Plate	10.1	30	$1.29\times10^{-2}\,Re^{0.865}$	0.988
B	Pinned	10.1	20	$5.13\times10^{-4}\,Re^{1.425}$	0.979
C	Modified	10.1	30	$1.44\times10^{-4}\,Re^{1.751}$	0.971

Testing carried out to investigate the relationship between heat transfer and frequency whilst maintaining the Reynolds number constant. The results obtained are shown in figure 4. As can be seen the derived equations for the case of pinned sink and flat surface hold true. As frequency is increased the heat transfer relative to the Reynolds number remains constant.

Discrepancies at low frequencies are omitted as the power supplied from the loudspeaker was insufficient to produce flow superior to natural convection. Similarly at high frequencies inconsistencies occurred due to overheating of the loudspeaker where the cavity temperature increased due to the temperature of the speaker. As such any data points requiring a power supply greater than 1.5W are omitted.

3.3. Effect of heat sink orientation with respect to gravity

In an investigation into the effects of gravity on cooling the testing for spacing is repeated with the sink in a vertical orientation. The heat sink is perpendicular to the ground and the jet impinges on its surface normally. Figure 4 compares the data obtained from both the horizontal and vertical orientations.

In order to correctly analyse how the orientation affected the heat transfer only the forced convection components of heat transfer are presented in the above data. Although convention would suggest that a vertical plate would achieve better heat dissipation the data obtained does not support this. Minor improvement occurs at some spacing but accounting for margin of error this change is insignificant. The results show also that the orientation has little effect on the optimum spacing. The optimum spacing is maintained for each heat sink irrespective of orientation.

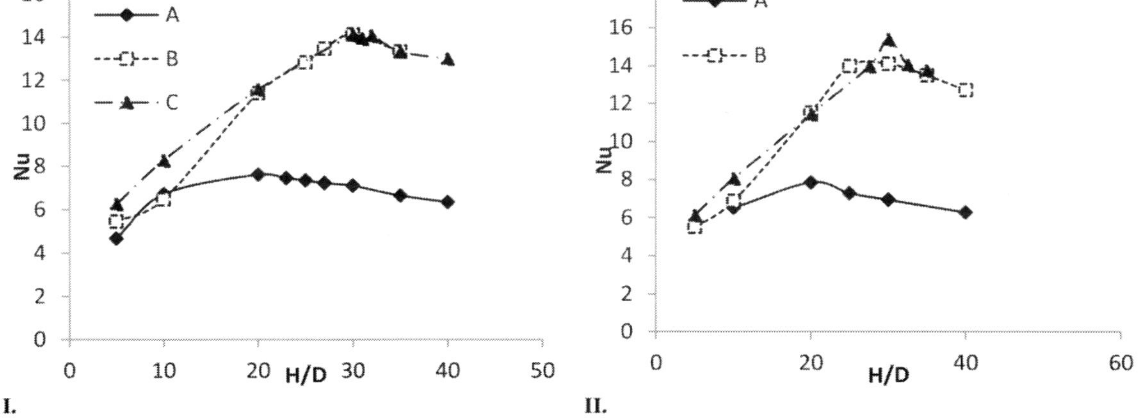

I. II.

Figure 3. Average Nusselt number as a funcion of nozzle-to-surface spacing H/D for three different heat sink geometries: (a) flat surface, (b) 14x14 Pins, (c) 14x14 array with central 6x6 pins removed. Re =1600, f =250Hz. (I. Horizontal sink base, II. Vertical sink base)

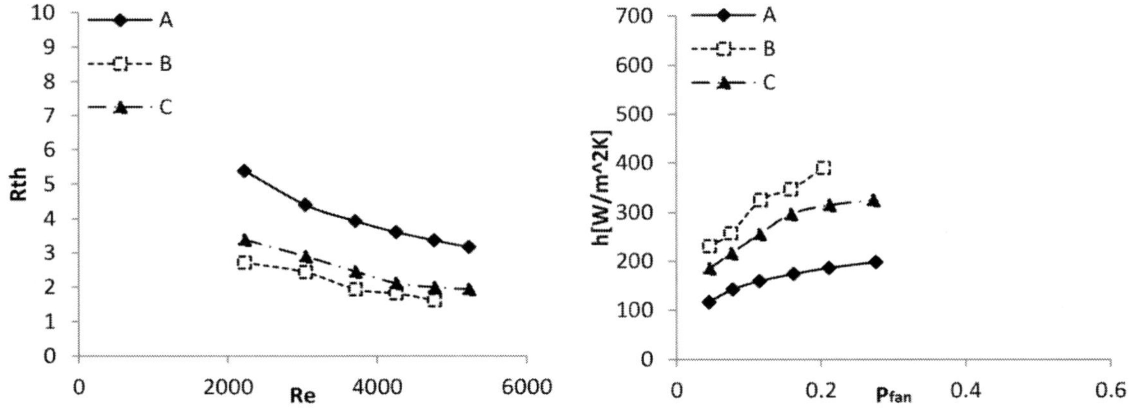

I. II.

Figure 4. I. Thermal resistance as a function of Reynolds for three different heat sink geometries for fan created flow. II. Heat transfer coefficient as a function of Power for fan created flow; (A) Flat plate, (B) 14x14 Pin array, (C) 14x14 Pin array with central 6x6 Pins removed

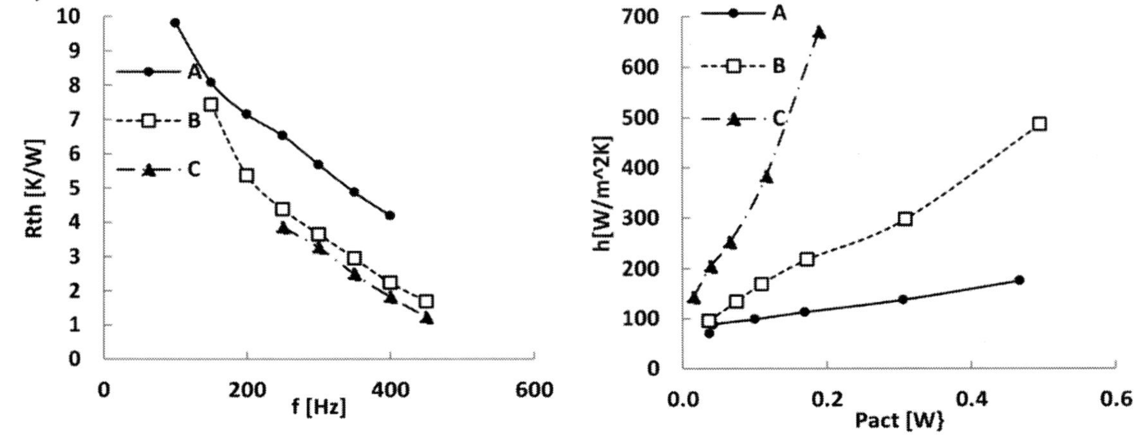

Figure 5. I. Electric power as a function of frequency for three different heat sink geometries ($L_0/D = 10$). II. Heat transfer coefficeint as a function of Power ($L_0/D = 10$); (A)Flat plate (H/D = 20), (B)14x14 Pin array (H/D = 30), (C) 14x14 Pin array with central 6x6 Pins removed(H/D = 30).

3.4. Power consumption

A comparison of the data in figure 4 and figure 5 shows that under certain conditions synthetic jet cooling devices outperform traditional fan cooling devices. Of particular interest is the relationship between heat transfer coefficient and input power to the cooling device. As previously stated, the modified pinned heat sink operates most effectively in the case of synthetic jet flows. However for a fan flow a pinned heat sink achieves the highest heat transfer coefficient. As expected as the turbulence of the flow increases the thermal resistance reduces. Although the synthetic jet flow has a much higher thermal resistance at low frequencies the rate of reduction is much greater and thus at greater operating frequencies the thermal resistance is comparable to that of the fan produced flow. Figure 4 II shows that the pinned heat sink has the maximum heat transfer coefficient for any fan power input. This heat transfer coefficient is almost twice as great as h for and equivalent power input to the jet actuator for the same sink. However, the implementation of the modified sink with jet actuation greatly improves the heat transfer coefficient over any of the sinks in use with a fan flow.

4. Conclusion

The study has investigated some of the main parameters pertaining to the heat transfer attainable with impinging synthetic jets on conventional extended surface heat sinks. Experimentation carried out on three different heat sink geometries has shown that any sink implemented will alter the heat transfer relative to its specific geometry independent of flow conditions. In addition to this each case has a unique optimum spacing between the jet actuator and sink where heat transfer is maximized for the given configuration. The study examined the heat transfer from sinks at a variety of flow conditions (in terms of jet Reynolds number $500<Re<2{,}500$ and dimensionless stroke length $5 < L_0/D < 12$) and found that the manner in which heat dissipation changed with spacing is repeated.

The comparison of the performance of different heat sink geometries in combination with the same synthetic jet flow shows that at any spacing a modified heat sink with pins surrounding a flat impingement zone provides superior cooling to either a flat surface or a fully pinned sink. The study finds that more research should carried out in the field of impingement surfaces for synthetic jets as flat surface impingement is inferior to pinned array sinks, particularly at larger spacing ($H/D > 10$).

Orientation of heat sinks in the vertical plane showed no increase in heat transfer capabilities over a sink lying in the horizontal plane. The orientation in the vertical plane further shows that the heat sink geometry affects heat transfer independent of flow conditions.

Investigation into the effects of frequency on the heat transfer shows that at large stroke lengths a change to frequency will have a greater effect on the heat dissipated from the surface. However, at smaller stroke lengths the effect of frequency is decreased.

The analysis of heat transfer coefficient in relation to power consumption indicates that by using the best available heat sink at optimum spacing a synthetic jet flow can enhance cooling.

Acknowledgements

The authors would like to thank Mr Michel Slusarz for previous work which contributed greatly to this research. All heat sink devices tested within the study are products provided by Malico Inc.

References

[1] Garimella, S.V., et al., *Technological drivers in data centers and telecom systems: Multiscale thermal, electrical, and energy management.* Applied Energy, 2013. **107**: p. 66-80.
[2] Glezer, A. and M. Amitay, *Synthetic jets.* Annual Review of Fluid Mechanics, 2002. **34**(1): p. 503-529.
[3] Smith, B.L. and A. Glezer, *The formation and evolution of synthetic jets.* Physics of Fluids, 1998. **10**(9): p. 2281-2297.

[4] Persoons, T., A. McGuinn, and D.B. Murray, *A general correlation for the stagnation point Nusselt number of an axisymmetric impinging synthetic jet.* International Journal of Heat and Mass Transfer, 2011. **54**(17–18): p. 3900-3908.

[5] Holman, R., et al., *Formation Criterion for Synthetic Jets.* AIAA Journal, 2005. **43**(10): p. 2110-2116.

[6] Chaudhari, M., B. Puranik, and A. Agrawal, *Heat transfer characteristics of synthetic jet impingement cooling.* International Journal of Heat and Mass Transfer, 2010. **53**(5–6): p. 1057-1069.

[7] Chaudhari, M., B. Puranik, and A. Agrawal, *Effect of orifice shape in synthetic jet based impingement cooling.* Experimental Thermal and Fluid Science, 2010. **34**(2): p. 246-256.

[8] Rylatt, D.I. and T.S. O'Donovan, *Heat transfer enhancement to a confined impinging synthetic air jet.* Applied Thermal Engineering, 2013. **51**(1–2): p. 468-475.

[9] Gillespie, M.B., et al., *Local Convective Heat Transfer From a Constant Heat Flux Flat Plate Cooled by Synthetic Air Jets.* Journal of Heat Transfer, 2006. **128**(10): p. 990-1000.

[10] Pavlova, A. and M. Amitay, *Electronic Cooling Using Synthetic Jet Impingement.* Journal of Heat Transfer, 2006. **128**(9): p. 897-907.

[11] Valiorgue, P., et al., *Heat transfer mechanisms in an impinging synthetic jet for a small jet-to-surface spacing.* Experimental Thermal and Fluid Science, 2009. **33**(4): p. 597-603.

[12] Gharib, M., E. Rambod, and K. Shariff, *A universal time scale for vortex ring formation.* Journal of Fluid Mechanics, 1998. **360**: p. 121-140.

[13]. Persoons, T., *General Reduced-Order Model to Design and Operate Synthetic Jet Actuators.* AIAA Journal, 2012. **50**(4): p. 916-927.

[14]. Zhang, J. and X. Tan, *Experimental study on flow and heat transfer characteristics of synthetic jet driven by piezoelectric actuator.* Science in China Series E: Technological Sciences, 2007. **50**(2): p. 221-229.

[15]. Farrelly, R., et al., *A Heat Transfer Comparison Between a Synthetic Jet and a Steady Jet at Low Reynolds Numbers*, in *ASME International Mechanical Engineering Congress and Expostion* 2008, ASME: Boston, Massachusetts.

[16]. Chaudhari, M., B. Puranik, and A. Agrawal, *Multiple orifice synthetic jet for improvement in impingement heat transfer.* International Journal of Heat and Mass Transfer, 2011. **54**(9–10): p. 2056-2065.

[17]. Persoons, T. and T.S. O'Donovan, *A pressure-based estimate of synthetic jet velocity.* Physics of Fluids, 2007. **19**(12): 128104.

[18]. Cengel, Y. A. and Ghajar, A. J. 2011. *Heat and Mass Transfer: Fundamentals & Applications,* New York, McGraw Hill.

AUTHOR INDEX

Institute of Physics
Dirac House, Temple Back
Bristol BS1 6BE UK

CURRAN ASSOCIATES INC
proceedings
.com